APPLIED MATHEMATICAL MODELING

A Multidisciplinary Approach

APPLIED MATHEMATICAL MODELING

A Multidisciplinary Approach

D. R. Shier
K. T. Wallenius

CHAPMAN & HALL/CRC

Boca Raton London New York Washington, D.C.

Library of Congress Cataloging-in-Publication Data

Shier, Douglas R.
    Applied mathematical modeling : a multidisciplinary approach /
Douglas R. Shier, K.T. Wallenius.
        p.   cm.
    Includes bibliographical references and index.
    ISBN 1-58488-048-1 (alk. paper)
    1. Mathematical models. I. Wallenius, K. T. II. Title.
QA401 .S465 1999
511′.8—dc21
                                                                99-050198
                                                                       CIP

© 1999 by Chapman & Hall/CRC

No claim to original U.S. Government works
International Standard Book Number 1-58488-048-1
Library of Congress Card Number 99-050198
Printed in the United States of America   2 3 4 5 6 7 8 9 0
Printed on acid-free paper

To Clayton Aucoin
for his visionary leadership in mathematical sciences education and for
the friendship he has bestowed

Contents

viii

Foreword

We are pleased to lend our support to this deserved appreciation of Clayton Aucoin for his efforts in curriculum reform in the mathematical sciences. Our involvement with the mathematical sciences at Clemson University began over twenty years ago as members of an "advisory committee" composed of academicians and representatives of business, industry, and government. Our task was to consult with a group of faculty members headed by Clayton Aucoin concerning their efforts to develop graduate programs based on breadth of training in mathematics, computing, operations research, and statistics. The central concern at the time was to establish a master's level curriculum with special attention to providing graduates with a broad knowledge of the mathematical sciences useful in applied settings. This goal was in sharp contrast to the traditional master's degree, typically viewed as a preliminary part of a doctoral program or even as a consolation prize for students who could not or did not choose to complete a (research-oriented) Ph.D. degree. Later, the focus of the Clemson program expanded to include doctoral work not limited to pure mathematics but featuring applied research, often sponsored by industries. The expansion included interdisciplinary doctoral programs worked out with and shared by other academic units within the university.

We found a set of enterprises all stimulated by Clayton's activities and all related to the roles of the mathematical sciences in the real world. A surprisingly comprehensive and well-developed undergraduate program provided a model for the structure of the proposed graduate programs. The entire Clemson program serves as a model nationally and has had a substantial impact on mathematical education at all levels. Clayton's central contribution in these many initiatives was to serve as a "guiding spirit" for all of them and to provide a long range vision which wove these separate strands into a cohesive fabric. This achievement required active support by the entire department. Individuals were encouraged to look beyond their own specialties as ends in themselves to see how their interests could fit into the larger picture. Clayton's steadying hand was ever present — not only in conceptual issues of development, but also in matters of redirecting conflicts between representatives of specialized mathematical areas into synergistic cooperation. Clayton recognized the significance of mathematical models, carefully formulated not only to be mathematically sound but also to illuminate important real world problems. To this end, he enlisted the cooperation of industrial managers both as advisers on curriculum and for their awareness of unsolved quantitative problems in manufacturing, business, and governmental operations. He kept a close scrutiny of national trends and managed to secure important and timely financial support for the department's innovative activities. We salute Clayton Aucoin for his masterful and creative leadership of mathematical science developments at Clemson over a period of many years. This volume of diverse models in the mathematical sciences is then a fitting tribute to one for whom modeling played such a key role in designing a modern mathematical sciences curriculum.

H. T. Banks

Don Gardner

Stu Hunter

Bill Lucas

Bob Lundegard

Don McArthur

Jim Ortega

Bob Thrall

Preface

Creation of this book was one of two related activities undertaken by a committee of colleagues to recognize Clayton Aucoin on his 65th birthday for his contributions to mathematical sciences education. A book on modeling seemed an appropriate tribute to one who insisted that modeling play a central role in the mathematical sciences curriculum. That viewpoint guided the development of the Clemson graduate program in the mathematical sciences, recognized as an exemplar of a successful program. Without doubt, it was a model-oriented multidisciplinary approach to the curriculum that earned this recognition. The second celebratory activity was the creation of the Clayton and Claire Aucoin Scholarship Fund at Clemson University. All royalties generated by sales of this volume will help support students in the mathematical sciences at Clemson University.

To create this modeling book, a five-person Editorial Committee was formed from department members whose areas of expertise span the mathematical sciences: applied analysis, computational mathematics, discrete mathematics, operations research, probability, and statistics. The Committee believes that modeling is an art form and the practice of modeling is best learned by students who, armed with fundamental methodologies, are exposed to a wide variety of modeling experiences. Ideally, this experience could be obtained through a consultative relation in which a team works on actual modeling problems and their results are subsequently applied. But such an arrangement is often difficult to achieve, given the time constraints of an academic program. This modeling volume therefore offers an alternative approach in which students can read about a certain model, solve problems related to the model or the methodologies employed, extend results through projects, and make presentations to their peers. Consequently, this volume provides a collection of models illustrating the power and richness of the mathematical sciences in providing insight into the operation of important real world systems.

Indeed, recent years have witnessed a dramatic increase in activity and urgency in restructuring mathematics education at the school and undergraduate levels. One manifestation of these efforts has been the introduction of mathematical models into early mathematics education. Not only do such models provide tangible evidence of the utility of mathematics, but the modeling process also invites the active participation of students, especially in the translation of observable phenomena into the language of mathematics. Consequently, it is not surprising that one can now find over fifty textbooks dealing with mathematical models or the modeling process.

Reform of graduate education in the mathematical sciences is no less important, and here too mathematical modeling plays an important role in suggested models of curricular restructuring. Whereas there are several excellent textbooks that provide an introduction to mathematical modeling for undergraduates, there are few books of sufficient breadth that focus on modeling at the advanced undergraduate or beginning graduate level. The intent of this book is to fill that void.

The volume is conceptually organized into two parts. Part I, comprising three chapters written by well-known experienced modelers, gives an overview of mathematical modeling and highlights the potentials (as well as pitfalls) of modeling in practice. Chapter 1 discusses the general components of the modeling process and makes a strong case for the importance of modeling in a modern mathematical sciences curriculum. Chapter 2, although not intended as a portmanteau of specific techniques, contains important ideas of a general nature on approaches to model building. It uses simple models of physics and more realistic models of efficient economic systems to drive home its points. Chapter 3 describes an experienced modeler's decade-long "odyssey" of modeling the AIDS epidemic. As in most journeys, the traveler often encounters new information as the trip evolves. Initial plans must be modified to meet changing conditions and to use new information in an intelligent manner as it becomes available. Road maps that were current at the beginning of the journey may not show the detours ahead. So too does this chapter illustrate the evolutionary nature of successful model building.

Part II is a compendium of sixteen papers, each a self-contained exposition on a specific model, complete with examples, exercises, and projects. Diverse subject matter and the breadth of methodologies employed reinforce the flexibility and power of the mathematical sciences. To avoid the appearance that one "correct" model has been formulated and analyzed, the treatment of most modeling situations in Part II is deliberately left open ended. A unique feature of many of these models is a reliance on more than one of the synergistic areas of the mathematical sciences. This multidisciplinary approach justifies use of that word in the book's title.

The level of presentation has been carefully chosen to make the material easily accessible to students with a solid undergraduate background in the mathematical sciences. Specific prerequisites are listed at the start of each chapter appearing in Part II. Included with each model is a set of exercises pertaining to the model as well as projects for modification and/or extension of results. The projects in particular are highly appropriate for group activities, making use of the reinforcing contributions of group members in a collaborative learning environment. A number of the chapters discuss computational aspects of implementing the studied model and suggest methods for carrying out requisite calculations using high-level, and widely available, computational packages (as Maple, Mathematica, and MATLAB).

This book may be viewed as a handbook of in-depth case studies that span the mathematical sciences, building upon a modest mathematical background (differential equations, discrete mathematics, linear algebra, numerical analysis, optimization, probability, and statistics). It makes the book suitable as a text in a course dedicated to modeling, in which students present the results of their efforts to a peer group. Alternatively, the models in this volume could be used as supplementary material in a more traditional methodology course to illustrate applications of that methodology and to point out the diversity of tools needed to analyze a given model. In either situation, since communication skills are so important for successful application of model results, it is recommended that students, working alone or in groups, read about a specific model, work the exercises, modify or extend the model along the suggested lines, and then present the results to the rest of the class. This volume will also be useful as a source book for models in other technical disciplines, particularly in many fields of engineering. It is believed that readers in other applied disciplines will benefit from seeing how various mathematical modeling philosophies and techniques can be brought to bear on problems arising in their disciplines. The models in this volume address real world situations studied in chemistry, physics, demography, economics, civil engineering, environmental engineering, industrial engineering, telecommunications, and other areas.

The multidisciplinary nature of this book is evident in the various disciplinary tools used and the wealth of application areas. Moreover, both continuous and discrete models are illustrated, as well as both stochastic and deterministic models. To provide readers with some initial road maps to chart their course through this volume, several tables are included in this preface. In keeping with the multidimensional nature of the models presented here, the chapters of Part II are listed in simple alphabetical order by the author's last name. Whereas in most mathematics texts, one must master the concepts of early chapters to prepare for subsequent material, this is clearly not the case here. One may start in Chapter 5 if cryptology catches your fancy or in Chapter 12 if bursty traffic behavior is your cup of tea.

Disciplinary Tools	Chapters
applied analysis	11,12
data analysis	3,16,19
data structures	13
differential equations	2,3,4,7,9,11,12,17,18,19
dynamical systems	4,11
graph theory	13,15
linear algebra	2,4,5,11,14,17
mathematical programming	2,10,16
modern algebra	6,13,14
number theory	5,6
numerical analysis	9,17
probability	8,10,12,13
queueing theory	8,12
scientific computing	9,17,19
statistics	8,10

Application Areas	Chapters
chemistry	19
civil engineering	18
communications	12,15
cryptography	5,6
demography	3,4
economics	2,16
environmental engineering	9,16
error-correcting codes	14
manufacturing	10,14
physics	7,11,17,18
public health	3,4
queueing systems	8,12

Model Types	Chapters
continuous	2,3,4,7,8,9,11,12,17,18,19
discrete	4,5,6,10,13,14,15,16
deterministic	2,3,4,5,6,7,9,10,11,14,15,16,17,18,19
stochastic	3,4,8,10,12,13

The book concludes with an appendix that provides an overview of the evolution and structure of graduate programs at Clemson University, programs that rely heavily on the pedagogical use of mathematical modeling. It recapitulates the fundamental importance of mathematical modeling as a driving force in curriculum reform, echoing the points made in Chapter 1 and concretely illustrating the need for a multidisciplinary approach.

The Editorial Committee has done its best to provide a sample of the wide range of modeling techniques and application areas. This book will be considered a success if it has whetted the reader's appetite for further study. Consequently, references to both printed materials and to websites are provided within the individual chapters. Supplementary material related to the models developed in this volume can be found at the website <`http://www.math.clemson.edu/modeling/`> for this book.

Acknowledgments

We are indebted to Dawn M. Rose for her unflagging and careful efforts in editing this unique volume.

The Editorial Committee,

Joel V. Brawley

T. G. Proctor

Douglas R. Shier

K. T. Wallenius

Daniel D. Warner

Contributors

W. Adams Department of Mathematical Sciences, Clemson University

Marc Artzrouni Department of Applied Mathematics, University of Pau

Joel Brawley Department of Mathematical Sciences, Clemson University

Philip B. Burt Department of Physics, Clemson University

Marie Coffin Department of Mathematical Sciences, Clemson University

Christopher L. Cox Department of Mathematical Sciences, Clemson University

P. M. Dearing Department of Mathematical Sciences, Clemson University

Jinqiao Duan Department of Mathematical Sciences, Clemson University

Robert E. Fennell Department of Mathematical Sciences, Clemson University

Mark Fitch Department of Mathematical Sciences, Clemson University

Shuhong Gao Department of Mathematical Sciences, Clemson University

Jean-Paul Gouteux Department of Biology (IRD), University of Pau

J. P. Jarvis Department of Mathematical Sciences, Clemson University

J. D. Key Department of Mathematical Sciences, Clemson University

Peter C. Kiessler Department of Mathematical Sciences, Clemson University

J. A. Knisely Department of Mathematics, Bob Jones University

R. Laskar Department of Mathematical Sciences, Clemson University

R. Lougee-Heimer Mathematical Sciences, IBM TJ Watson Research Center

William F. Lucas Department of Mathematics, Claremont Graduate University

William F. Moss Department of Mathematical Sciences, Clemson University

Tamra H. Payne Department of Mathematical Sciences, Clemson University

T. G. Proctor Department of Mathematical Sciences, Clemson University

Herman Senter Department of Mathematical Sciences, Clemson University

D. R. Shier Department of Mathematical Sciences, Clemson University

James R. Thompson Department of Statistics, Rice University

Robert M. Thrall Graduate School of Administration, Rice University

Yuan Tian Polymer Technology Group, Berkeley, CA

K. T. Wallenius Department of Mathematical Sciences, Clemson University

James M. Westall Department of Computer Science, Clemson University

Christian J. Wypasek GE Capital Mortgage Corporation, Raleigh, NC

Hong Zhang Department of Mathematics, Louisiana State University

Chapter 1

The Impact and Benefits of Mathematical Modeling

William F. Lucas

1.1 Introduction

This chapter discusses the nature of mathematical modeling and why we are including more of it in our curriculum. Modeling is being inserted as parts of our courses, is often the primary focus of whole courses, and is frequently covered now in various non-traditional teaching formats. The resulting benefits for our students range from the very practical (being better prepared to earn a good living) to improving education overall. Some benefits also extend to the faculty, to the colleges, and to the whole mathematical community in either academic or industrial settings. Problem solving in applied mathematics also impacts research and education in pure mathematics, although some methodologies such as the axiomatic method may play somewhat different roles. There still remain formidable challenges in realizing more and better modeling experiences.

1.2 Mathematical Aspects, Alternatives, Attitudes

People are motivated to study mathematics for two reasons. One is to enjoy the sheer intellectual beauty of its many abstract structures as well as the challenge of understanding its deep and far-reaching mental constructs. The other purpose is to be able to employ the great power of mathematics as a language and tool to understand and explain science and technology. These are referred to as *pure* and *applied* mathematics, respectively. This first attitude is described well in the often cited "greatbook" *A Mathematician's Apology* (1940) by the outstanding Englishman G. H. Hardy and more recently in *I Want to be a Mathematician* (1985) by P. R. Halmos of the University of Santa Clara. The latter aspect about mathematics' extensive interaction with many other subjects and its interrelationship to the more general culture of its time is presented in topologist R. L. Wilder's highly readable book *Evolution of Mathematical Concepts: An Elementary Study* (1968). There is a long history to demonstrate that what arises originally as pure mathematics often becomes most useful in applications. Also, what begins as a mathematical attempt to solve a very real and concrete problem may in time develop into a rich and major branch of pure mathematics. Beauty and utility need not be in conflict, and one can reap the rewards of both while studying this discipline. The dreamer and pragmatist can be at home in mathematics.

Tales are legend about rather eccentric scientists and artists devoting their entire lives almost exclusively to the pursuit of knowledge or art purely for their own sake. One week in the Fall of 1987 I saw or read rather accurate accounts of this sort about three of the greatest pure mathematicians of this century: Alan Turing (1912–1954) in the popular play *Breaking the Code*, Paul Erdös (1913–1996) in *The Atlantic Monthly* (November, 1987), and Kurt Gödel (1906–1978) in Chapter 3 of the book *Who Got Einstein's Office?* (1987) by Ed Regis. On the other hand, many credit logician Turing as the person most responsible for winning World War II by breaking the German code, while the deep "undecidable propositions" discovery by logician Gödel has had a profound influence on philosophy and caused havoc with works like Whitehead and Russell's *Principia Mathematica*. Modern mathematicians are said to have an "Erdös number" — how many joint publications they are "removed" from being a co-author with him on one of his great number of papers.

Most mathematicians also divided their subject into continuous and discrete, or analysis and algebra to some. Continuous mathematics begins with calculus and its study of the "solid" real number line, smoothly bending functions of real numbers (curves), and its many physical applications such as continuous dynamics and infinitesimal changes in time. Calculus and its extensions (called *analysis* more generally) have, for very good reasons, dominated both pure and applied mathematics for the past three centuries, and the main goal of precollege education has been to prepare one for calculus as rapidly as possible. Prior to 1700 most of mathematics was discrete in nature. That is, it was concerned primarily with properties of finite sets or "countable" sets like the integers. Moreover, there has been an explosion of advances in discrete mathematics over the past 50 years. Many of the new and traditional areas making use of mathematics in a crucial way today employ the methods of discrete mathematics. Furthermore, the revolution in digital computers involves discrete mathematics in nearly all of its aspects, as do most of the developments in the newer systems and decision sciences.

Many physical objects, such as metal bars, are often very discrete in nature, being composed of crystals, molecules, atoms, and/or elementary particles. Nevertheless, many of their properties can be discovered theoretically by modeling these as solid objects and using the mathematical methods of continuous analysis. On the other hand, many large-scale problems arising in logistics, decisionmaking, and social science are inherently multivariable and yet typically cannot be well described in terms of continuous mathematical representations. Thus the serious need for discrete models and often solutions via computers.

The greatest success story of applied mathematics (and perhaps all of science) over the last three centuries has been its keen ability to model the *laws of nature* and to use these constraints to assist in traditional engineering problem solving. In the last 50 years mathematics has, in addition, been employed increasingly to study the workings of large systems and institutions *invented by humans*. This gave rise to modern subjects to study the workings of large systems and institutions such as system analysis, operations research, and combinatorial optimization. In more recent decades mathematics has spread even further into the investigation of *human actions* themselves. This includes a host of new mathematical specialties within the social, behavioral, managerial, and decisional sciences. There are today whole new quantitative fields concerned with topics like one's utility for outcomes, the structure and nature of democracies, bargaining and conflict resolution, practical ways to arrive at a fair division, and many others. A major fraction of these new directions in applied mathematics are consumers of the parallel advances being made in discrete mathematics.

The recent developments in mathematics provide the potential for profound and highly beneficial changes in mathematics education at all levels. One need no longer master calculus and mechanics in order to be able to appreciate the great modeling ability of this subject. There can be valid alternate tracks and strands which can lead one to appreciate the beauty and power of mathematics, as well as to understand serious applications of the subject. Mathematics no longer needs to be viewed by most students as a rather old and completed field. It need not appear devoid of intellectual excitement. One also can easily show really convincing and relevant applications to one's routine interactions with other people as well as one's physical surroundings. Furthermore, the totality of mathematical knowledge doubles each decade. Yet, few non-professionals know any part of it that is less old than themselves. The author [4, 8] has written elsewhere about the great need to "modernize" the image of mathematics.

This chapter is concerned primarily with applied mathematics and how to teach it most effectively. Since the 1960s there have been, for various reasons, attempts to include more and current applications in the curriculum. The term "applicable mathematics" was introduced at the time to distinguish topics intended for immediate use to solve specific types of relevant problems, in contrast to mathematics that originally arose from applications but has become over time highly theoretical and rarely used in practice in its abstract format. A colleague suggested that we really need four expressions (or a 2 × 2 array) with PP, PA, AP, and AA, where P is for pure and A is for applied. For instance, AP would include subjects that originally arose in pure mathematics and are now being used in applications, such as number theory in cryptanalysis, whereas PA would stand for much of the work they do so well at places like the Courant Institute at NYU and The University of Minnesota.

In the 1970s some mathematics departments, led by Clemson University, even changed their name to "mathematical sciences," etc., to reflect their broader interests, or perhaps to ward off potential intruders on their boundaries. In this article I will use the term "mathematics" in its broadest and most inclusive sense, and to embrace all various sectors of applications as well. This contrasts with the beautiful writings by the late Morris Kline on the great need for applications in the curriculum. On a couple of occasions, in response to my inquiries, he made it very clear that he was referring to the great tradition of using mathematics in the physical sciences and classical engineering, and his aim was not met by many of the newer directions of applications. Clearly the time has come for mathematics programs to extend themselves beyond their traditional frontiers in all sorts of newer directions.

Mathematical education has been seriously damaged over the years by certain elitist attitudes and misconceptions. These include: "real researchers don't waste time in doing educational innovations," "pure math is better than applied," "some directions of applications are better than others or better for teaching purposes," etc. Some recent statements by math education administrators even seem to imply that research mathematicians are not able to make worthy contributions to education. This entire situation is absolutely ridiculous, and should be strongly condemned. These attitudes must be replaced by a code. Among its tenets would be that no mathematician has any right not to contribute something back to the system that made his or her career possible. It is true that each person has his or her own individual "utility function." We say "to each his own" or, in economics, "one should not make interpersonal comparisons of utility." However our community as a whole is only hurt by those who "bad mouth" others. The challenges we face in mathematics education will certainly require the full collaboration of everyone if we are to have any hope of solving our very serious problems.

Mathematics is often viewed like a great oak tree with a massive trunk supporting huge branches, leading in turn to successively smaller ones, and eventually to little twigs holding leaves. To add a beautiful new leaf one must necessarily climb high up through this long network in order to reach some elevated point. Although this analogy is a good one for many purposes, it can also be misleading and discouraging for some students. On the other hand, there are very recently created subjects and new types of applications that provide reachable branches on the tree, and even some suckers sprouting from the roots. There are also ladders, such as computers, that make some high points easily accessible. Furthermore, teaching methods based upon discovery and modeling can give the rewarding feeling of creating a new leaf without an exhausting or unreasonable climb through much of the tree of mathematics.

One excellent general discussion about the role of applications of mathematics in the undergraduate curriculum that emphasizes the unity of mathematics is given in the NRC's 25-page Hilton report [2] from 1979. This modeling book and former Clemson University reports address applications in graduate education. Very useful information about modeling courses, texts, and workshops can be found on the website <http://www-unix.mcs.anl.gov/mathmodeling/>.

1.3 *Mathematical Modeling*

The scientific method has had revolutionary impact on humankind for over 300 years. Mathematical modeling illustrates the more theoretical and intellectual aspects of this general scientific approach, and is a primary ingredient of applied mathematics. A *model* is a simpler realization or idealization of some more complex reality. It is created with the goal to gain new knowledge about the real world by investigating properties and implications of the model. A model can take various forms. It may be a physical miniature such as a model airplane in a wind tunnel, or a topographical representation of a local watershed used to gain insights into water flow and possible flooding. A model may be a gaming exercise or a computer simulation. Such models may well provide improved understanding and, better yet, new information about the real world phenomena. It may be more feasible and cost effective to investigate the model rather than the more complex reality itself. For example, one may gain significant additional insight into natural disasters via gaming exercises or simulations of potential events.

Much of the modeling done in mathematics is of a purely mental nature. It is done "inside our heads," perhaps with the aid of pencil and paper. We observe some aspect about the natural world or human actions, and use our inner faculties to create a mental construct with many similarities (and some differences). Reality is perceived by our senses, assimilated in our mind, and mimicked in some abstract manner. The detailed problem solving process via mathematical modeling can be illustrated by Figure 1.1. The top arrow in the diagram corresponds to the initial part of the modeling process. One replaces the real world phenomena by an abstract model for the purpose of employing the tools of mathematical analysis. This model typically takes a mathematical formulation involving basic variables and relationships corresponding to the entities and the laws of nature or behavior being observed.

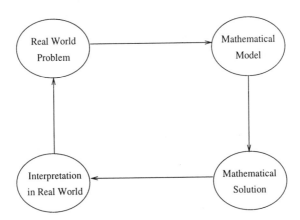

FIGURE 1.1   **The modeling process**

The right-hand arrow represents the solving of some resulting well-defined mathematical problem. The subsequent solution is usually in a mathematical form and must in turn be reinterpreted back in the original real world setting. Finally, in the left-hand arrow, the interpretation must be checked against the original reality. If this process does not result in the additional knowledge desired, then one may repeat the full cycle over and over again with alternate assumptions and tools. In practice, it is common to proceed around this modeling cycle many times. One repeatedly attempts to refine the model. One also goes back and forth between different aspects of the cycles before arriving at a satisfactory "solution" to the original problem. We will go over each of the four arrows shown in the diagram in more detail. Additional discussion on the latter arrows concerning interpretation and validation appear in Chapter 2 by R. M. Thrall.

The downward arrow on the right side of our diagram is the deductive part of the modeling activity. Given a well-formulated mathematical problem, e.g., a differential equation with boundary conditions, one attempts to find *all* the mathematical solutions. This task is rather straightforward in theory, but the degree of difficulty may vary greatly. We spend most of our time in traditional applied mathematics courses on this side of our modeling square. It is the best defined, most elegant, and most mathematical part of the modeling process. Unfortunately, we often overly stress such mathematical techniques in standard courses to the neglect of the other three arrows in the diagram.

It is the top arrow in this diagram that often involves the most creative and original input. It requires keen observation, or the use of other human senses, as well as communication skills and knowledge of other disciplines. This step is the crucial one in doing theoretical science or real applied mathematics. It uses more inductive type reasoning. One must translate the real phenomena into mathematics. It is essential for one to maintain the critical ingredients while filtering out the non-crucial elements in order to arrive at a realistic and still tractable mathematical problem. One must clearly identify the basic variables and characterize the fundamental laws or constraints involved. Many elements of such identifications, conceptualizations, and idealizations needed to arrive at an abstract formulation are difficult to teach. One often must forgo excessive details in order to keep the model manageable. What delicate tradeoffs one should make are traditionally learned from experience. In many ways, model building is an art form rather than a science per se. Eddington once said: "I regard the introductory part of a theory as the most difficult, because we have to use our brains all the time. Afterwards we can use mathematics." The person who can do this initial step well makes a most valuable contribution.

The bottom arrow in our modeling figure is concerned with how the essentially mathematical solution relates back to the original real world problem. One must accurately translate the mathematical results back into the initial setting. The purpose of modeling is to draw valid conclusions about what we did not know beforehand from things we could see in advance. It is important to communicate these interpretations in a precise, concise, and very lucid manner. Many good ideas are never implemented, or they are long delayed in being accepted, because of a poor job in this regard. In Tucker's MAA-CUPM report [11, p. 75] the author wrote the following paragraph regarding the potential ethical, political, and educational aspects of this bottom arrow in the modeling box.

> In describing the meaning of a mathematical solution one must take great care to be complete and honest. It is dangerous to discard quickly some mathematical solutions to a physical problem as extraneous or having no physical meaning; there have been too many historical incidences where "extraneous" solutions were of fundamental importance. Likewise, one should not select out just the one preconceived answer which the "boss" is looking for to support his or her position. A decisionmaker frequently does not want just one optimal solution, but desires to know a variety of "good" solutions and the range of reasonable options available from which to select. There is an old adage to the effect that bosses do not act on quantitative recommendations unless they are communicated in a manner which makes them understandable to such decisionmakers. This communication can often be a difficult task because of the technical nature of the formulation and solution, and also because large quantities of data and extensive computation may need to be compressed to a manageable size for the layman to understand in a relatively limited time. If mathematical education gave more attention to this aspect of mathematical modeling, there might be wider recognition and visibility of mathematicians in society beyond the academic world!

The left-hand (upward) arrow in our diagram is the validation part of modeling. One must check whether the model is reasonable. Does it agree with previously known aspects of the original problem? One must compare and verify the interpretation of the solution against reality. Does the solution provide accurate and reliable predictions? This may involve experiments and sensitivity analysis regarding the limits on, or the range of, variables for which the answers are dependable. The goal of modeling is to obtain new knowledge, but it should also be in agreement with previously known facts. It is not unusual for students to present the results of a modeling project to a lay expert only to have them reply that: "It is obvious. Anyone in this 'business' knows that it happens." Modelers should not be discouraged by such remarks. Such confirmation of fully known facts is a major part of validation. On the other hand, real gains are obtained when the model predicts something new that was essentially unknown before and which does check out when tested.

Most mathematics courses do not get involved to a significant degree with this left-hand arrow in our modeling box. Avoiding this aspect of modeling is to miss an important element of problem solving, as well as lose an opportunity to be involved with the joy of seeing the results communicated to others and perhaps implemented in society. In order to teach all aspects of applied mathematics one must cover all four of the "arrows" in the modeling diagram. It is not sufficient to overlook all but the right-hand one.

1.4   *Teaching Modeling*

There are a good number of different ways to include aspects of modeling in the mathematics courses. Traditionally, much of this has been left for courses taken outside of mathematics departments. In the "old days" most undergraduate math majors had at least a minor in physics. Many graduate degree requirements also required one or more courses from outside of the mathematics department. In the 1970s it again became common to require math degree students to take one or more courses from another program that emphasized some concrete applications of mathematics. These now may permit appropriate cross-discipline courses in the social, behavioral, systems, and decision sciences, as well as those in the natural sciences and engineering. Of course, separate mathematical sciences departments outside of the math department itself, such as computer science and operations research, often provide several appropriate courses containing modeling. In addition, more examples of contemporary modeling are finding their way into our more classical applied mathematics courses in recent years.

Furthermore, it has become popular in the last twenty years to introduce whole new courses devoted primarily to modeling. Several projects run by the MAA and others have prepared current teaching material for such courses. Nevertheless, many schools were slow to get *directly* involved. For example, Cornell University introduced a course called "Mathematics in the Real World" and made it a requirement for all undergraduate majors. However, it was not taught for the first few years, and students were allowed to take a suitable course in another department instead. For the first ten years that it was offered, it was taught by three professors whose primary appointments were outside of the mathematics department. The author has taught such modeling courses, and directed projects and institutes to prepare new materials for them. Nonetheless, he feels that it would be better instead to have ample modeling sprinkled at suitable spots throughout most math courses in the curriculum. This could be combined with efforts to also include more of other things like modern history and the culture aspects of mathematics in all of our courses.

In other subjects, much of the course exposure to modeling takes place in the case studies format. This approach has become well developed in some of our more professional programs such as in business and management science. Although this is often done well in some mathematical science areas such as operations research (and industrial engineering, one of OR's predecessors), it seems to be not very popular in math courses as such. To have the students obtain real hands-on experience with modeling, it appears that it is necessary to do more than merely read about how others did it. Consequently, many of the formats for teaching modeling are done as independent projects within a class, or else outside of the routine class setting. This contrasts to some extent with subjects like physics where they do prepare good modelers by mostly going over great historical models that form the basis of the subject.

Experience in modeling can be gained from projects done by whole classes, individuals, or teams of students as an adjunct to, or as an integral part of, a formal course without any outside funding. This experience may come as a part of thesis research at the B.S., M.S., or Ph.D. level. Some schools replace the senior or M.S. thesis by a team project. Some departments involve students in various ways in real consulting projects. This can be done through summer jobs, part-time work, co-op programs, or consulting centers as is common in statistics programs. Team projects funded by some outside sources such as industry, business, or government have been highly successful at places like Clemson University and The Claremont Colleges. The National Science Foundation (NSF) has funded Research Experiences for Undergraduates (REUs) for research by individuals or groups. These grants often cover several students at some center for a summer period. However, funds may come for individual student research and as "add ons" to regular NSF research grants, as is done in a big way at The University of Michigan during the academic year as well as summers.

The purpose of teaching modeling is to involve students in a hands-on manner on a more-or-less open-ended, real world problem. It should allow the students to collect data, make observations, apply empirical laws, discover relationships, and use their own independent abilities to create. The students should have some chance to experience all aspects of the modeling process discussed above. This normally involves some direct contact with people from outside of the purely academic setting. Projects can vary in length. However, they must be doable in the given time frame with the background the particular students possess. Modeling can easily be overwhelming and discouraging. The choice of problem should not be overly ambitious. It must be reasonable and possible to finish, given the participant's knowledge level. It is better to do well on a problem of modest scope than fail to arrive at a reasonable conclusion. One must give very careful attention to the possible prerequisites. Will the problem likely involve differential equations, statistics, algorithms, or numerical methods? Is the solution likely to make use of continuous or discrete methodologies? Do the students have appropriate experience, maturity, and sufficient time? What will the effort level be? One must ask whether all the needed resources are available. (These may include library, data banks, and computation skills and facilities.) Will there be help from some more advanced students, professors, or industrial liaisons? If at all possible, modeling projects should result in some formal presentation. This can vary from an oral class report, talk at "clinic day," or presentation at some regional or national meeting of a professional society. Any written report should be available in a college or departmental library. Such presentations and reports improve the exposition elements of the project and are valuable skills to develop. Students doing modeling should have access to write-ups of previously done projects. Publication in local "journals" may be a possibility. Above all, modeling projects should allow students to use their own ingenuity and feel the rewards of discovery and telling about it.

1.5   *Benefits of Modeling*

These are some obvious and immediate benefits that should follow from including more and better applied mathematics and modeling in our curriculum. Primarily, our graduates will improve significantly their chances of obtaining appropriate employment. There is clearly a major mismatch now between our graduates' training and the job market. On a more global scale their preparations for careers are not particularly in line with changing national needs. The job situation in recent years is reminiscent of the early 1970s, but the current decline may well persist for a longer period than on the prior round. In hindsight, there is ample evidence that the mathematics community has missed several golden opportunities in the past. Their long delay in taking computer science seriously is merely one example of the community's slowness in embracing other areas in their initial stages of mathematizations. We know that the 1950s and 1960s witnessed a great period of discovery and growth in pure mathematics. This was probably the first time in modern history when mathematics departments viewed themselves as more than just service departments. They could justify preparing math students to go on for Ph.D.s and do research in pure mathematics. Applied mathematics and the newer mathematical sciences were thus somewhat "squeezed" into the curriculum. Hopefully, our community will learn from the past and not overlook again many recent and future opportunities.

There have been many letters and articles in places like *The Notices of the AMS* and reports by NRC/NAS on problems in mathematics, and additional reports will likely produce diminishing returns. It would be better to take a closer look at those schools that have innovative applied programs and whose students are still obtaining good jobs. Many of these positions are in various fresh areas such as systems, finance, insurance, instrumentation, etc. On the other hand, a good fraction of the postdoctoral awards *for applications* appear still to be going to the "more respectable" areas such as aerospace, where employment possibilities will likely continue to decline.

One can continue to espouse the great value of a good pure math training in preparing students for applications, or quote distinguished applied mathematicians who claim that real analysis is "the best course." There are also ample statements in recent reports that can provide one with excuses for not seriously changing what we are now doing. Instead, we should examine our consciences, broaden our outlook, communicate more widely, undertake greater outreach, find out what others really desire from mathematicians, and thus better prepare our students for the wider range of employment opportunities awaiting the suitably prepared mathematical scientist. Leadership will be a very crucial element in these attempts. We also must improve the public image of our subject, but this must be earned by undertaking real change. In addition to the practical benefits of more and better applied mathematics education, there are great many more internal gains to be achieved as well for both mathematics itself and its practitioners.

1.6   Educational Benefits

Learning in mathematics seems to be a rather personal individualized endeavor. Students appear to need to rediscover and internalize new mathematical concepts. Lecture classes of 200 students sometimes work all right in the physical sciences where ideas often appeal to some more common human experiences, whereas they usually fail badly to communicate much in mathematics. Doing mathematics involves more than examples, factual knowledge, and analogy. Elements of imagination, intuition, and creativity are involved. Learning mathematics typically involves both very private aspects, although it can benefit from more group interactions as well. In their 1995 calendar The RAND Corporation quotes Robert Louis Stevenson in "An Apology for Idlers" (1876):

> Extreme busyness, whether at school or college ... is a symp-
> tom of deficient vitality; and a faculty for idleness implies a
> catholic appetite and a strong sense of personal identity. ...
> they have no curiosity ... they do not take pleasure in exer-
> cise of their faculties for its own sake ...

and, on the other hand, Samuel Johnson in "The Adventurer, No. 126"
(1754):

> Though learning may be conferred by solitude, its applica-
> tion must be attained by general converse ... he that never
> compares his notions with those of others ... very seldom dis-
> covers the objections which may be raised against his opin-
> ions ... he is only fondling an error ...

Whether learning is done in an individual or collaborative setting, motivation of some sort is a crucial prerequisite.

It is frequently asserted that we learn from experience. (It is also said
that experience is what we obtain when we don't get what we want.)
One typically learns more about being an auto mechanic by working at
a service station than from reading manuals. If you were badly hurt
in an auto accident would you prefer to have a young trauma surgeon
who had his or her residency at a much-maligned big city trauma center
where one obtains significant hands-on experience with many cases, or
else at a very prestigious suburban hospital where he or she "watched"
well-known surgeons perform? Mistakes in these illustrations may be
very costly. Nevertheless, we do learn from our mistakes and, if the
damage is not too great, individuals should be allowed to make mistakes
in their educational activities.

Howard Gardner of Harvard University divides learning styles or domains of intelligence into seven types:

(a) Linguistic (word players)

(b) Logical/Mathematical (questioners)

(c) Spatial (visualizers)

(d) Musical (music lovers)

(e) Bodily/Kinesthetic (movers)

(f) Interpersonal (socializers)

(g) Intrapersonal (individuals)

The author feels that the first three of these are more basic and internal to the learner, whereas the latter four are more about the context or environment which stimulates the learning to take place. The psychologist may relate (a) and (b) to "left brain" and (c) to "right brain," in the normal right-handed person. What is clear, however, is that different people learn more effectively in different ways and in different situations. Moreover, the teaching of mathematics should involve some elements of each possible learning style as well as each basic mathematical strand. Mathematical modeling is surely one approach that can do this in a relatively workable manner.

The past two decades have witnessed major changes in the way we teach mathematics at all levels. The dominance of lectures is rapidly declining. More and more teachers relate how they are throwing out their lecture notes. Over this period we have seen movements toward more interaction, discovery approaches, team efforts, use of calculators, and experimentation in the classroom. Outside of class the students often work together in groups and on problems with more open-ended or nonunique solutions (whose answers do not appear in the back of the book). These approaches build motivation, questioning attitudes, experimental attitudes, confidence, improved vocabularies, and better listening and communication skills. These are the attributes that are so important for lasting success after graduation. The move toward more modeling in the curriculum was one major change that is an extremely efficient and effective way to embed most of these new emphases into the curriculum, as well as a good way to cover more applied mathematics and better prepare our students for a broader range of employment opportunities. The modeling approach in applied mathematics has much in common with the discovery methods used in pure mathematics, such as the famous R. L. Moore approach.

It may be helpful to review major reforms in math education over a longer time period in order to assess the role of applications and modeling in a more long-term framework. The author's perspective on this and some of his own involvement in over 40 years of teaching follows in brief. The first major change in this period came in the 1950s and 1960s. This change involved a great increase in abstraction, use of logic, and concern with fundamental structures. It brought together the great successes in logic and abstract algebra starting a century earlier, as illustrated in the work of N. Bourbaki. This was somewhat essential because of the explosion of new mathematical discoveries and fields, and a need to isolate what is basic. It was accompanied by a golden age of great mathematical advances in and appreciation of pure mathematics. As related above, academic mathematics departments felt that they could justify spending much of their efforts in training mathematicians through the Ph.D. in contrast to their former stress on mostly service to others. These changes were of a top-down nature. More abstraction went from graduate schools to elementary schools (the "new math") in less than two decades. Most feel that at the precollege level the new math was a failure. I personally feel that the biggest problem was in stressing some two or three basic and potentially interesting aspects of mathematics while overlooking some twenty other interesting and yet fundamental strands in the subject. Two other revolutionary changes were emerging in this period: (a) the rapid increase of exciting new topics in pure and applied discrete mathematics (along with parallel advances in the digital computer), and (b) the broadening of the scope of applied mathematics to include the life, decisional, managerial, and social sciences. The educational part of these latter changes can be glimpsed from the many booklets put out by the MAA's CUPM in the 1960s and early 1970s. Several major changes in the teaching of calculus have been ongoing for some time. These include the addition of more modeling and realistic applications, as well as the reordering of many topics within this subject. Calculus was also moving down to the freshman and high school levels.

The second major change began around the early 1970s, and was mostly concerned with applied mathematics. It was more "bottom-up" in nature. It began with the employment situation. New Ph.D.s, especially in pure math, were not obtaining the types of jobs that had been anticipated from past experience. Many instead took nonacademic positions in applications, or else in teaching at "small" colleges where they often replaced teachers without the doctorate. They were often in small departments with few math majors. They soon recognized there was a significant need to revitalize undergraduate math as well as to present more applications. Things had to change compared to what they themselves went through. Many of these well-trained and bright young faculty made their own changes in how and what they taught, but were mostly working in isolation. Before long they met others at conferences and saw that they were not alone. This general restlessness began to spread. Clearly, something was seriously wrong. At the same time there was great concern about preparing Ph.D.s for nonacademic positions and the fact that applied mathematics had been terribly squeezed into math programs as a result of the many great breakthroughs in pure math. More applications, and more contemporary and relevant ones, needed to be injected into the curriculum at both the graduate and undergraduate levels. The question was how to best teach more interesting and contemporary applications and newer methodologies in the most effective way. There had, in addition, already been an earlier wave at many places like Berkeley of students working in groups, special activities for minorities, professors visiting high school classes, help sessions outside of class, collaborative and interactive approaches in class, etc. It became obvious to many that the modeling approach was one that would meet many of the new goals.

The introduction of more applications and modeling in university level mathematics courses over the past twenty years was not merely one part of a major reform taking place in math education more generally. One could argue that it was a basic source of many changes in pure as well as applied courses. Some of the things that are so naturally associated with modeling are: students working in groups, students consulting with various people (or computers) beyond the classrooms, discovery approaches on concrete or intuitive-rich problems, open-ended and practical problems with nonunique solutions, broad class discussions, laboratory type settings, experimental methodologies, and oral or written reports. These have much in common with the changes taking place in recent reforms in more fundamental pure math courses. These often occurred initially in the applied context and before recent topics like the calculus reform. There is much in common in the newer adaptations in both pure and applied courses and alternate teaching methods. The fundamental role of modeling in math educational reform is made in the article "Modeling as a Precursor and Beneficiary of Mathematics Reform" by Ricardo [9].

Furthermore, one could even argue that calculus reform over the last decade was a reaction to the "threat" from applied and discrete mathematics, in addition to the clear need to improve what was actually happening in calculus itself at some schools. About thirty years ago there were attempts to introduce more (pure and applied) discrete math and at lower levels, as well as more applications of a discrete (or continuous) nature. The advances in fields like computer science, operations research, and mathematical social science called out for more discrete topics. The popularity of low level discrete courses at many schools was clear. So the discrete and applications increasingly competed for the space traditionally held by calculus and its normal successors. On the other hand, there are individuals, and even some whole university departments, who held that analysis and the pure are "better" than discrete or applied. There were even scheduled debates at conferences (e.g., [10]) about whether discrete math should be covered in lieu of a little less calculus at the lower divisional level, or whether several discrete topics should even be integrated into the calculus. The calculus supporters argue that their subject was more well defined and had proved itself over time. One of several observations by the opponents was that calculus was being done very badly for the majority of students at many schools. This was very true and hit home to many continuous supporters. Such "negatives" and "reactions" may well be a primary cause of calculus reform. The first Sloan funded conference in 1986 required the support of a distinguished discrete mathematician in order to come about. This in turn led to mathematics being well positioned to obtain funding for calculus when the NSF returned to supporting college level science education in 1987. The resulting calculus reform had much in common with modeling and other ongoing changes in the teaching of mathematics.

There now exist several successful educational programs with alternate ways to experience current applications of mathematics. In addition, there is an ample supply of teaching materials to organize modeling courses of various sorts. Two MAA projects in the 1970s led to the CUPM report on *Case Studies in Applied Mathematics*, edited by Maynard D. Thompson in 1976, and to the four Committee on Institutes and Workshops (CI&W) volumes on *Modules in Applied Mathematics*, edited by William Lucas (Springer-Verlag, 1983) and containing most of the 60 modules started several years before at the 1976 workshop at Cornell University. Modeling has been a popular topic in the MAA minicourses and AMS short courses at the annual meetings for about twenty years now. There now exist the *UMAP Journal* and many other UMAP modules (some in pure math) as well as other products from COMAP, Inc. There are also dozens of good texts on modeling. It is not difficult today for a nonexpert to become involved in the teaching of modeling.

1.7 Modeling and Group Competition

The field of mathematics has for many years had competitions at various student levels. These usually score the results of individual contestants and cover more pure math sorts of problems, although some less traditional topics like combinatorics are popular for various reasons. The most famous competition in the United States is the Putnam exam. As in several competitions there are "team scores," but these are merely the sums for the individuals on the team. There has also been a noticeable shortage of many women's names in the top scorers on the Putnam, although this seems to be improving somewhat in recent years.

Among other things, the series of educational reforms begun in the 1970s included two objectives in particular: a desire for more applied mathematics and experience in modeling, and to obtain the potential benefits of more student interactions and collaborative learning. This led to the idea of real team competitions on more open-ended applied math problems done over a somewhat longer time frame. The nature of such competitions and how to bring them about was a topic of discussion at several meetings held by the MAA and others from the mid-1970s to the early 1980s. For example, this idea was considered at some CUPM meetings at the time. The MAA committees concerned with continuing education (which succeeded the MAA's CI&W) discussed this, and two of the first three MAA minicourses in 1979 were on modeling. Meetings for various UMAP projects and the boards at its successor, COMAP, Inc., talked about having such competitions. The author recalls a dinner he had with Bernard A. Fusaro at his house on June 8, 1977, where he summarized these considerations and spelled out what was to later become the COMAP Mathematical Competition [Contest later] in Modeling (MCM). Since 1985 this has become a highly successful activity where groups of three students work on one problem over a long weekend. This was accomplished under Ben Fusaro's leadership for its first seven years, and has been led by Frank R. Giordano since then.

In the late 1970s we referred to the need for an "applied math Putnam," although a greatly different format from the Putnam exam was visualized. It is an interesting aside that the brilliant think-tank guru, Herman Kahn, used to say that it takes about seven years from the time that a new idea is clearly formulated at a known research organization until government or business gets around to implementing it — which was the case with the MCM. On the other hand, it was only about three years from the time that MAA committees recommended minicourses until most of the MAA leadership very reluctantly allowed them to be tried; and they have since become the most popular new aspect of national meetings in decades.

COMAP's MCM is now a very successful modeling competition at the undergraduate level, and is documented in the special edition of the *UMAP Journal (Tools for Teaching)*, 1994. The participation of women students in MCM appears to be at least 30%. Some successful modeling experiences for graduate students are also available now. These last up to a couple of weeks in time, prior to write-up. Ellis Cumberbatch at The Claremont Graduate School has been a leader in these efforts, and a related article appeared in the *SIAM NEWS*, November, 1993. One typically does not attempt to rank order teams at this level. The international consortium for SIAM-type organizations in Europe, ECMI, runs a modeling activity for graduate students as well.

For some dozen years there have been many discussions about a major modeling competition for groups of high school students. Some smaller local activity already appears to exist, but the size of a fully national contest seems forbidding and has not yet been realized. It should perhaps be started by organizing it around some structure like the MAA, AMATYC, or SIAM local sections where this may well be feasible. The excellent experiences that students obtain from such structured interactive competitions is usually highly beneficial to them and our profession.

It seems to be one of the good side effects of concerns for more modeling in the mathematics curriculum.

1.8   Other Benefits of Modeling

The introduction of modeling courses and other alternative formats, such as math clinics, for providing meaningful training in modern applied mathematics can provide a host of benefits that extend far beyond the students themselves. Individual faculty, whole mathematics departments, and various people outside of these departments can benefit as well. Moreover, improved teaching of contemporary applications can also have profound effects on developments within the realm of pure mathematics.

Some highly disturbing things about math education are: the small percent of math majors at all but some dozen or two colleges in the country, the significant dropout rate early in Ph.D. programs, and the extremely high "research dropout rate" that occurs soon after the Ph.D. is awarded. Above all, mathematicians rarely do research in areas significantly far away from the topics of their Ph.D. theses. This is presumably due in part to the fact that the thesis is usually on a very deep and narrow problem, and it is difficult to again achieve this depth in a rather different field. The consequence of specialization may well be a lack of breadth. Teaching modeling or directing a serious student project can prove to be just the "medicine" needed to reinvigorate an individual or group of faculty members, as well as to keep students longer in exciting and relevant math studies. I know several mathematicians who re-entered research in new areas as a result of rather mild doses of modeling obtained at a workshop, by directing a project while on sabbatical, or suchlike. If the United States really desires more research mathematicians, the most cost-effective way would be to decrease the large numbers of those who quit serious research soon after receiving their doctorates. This last sentence is in no way intended to assign lesser value to what most of the "research dropouts" contribute in other ways. Our professional societies should help provide better outlets for some of these activities as well.

Many advantages accrue when mathematical sciences departments become involved in applications that require interdisciplinary efforts, nonuniversity contacts, and learning new problem areas or solution techniques. Simply the group nature and planning aspects are of benefit. On both the individual and departmental levels this can result in additional funding for faculty as well as students. So in addition to educational research stimulus, the rewards can include new grants, consulting arrangements, or summer jobs from new or traditional sources for both students and faculty.

As suggested at the start of this chapter it is well recognized that what starts as applied mathematics often leads to beautiful pure mathematics. Likewise the physicists [1, 13] have expressed well how surprising it is that some existing pure mathematics proves to be the perfect tool to solve problems arising in the natural sciences and elsewhere. However, it should furthermore be acknowledged that applications and the way they are taught has had profound effects on bringing new fundamentally pure mathematical topics into the main stream of the field. Let us consider four illustrations of this that have occurred over the last 50 years.

Mathematical Programming. One can argue that linear programming (LP) is just as important (as an applied or pure subject) as is linear algebra. Polyhedra and cones are hardly less basic things than vector spaces. Yet it was the great success of LP in applications that brought it such widespread attention. LP was also the obvious tool for solving a great variety of different sorts of problems. It thus became a very popular topic in many (less open-ended) modeling courses. Even then, it took several decades before many mathematics departments took it seriously. The abstractions and generalizations arising out of LP have a great number of followers, but the field could still benefit from additional input from deep pure mathematicians.

The New Modern Algebra. Some aspects of combinatorics have always been considered part of pure mathematics. Recent uses of mathematics beyond the natural sciences and engineering, however, have created a great variety of new discrete subjects. It is clear that the recognition of common elements in these concepts is leading to another round of abstract algebra, no less fundamental and potentially more useful than the traditional modern algebra which has been one of the leading stars of twentieth century pure mathematics. A few people with deep knowledge in traditional algebraic areas and knowledgeable in contemporary discrete math are really "cleaning up" in terms of new discoveries. Nevertheless, there remain many algebraists seemingly unaware that they are superbly prepared to research in these newer directions of combinatorial and discrete mathematics.

Higher-Dimensional Geometry. Many problems in the social and system sciences are naturally multivariate and not easily approximated by continuous or statistical methods. Subjects like multiattribute utility, multiobjective programming, multiperson game theory, etc., could greatly benefit from experts in many parts of higher-dimensional geometry. There remains a large gulf between these practitioners.

The "Missing Strand." In the fall of 1987, while serving briefly as
a program officer in USEME at the NSF, I attended a meeting of the
advisory committee to the mathematical sciences division at NSF. I discussed the fine reports on mathematical education appearing at that
time with Kenneth Hoffman. I expressed the reservation, however, that
a structure was being set up that well may result in a great number of
reports and that they would likely have diminishing value. On the other
hand, I did say that there should be one more report which described
what was really fundamental in mathematics. When requested, I wrote
such a list: (a) numbers (algebra), (b) space (geometry), (c) randomness
and uncertainty (probability and statistics), and (d) continuous change
(calculus, etc.) and *discrete change*. In early 1988 I heard that there
was a project on "strands" in mathematics and I wrote a letter to Mar-
cia Sward in more detail about my list. In 1990 there appeared the
NRC/NAS book *On the Shoulders of Giants* edited by Lynn A. Steen.
It offered five examples of the developmental power of deep mathematical ideas: (a) dimension, (b) quantity, (c) uncertainty, (d) shape, and
(e) change. The wonderful chapter on change by Ian Stewart contained
mostly continuous illustrations, and nothing on what I had in mind when
I listed "discrete change." I had in mind things from modern discrete
mathematics: directed networks, graph theory (road maps or elementary
topology), edges on polytopes, path-following algorithms, system analysis type thinking, deterministic operations research, and many others.
Of course, I had a secret agenda behind suggesting discrete change as a
fundamental notion. If this idea were to become accepted, then people
may consider introducing more such topics at lower levels in the schools.
Surely, many basic topics in discrete math could be covered at very low
levels, and topics from applications like OR could allow students to do
good modeling in grade school mathematics. I believe most strongly that
a golden opportunity in math education at all levels is being neglected
in this regard. I now refer to "discrete change" (or the wider pure and
applied developments in discrete mathematics) as "the missing strand."

The point is that new applications bring many fresh and basic notions
into pure mathematics as well. This contrasts somewhat with the popular viewpoint that mathematics as one of the humanities is driven more
by internal needs and aesthetical considerations. It is said that John von
Neumann often expressed concern about how the directions for future
mathematical research were determined. The following frequently given
quote [12] by him on empirical sources for new mathematics is worth
stating again.

As a mathematical discipline travels far from its empirical source, or still more, if it is a second and third generation only indirectly inspired by ideas coming from "reality," it is beset with very grave dangers. It becomes more and more purely aestheticizing, more and more purely *l'art pour l'art.* This need not be bad, if the field is surrounded by correlated subjects, which still have closer empirical connections, or if the discipline is under the influence of men with an exceptionally well-developed taste. But there is a grave danger that the subject will develop along the line of least resistance, that the stream, so far from its source, will separate into a multitude of insignificant branches, and that the discipline will become a disorganized mass of details and complexities. In other words, at a great distance from its empirical source, or after much "abstract" inbreeding, a mathematical subject is in danger of degeneration. At the inception the style is usually classical; when it shows signs of becoming baroque, then the danger signal is up. ... In any event, whenever this stage is reached, the only remedy seems to me to be the rejuvenating return to the source: the reinjection of more or less directly empirical ideas. I am convinced that this was a necessary condition to conserve the freshness and the vitality of the subject and that this will remain equally true in the future.

Similar views have been expressed by others who excelled in both pure and applied mathematics (such as Hermann Weyl and Richard Courant).

1.9   The Role of Axioms in Modeling

The axiomatic method has been a hallmark of twentieth century mathematics. Axioms may, however, play a much different role in research versus education, and even more so in pure versus different sorts of applied mathematics. There is little question about the value and success of the axiomatic approach in pure math research and higher education. However, the way they are used in school mathematics can cause one to wonder. They often take the form of the obvious, and lead children to ask: Why would not: "$a + b = b + a$" or "two different nonparallel lines cross at one point." Often the teacher cannot give a case where the associative law fails. In an interview the great geometer H. S. M. Coxeter [3] spoke of "... dull teaching: perhaps too much emphasis on axiomatics went on for too long a time ... That did a lot of harm." As we know, the "new math" did not go over well (for several reasons).

In the realm of applied mathematics, axioms often play a different role and their use can be more controversial. In 1973 L. C. Wood [14] entitled an article "Beware of Axiomatics in Applied Mathematics" and warns of the danger of their excessive use in classical applications and their misuse in teaching the physical sciences. The stress on axioms in these cases is often put in the wrong place. They often come after the really creative activity is done. They get formulated toward the end of a theory when one attempts to refine it into a precise, concise, and compact form. This does have the advantage of allowing one to teach a subject more rapidly so as to get on sooner to newer and more advanced topics. It is also true, on the other hand, that many great applied mathematicians, like John von Neumann, had a solid grounding in pure mathematics and the axiomatic approach was integral to his applied work, too. He always wanted to "get to the point" in the sense of what it is that is really fundamental. Nevertheless, observation, intuition, and experimentation, etc., are crucial in doing natural science as well as just listing basic assumptions and drawing logical deductions.

However, there can be a major difference between the way axioms are employed in classical applied mathematics and how they are frequently used in some of the newer directions of applications. The greatest success story of applied mathematics over three centuries has been its use to explain how Mother Nature behaves, and this assisted in the industrial and technological revolutions. One of the megatrends in applied mathematics in the past 50 years has been its use in explaining how human beings interact with each other and so develop systems and structures in a society. In modeling in the social sciences, axioms often come in at the earliest stage of the quantitative model formulation process. Consider, for example, a subject like social choice theory or Kenneth Arrow's famous impossibility theorem, or the development of some fair division scheme. These topics begin with axioms in the form of basic principles or desirable properties you wish to be satisfied. One may move almost immediately to proving existence and uniqueness theorems from some subset of the axioms. Some mathematics for the social scientists is, in this sense, more like synthetic (axiomatic) geometry than most people realize. The discovery of axioms and selecting consistent subsets of them is part of the early modeling process itself. For example, what is being assumed when one asserts that some fair division method is in fact "fair"? Since these newer directions of applicable mathematics typically assume fewer nonmath prerequisites, they can also be done at much lower grade levels. We can easily teach nontrivial modeling, axiomatics, and aspects of the "missing strand" in grade school courses! We are certainly overlooking a wonderful opportunity to greatly improve mathematical education at all levels.

1.10   The Challenge

A main problem with university level education in mathematics is that there is just too much to cover. The sheer volume of material to prepare a student in those precious aspects of pure math as well as to develop some techniques of applied mathematics is too excessive to fit into our programs — not to mention the need for skills in computing and some exposure to science. We face the problem of how to select from a variety of tradeoffs. This includes the basic question of breadth versus depth. In order to squeeze everything in, one risks watering down the contents. It is true that there is a sort of "trivialization of knowledge over time." Clearly some old topics can be deleted sooner (determinants) and prerequisite chains can often be shortened. But one cannot shortchange that valuable commodity known as "mathematical maturity" in order to just acquire more factual knowledge. Nevertheless, some believe that there are potential solutions. A modeling approach (or perhaps parallel discovery methods in pure math) appears to be one way to realize certain efficiencies in terms of mathematical content. Other educational techniques may also help.

There are now many superb modeling courses and successful alternate formats for students to gain such skills in applications that are ongoing across the country. There are also ample materials available to assist anyone willing to do the "homework" and desiring to introduce new courses or project activities. Schools should make use of their own particular local talent and resources and survey the needs of their region in order to make suitable connections.

Admittedly, there are limiting factors to how far mathematics can extend its frontiers. I do not agree with the pessimistic view that mathematicians must become "engineers" in order to find employment. They must, however, look for new uses of mathematics in fields that had little need for this subject in the past. One would hardly assert that physicists have to become computer scientists because they developed the new area of computational physics, or that economists are now psychologists when they advance the field of experimental economics.

I believe firmly that there are many situations where "a little bit goes a long way," and that this is true for "fun parts" and modeling aspects in mathematics courses. Another observation relates to my 15 years as a member of the operations research department at Cornell University. The OR undergraduate B.S. degree in the College of Engineering was a highly mathematical one, except that it replaced things like algebra and analysis by somewhat similar subjects like mathematical programming and stochastic processes. But most students also took a course in accounting and cost analysis, and a second one-semester course covering some 45 "tricks" that everyone should know and use, to make up most of the old industrial engineering curriculum (e.g., PERT/CPM). These two courses allowed students to communicate with people in industry and business, and thus acquire jobs. It was the deeper math and their modeling experiences that proved of far greater value in the long run.

It remains for the leadership of the mathematics community to encourage experiments with a broader array of mathematics programs, and for individual faculty or departments to have the courage to dare to be different. We should not get bogged down with the idea that our subject has some large unchangeable core of knowledge that every mathematics student must know. The problems caused by the seemingly endless explosion of new mathematical advances are hardly insurmountable. There are still many different ways that a mathematics student can prepare for an enjoyable and prosperous career in this field. The faculty may, however, have to do some things differently to bring this about.

1.11 References

[1] F. J. Dyson, "Missed opportunities," *Bulletin of the American Mathematical Society* **78**, 635–652 (1972).

[2] P. Hilton, et al., *The Role of Applications in the Undergraduate Mathematics Curriculum*, National Research Council, National Academy of Sciences, Washington, D.C., 1979, 25 pages.

[3] D. Logothetti, "An interview with H. S. M. Coxeter, the king of geometry," *The Two-Year College Mathematics Journal* **11**, 2–19 (1980).

[4] W. F. Lucas, "Growth and new intuitions: can we meet the challenge?" in *Mathematics Tomorrow*, L. A. Steen (Ed.), Springer-Verlag, New York, 1981, 55–69.

[5] W. F. Lucas, "Operations research: a rich source of materials to revitalize school level mathematics," in *Proceedings of the Fourth International Congress on Mathematical Education*, Berkeley, 1980, M. Zweng, T. Green, J. Kilpatrick, H. Pollak, and M. Suydam (Eds.), Birkhäuser, Boston, 1983, 330–336.

[6] W. F. Lucas, "Problem solving and modeling in the first two years," in *The Future of College Mathematics*, A. Ralston and G. S. Young (Eds.), Springer-Verlag, New York, 1983, 43–54.

[7] W. F. Lucas, "Discrete mathematics courses have constituents besides computer scientists," *College Mathematics Journal* **15**, 386–388 (1984).

[8]   W. F. Lucas, "Can we modernize what we teach?" *UME Trends*
      1, 7–8 (1989).

[9]   H. J. Ricardo, "Modeling as a precursor and beneficiary of math-
      ematics reform," *UMAP Models: Tools for Teaching*, COMAP,
      Lexington, MA, 1994, 25–28.

[10] F. S. Roberts, "Is calculus necessary?" in *Proceedings of the
      Fourth International Congress on Mathematical Education*, Berke-
      ley, 1980, M. Zweng, T. Green, J. Kilpatrick, H. Pollak, and
      M. Suydam (Eds.), Birkhäuser, Boston, 1983, 50–53.

[11] A. C. Tucker (Ed.), *Recommendation for a General Mathematical
      Sciences Program*, Committee on the Undergraduate Program in
      Mathematics, Mathematical Association of America, Washington,
      DC, 1981, 102 pages. (A revised version appears as the first six
      chapters in *Reshaping College Mathematics*, edited by L. A. Steen,
      MAA Notes, No. 13, 1989.)

[12] J. von Neumann, "The mathematician," in *The Works of the Mind*,
      R. B. Heywood (Ed.), University of Chicago Press, Chicago, 1966,
      180–196.

[13] E. P. Wigner, "The unreasonable effectiveness of mathematics in
      the natural sciences," *Comm. Pure Appl. Math.* **13**, 1–14 (1960).

[14] L. C. Wood, "Beware of axiomatics in applied mathematics," *Bul-
      letin of the Institute of Mathematics and Its Applications* **9**, 25–40
      (1973).

Chapter 2

Remarks on Mathematical Model Building

Robert M. Thrall

2.1 Introduction

I was asked by the Editorial Committee of this modeling volume to provide an overview of "the processes of model building, model fitting, and model evaluation." Since I am certainly in no position to undertake such a monumental task, I have employed the less pretentious title "Remarks on Mathematical Model Building." Most of these remarks are based on my own personal experiences with model construction and application. I will rely especially on two of my papers [8] and [11]. The first model considered in Section 2.2 is a classic, simple model of physics, yet it contains germs of modeling truth that are applicable to more complex models, such as the DEA model discussed in Section 2.3.

2.2 An Example of Mathematical Modeling

A familiar mathematical model uses the equation

$$y = y_0 - \frac{gt^2}{2} \tag{2.1}$$

to relate the height y in feet at time t in seconds to an initial height y_0 from which a ball is dropped at time $t = 0$. Here g, the acceleration due to gravity, has the units "feet per second squared."

A question related to (2.1) is what is the *striking time* $t(y_0)$ for a ball to reach the ground ($y = 0$) from an initial height y_0. Solving (2.1) for this time t produces

$$t(y_0) = \sqrt{\frac{2y_0}{g}}. \tag{2.2}$$

One can test the relation (2.2) experimentally by (a) graphing a selection of (initial height, striking time) pairs, (b) noting that the plotted points have a parabolic shape, and (c) fitting these points to an algebraic curve of the form

$$at^2 = y.$$

Here a turns out to be approximately 16.

A more sophisticated second approach starts with the theoretical assumption that the downward acceleration $\ddot{y}$ of y is a constant $-g$, so that

$$\ddot{y} = -g. \tag{2.3}$$

Integrating this equation gives an equation for the velocity $\dot{y}$:

$$\dot{y} = -gt + c_1, \tag{2.4}$$

where $c_1 = 0$ if t is measured from the instant the ball is dropped. Now (2.1) and (2.4) can be checked experimentally, although measuring $\dot{y}$ requires fairly sophisticated equipment.

A closer look at the behavior of falling bodies discloses that (2.1) is not valid when a feather or parachute is substituted for the ball, since now a resistance force enters the picture, so that (2.3) is replaced by

$$\ddot{y} = -R\dot{y} - g. \tag{2.5}$$

Here we assume that resistance is proportional to the speed $-\dot{y}$ (since the downward motion yields a negative velocity $\dot{y}$).

From (2.5), the standard differential equation solution is

$$y = -\frac{g}{R^2}e^{-Rt} - \frac{g}{R}t + y_0 + \frac{g}{R^2}, \tag{2.6}$$

where the constants of integration have been determined by the initial conditions $y = y_0$ and $\dot{y} = 0$. Then the velocity satisfies

$$\dot{y} = \frac{g}{R}e^{-Rt} - \frac{g}{R}. \tag{2.7}$$

The fact that $\dot{y}$ approaches $-\frac{g}{R}$ as a limit with the passage of time explains why parachute jumps are not fatal!

This sketchy description of falling body models is provided to illustrate how theories develop by interaction with observation and experiment.

Clearly model (2.5) is theoretically superior to model (2.3), but when resistance is negligible, as in Galileo's falling ball experiments, (2.3) gives much more useful results than (2.5).

This illustrates a frequently encountered phenomenon — namely, that simplifying assumptions may lead to important results. The genius of a scientific model builder lies in sensing which simplifications will lead to viable theories. Here we also get some guidance as to the proper role of the mathematician in model building. For example, both (2.3) and (2.5) come from physics, as do the initial conditions and the values of the constants g and R. Then, deriving (2.4), (2.1), (2.6), and (2.7) is a purely mathematical task. Interpretations of the derived formulas again are in the physicist's domain, although these models fall well within the grasp of a mathematician.

In more general quantitative problems, the separate roles of the mathematician and the physicist, economist, biologist, or other specialist in the real world arena being modeled need to be recognized. An applied mathematician may well be sufficiently knowledgeable in some arenas to "go it alone," and vice versa for the real world specialist with a strong mathematics background. However, for many problems an interdisciplinary approach is called for, and building a successful model may require a great deal of interaction between the arena specialist and the mathematician.

One important feature of the mathematics curriculum at Clemson is its recognition of the importance of supplementing the standard mathematics offerings with courses and seminars that help bridge the gap between mathematics and real world problems. These supplements cannot be expected to provide mastery of these arenas, but they can facilitate the intelligent participation of their graduates in the interdisciplinary teams needed to deal with complex quantitative problems in the physical, biological, and social sciences.

2.3   Model Construction and Validation

I now turn to real world model building per se. It is convenient to divide the process into separate phases, bearing in mind that each phase may occur more than once as the construction proceeds.

Problem Identification. This phase entails describing the problem to be solved. Unless a solution would make a noticeable contribution in its arena, model construction has little purpose and the process stops here.

Problem Definition. This phase begins with some (at least tentative) formal statement of the identified problem. I assume that the modeling team either has members with expertise in the arena or at least has ready access to appropriate specialists. The significant variables including inputs and outputs need to be identified and precisely defined. In many situations, the initial choice of variables may turn out to require some augmentation or pruning or both. For example, if an agricultural state has one snake farm, its special inputs and outputs are unlikely to be useful in a statewide model of farm operation.

For a variable to be useful it must be measurable, either directly or indirectly. Choice of variables and methods to measure them fall in the domain of the arena specialists and specialists in data handling.

One of my model building experiences involved a study of the efficiency of farms in the State of Kansas [7]. One member of our team had wide experience both with state-operated agricultural agencies and with "trade" associations such as the Farm Bureau and labor unions. Another member was an agricultural economist with considerable expertise in statistics and a strong mathematics background. Two junior members of the team carried out most of the details of computation and data organization, and I was the team's mathematician. The model we built considered four *inputs*:

1. land (in acres)

2. operating capital (in dollars)

3. labor (in hours)

4. capital investment (in dollars)

We used three *outputs* O1, O2, and O3, each measured in bushels and consisting of one or more types of grain. The types of grain are distinguished by the amounts of water required for growth. Farms fell into three classes: dry land (i.e., with very low rainfall), land with enough rainfall for sorghum and corn, and irrigated land. At this stage for each farm j we must split the first input (land) x_{1j} into three parts $x_{1j}(1)$, $x_{1j}(2)$, and $x_{1j}(3)$, according to the number of acres used in the respective land types: (1) dry land, (2) moderately moist land, and (3) irrigated land. Since each farm has a unique land type, two of the three land inputs for each farm j are zero.

The selection of three (composite) outputs required a variety of decisions as to how to group the many varieties of grain. Some very minor varieties were dropped from the study. In these decisions, our agricultural expert played the central role. He selected 83 farms in the state and provided input-output data for each of these. Thus at this stage we had a data matrix with $9 = 3 + 6$ rows and 83 columns. The nine numbers in each column record the amounts of each output and input used or supplied by the corresponding farm; they constitute an input-output vector whose first three components constitute an output vector and whose six remaining components constitute an input vector.

We are now ready to pose the problem of determining the relative efficiencies of the 83 farms in terms of the ratio of some measures of the output and input vectors.

The Mathematical Model. In the past dozen years, my own research in model building has centered on productive efficiency and in particular on a model called *Data Envelopment Analysis* (known as DEA), which was formally introduced in the seminal 1978 paper of Charnes, Cooper, and Rhodes [1]. This model spawned a variety of additional DEA models and research papers. L. H. Seiford has issued a bibliography with over 800 published papers in the years 1978–1996 [6]. I will base my discussion of model building on DEA models. Additional bibliographic information as well as a classification of DEA models can be found on the website <http://www.warwick.ac.uk/~bsrlu/findex.htm>.

Suppose a model building team is asked by some potential user to study the relative efficiencies of n productive entities, called here *Decision Making Units*: $\mathrm{DMU}_1, \ldots, \mathrm{DMU}_n$. Suppose also that the team and the user, following a careful analysis of the structures and processes of the DMUs, have agreed on a set of s outputs and m inputs, and have measured (for $r = 1, \ldots, s$, $i = 1, \ldots, m$, and $j = 1, \ldots, n$) the amounts y_{rj} of output r produced and x_{ij} of input i used by DMU_j. These numbers are collected in an $(s + m) \times n$ matrix (see [5])

$$P = \begin{bmatrix} Y \\ -X \end{bmatrix} = [p_1 \ldots p_n] = \begin{bmatrix} y_1 & \cdots & y_n \\ -x_1 & \cdots & -x_n \end{bmatrix}, \qquad (2.8)$$

where

$$p_j = \begin{bmatrix} y_j \\ -x_j \end{bmatrix}, \quad y_j = \begin{bmatrix} y_{1j} \\ \vdots \\ y_{sj} \end{bmatrix}, \quad \text{and} \quad x_j = \begin{bmatrix} x_{1j} \\ \vdots \\ x_{mj} \end{bmatrix}. \qquad (2.9)$$

It is assumed that the vectors satisfy

$$y_j \geq 0, \quad x_j \geq 0, \quad y_j \neq 0, \quad x_j \neq 0 \qquad (2.10)$$

for all $j = 1, \ldots, n$. This is based on the concepts (a) that each DMU_j must use a positive amount of at least one input and produce a positive amount of at least one output, and (b) that there are no negative inputs or outputs.

Since this chapter is part of a volume on construction and validation of mathematical models, I will give some special attention to the mathematical scientists' role in model building.

One of these roles is the selection of notation and concepts for organizing data. In the transition from functions of a single variable to relations on a set of variables, the mathematical concepts of and notation for vectors and matrices are developed. See (2.8) and (2.9). The concept of a vector arose in physics, but the description of a vector by a column of numbers and the organizing concept of a matrix as a set of column vectors represent the mathematicians' touch.

A second role is in generalizing the concept of "greater than" from numbers to vectors. If z' and z'' are vectors with ℓ components, we say that z' is greater than z'' if

$$z'_1 > z''_1, \ z'_2 > z''_2, \ \ldots, \ z'_\ell > z''_\ell \tag{2.11}$$

and write

$$z' > z'' \tag{2.12}$$

as a simple notation for (2.11).

In many situations in mathematics as well as generally in the real world, when we wish to optimize (e.g., maximize or minimize) among numbers in some set we find that $>$ is not a satisfactory concept. For example, there is no real number z which is minimal among all numbers x for which $x > 2$. This led to the concept of "greater than or equal to." Similarly, with vectors we replace (2.11) with

$$z'_1 \geq z''_1, \ z'_2 \geq z''_2, \ \ldots, \ z'_\ell \geq z''_\ell \tag{2.13}$$

and (2.12) with (2.14)

$$z' \geq z'', \tag{2.14}$$

read as z' is greater than or equal to z''.

However, whereas for two real numbers a and b we have either $a > b$, $a < b$, or $a = b$, there is no such trichotomy for vectors. However, there is an important intermediate stage

$$z' \geq z'', \ z' \neq z'', \tag{2.15}$$

which means that in addition to (2.13) there is at least one j for which $z'_j > z''_j$. Clearly, (2.15) is more suitable than (2.14) as a criterion for dominance, and if (2.15) holds, we say that z' *dominates* z''. An element z'' in a set is said to be *maximal*, or *undominated*, if (2.15) holds for no z': that is, no vector dominates it.

Related to these considerations is the reason for the minus signs occurring in P and p_j. If we assume that larger outputs and lesser inputs represent improved efficiency of a DMU$_j$ for "dominance" of DMU$_j$ over DMU$_k$, we wish to have both $y_j \geq y_k$ and $x_j \leq x_k$ along with either $y_j \neq y_k$ or $x_j \neq x_k$. The simple device of inserting minus signs in p_j and p_k allows us to use (2.15) as a concise description of dominance among these vectors. This device arose in economics and enables us to use vector dominance even when both increasing some components and decreasing some others are desirable.

I believe that the lack of attention to inequalities in most elementary or intermediate mathematics courses is a weakness that should be addressed by curriculum planners.

We should be aware that the entries y_{rj}, x_{ij} are not purely numbers since they also have dimensions. For example, in the Kansas farm data there are four different input dimensions. If land were measured in hectares rather than in acres and labor in days rather than in hours, all of the x_{1j} and x_{3j} would have to be adjusted (by a cogredient transformation). Thus if, say, x_{3j} is 16 then the adjusted x_{3j} is 2 (assuming an 8-hour day). The important point for now is that summing the components of any x_j or y_j is dimensionally and hence scientifically unsound, a point that has been overlooked by many modelers in (and editors of) their research papers.

Given a data matrix, how do we compare efficiencies of the DMUs? The classic definition of efficiency is output divided by input and this works well when output and input have a common unit of measurement, say, both energy or both dollars. More generally, if every unit of each input and of each output has a well-defined dollar value, we can express the efficiency of DMU$_j$ as

$$\text{Eff(DMU}_j) = \frac{u^T y_j}{v^T x_j}, \qquad (2.16)$$

where u and v are vectors whose components are the dollar values of units of the respective outputs and inputs. But, suppose that one input is gallons of regular gasoline. What value do I use, when on my way to work I see prices ranging from 99.9 to 110.9? A few months ago these prices were much higher. If fuel economy is one factor in choosing a new car that I expect to last for several years, what price do I use for gasoline? How do I price labor hours to account for different wage rates for various classes of workers: e.g., for owners (or managers) as compared with hired hands?

Faced with this difficulty in formulating their Data Envelopment Analysis (DEA) model, the heart of the Charnes, Cooper, and Rhodes (CCR) model was to use the right-hand side of (2.16) in an ingenious way. Consider the n functions

$$h_j(u, v) = \frac{u^T y_j}{v^T x_j}, \quad j = 1, \ldots, n. \tag{2.17}$$

For each choice of nonnegative nonzero multiplier vectors u, v let

$$h_k(u, v) = \max\{h_1(u, v), \ldots, h_n(u, v)\}. \tag{2.18}$$

Then define

$$e_j(u, v) = \frac{h_j(u, v)}{h_k(u, v)} \tag{2.19}$$

as the *efficiency* of DMU$_j$ *relative to* the multipliers u, v. Clearly,

$$0 < e_j(u, v) \leq 1 = e_k(u, v). \tag{2.20}$$

Next, define

$$e_j = \max_{u, v}\{e_j(u, v)\}, \tag{2.21}$$

which we call the *DEA radial efficiency* of DMU$_j$. If $e_j = 1$ we say that DMU$_j$ is *DEA radial efficient*. As the qualifier "radial" suggests, there may be other stronger types of efficiency and we will consider this below.

The formulation of $e_j(u, v)$ and of e_j is a typical mathematical construct. That these quantities always exist and that they are uniquely defined require proofs which are omitted since they are readily available in Charnes, Cooper, and Thrall [5].

Computation of $e_j(u, v)$ is simple but, as formulated in (2.21), computation of e_j involves nonlinear optimization. Fortunately, as was shown in [1], there is an equivalent formulation involving linear programming. Here, I follow the alternative approach introduced in [5].

First, observe that for every u, v and positive numbers α, β we have

$$h_j(\alpha u, \beta v) = \frac{\alpha}{\beta} h_j(u, v)$$

so that

$$e_j(u, v) = e_j(\alpha u, \beta v). \tag{2.22}$$

Early DEA models required

$$u > 0, \quad v > 0, \tag{2.23}$$

but in [5] the authors were able to generalize the applicability of DEA by weakening the admissibility condition (2.23) as follows. A pair u, v

is *admissible* if it satisfies the following conditions:

$$u \geq 0, \;\; v \geq 0, \;\; u \neq 0, \;\; v \neq 0$$
$$u^T y_j > 0 \;\; \text{for at least one } j \tag{2.24}$$
$$\text{for no } j \text{ with } \;\; v^T x_j = 0 \;\; \text{is } u^T v_j > 0.$$

Here the first two conditions of (2.24) imply (2.20). We leave $h_j(u,v)$ undefined if both $v^T x_j = 0$ and $u^T y_j = 0$ hold. Then (2.19), (2.24), and (2.10) guarantee the existence of e_j in (2.21) with the understanding that $e_j(u,v)$ is only defined when $h_j(u,v)$ is, and undefined terms are omitted in maximizations.

Let DMU_0 be any selected DMU_j, and we focus on obtaining a linear program for obtaining its radial efficiency e_0. Using (2.22) we may limit our choice to multiplier pairs u,v for which

$$v^T x_0 = 1, \;\; u^T y_k = v^T x_k, \tag{2.25}$$

so that

$$e_k(u,v) = 1 \text{ and } e_0(u,v) = u^T y_0. \tag{2.26}$$

Then, (2.18) is equivalent to

$$h_j(u,v) \leq 1 \;\; \text{or} \;\; u^T y_j - v^T x_j \leq 0, \;\; j = 1,\ldots,n. \tag{2.27}$$

Now from (2.21), (2.25), (2.24), and (2.27), we get the linear program

$$\text{maximize} \;\; z_0 = u^T y_0$$

subject to

$$v^T x_0 = 1$$
$$u^T y_j - v^T x_j \leq 0, \;\; j = 1,\ldots,n$$
$$u_r \geq 0, \;\; r = 1,\ldots,s$$
$$v_i \geq 0, \;\; i = 1,\ldots,m.$$

This is called the *CCR multiplier program*. By the duality principles for linear programming we also get what is called the *CCR envelopment program*.

$$\text{minimize} \;\; \theta_0$$

subject to

$$\theta_0 \text{ sign free}$$
$$\lambda_j \geq 0, \;\; j = 1,\ldots,n$$
$$\sum_{j=1}^{n} y_{rj}\lambda_j \geq y_{r0}, \;\; r = 1,\ldots,s$$

$$\theta_0 x_{i0} - \sum_{j=1}^{n} x_{ij}\lambda_j \geq 0, \quad i = 1, \dots, m.$$

(Here e_0 is replaced by θ_0.)

These programs can be written more concisely as the primal-dual pair:

Envelopment (primal)

$$\text{minimize} \ \ \theta_0$$

subject to

$$\theta_0 \ \text{sign free}$$

$$P\lambda - p_0(\theta_0) \geq 0$$

$$\lambda \geq 0$$

(2.28)

Multiplier (dual)

$$\text{maximize} \ \ z_0 = u^T y_0$$

subject to

$$v^T x_0 = 1$$

$$w = \begin{bmatrix} u \\ v \end{bmatrix} \geq 0$$

$$w^T P \leq 0$$

where $p_0(\theta_0) = \begin{bmatrix} y_0 \\ -\theta_0 x_0 \end{bmatrix}$.

The passage from the nonlinear system (2.17)–(2.21) to the dual linear programs (2.28) is purely mathematical and involves typical mathematical concepts and constructions. Note that conciseness is a consequence of (or reward for) the minus signs in P and p_j. Note also that the duality is illuminated by comparison of the primal and dual portions of each row. Thus, "sign free" in the primal corresponds to "=" in the dual. These again are mathematical niceties which deepen understanding and appreciation of the role and significance of duality.

2.4   Model Analysis

It is not my purpose to go deeper into details of DEA. However, I will continue to refer to DEA developments to illustrate what happens after a first model is achieved.

Generally, in analyzing a model a researcher looks for strengths and weaknesses as well as for potential and actual applications. Eliminating or ameliorating a weakness usually involves some modification of the model. I will discuss two examples from DEA.

Given a feasible primal pair θ_j, λ^j for DMU$_j$ in (2.28), let

$$S^j = P\lambda^j - p_j(\theta_j).$$

Since θ_j and λ^j are feasible,

$$S^j \geq 0.$$

We call S^j a (primal) *slack vector*. We use asterisks to designate optimal solutions. If $\theta_j^* = 1$, we say that DMU$_j$ is *technically* or *DEA efficient* when $S^{j^*} = 0$ for every optimal radial solution. Suppose that $p_j \geq p_k$ and $p_j \neq p_k$. Clearly, any optimal radial solution λ^{j^*} for DMU$_j$ is also a feasible radial solution for DMU$_k$. From this it can be shown that $S^{k^*} \neq 0$ (unless $p_j = p_k(\theta)$ for some θ).

Thus nonzero (primal) slacks indicate that DMU$_k$ is less efficient than DMU$_j$. Moreover, if, say, the first input slack is positive for some optimal solution for DMU$_k$, then by the well-known complementary slackness property of linear programs, the first input optimal multiplier v_1 must be zero which shows that (2.23) cannot hold for DMU$_k$. See [1, 2, 3, 12] for discussions of slacks.

For a single inequality $a \geq b$, the difference $a-b$ is called its *slack*. The DEA primal slacks correspond to differences between $P\lambda^{j^*}$ and $p_j(\theta_j^*)$. In the multiplier program the dual slacks are defined by

$$T = -w^T P.$$

Although the dual slacks have not played a major role in most of the DEA literature, in more advanced DEA theory questions about them such as how many components of T^* are zero are important.

A natural question arising about DEA models is "how sensitive are the radial efficiencies and slack values to small changes (or errors) in the input output data?" This question was treated by Charnes and various collaborators in a series of papers including [4], which treats changes in a single data vector. Simultaneous changes in all $n(s + m)$ elements of the data matrix P are discussed in [9].

2.5 Some Pitfalls

As a referee for and also in reading technical papers, I have encountered numerous erroneous (or partly erroneous) theorems and proofs as well as shortcomings in notation. A few types of such errors, or pitfalls, are briefly illustrated here.

One frequently occurring class of examples is illustrated by equations such as

$$x_j = \max\{x_j : j = 1, \ldots, n\} \tag{2.29}$$

and

$$\theta x_{ij} = \sum_{j=1}^{n} x_{ij} \lambda_j. \tag{2.30}$$

In both of these examples the symbol j is "quantified out" on the right-hand side by the "$\max_j$" or "$\sum_j$" operation, and thus cannot properly appear also on the left-hand side. Here, both (2.29) and (2.30) can be corrected by replacing the (quantified) j on the left by a new (unquantified) symbol h. Having a mathematical scientist in the research team should help protect against such errors.

A more serious error occurs in dealing with a vector V whose components v_j measure separate undesirable factors, each with its own unit of measurement. The slack vectors S in DEA illustrate this point. To illustrate the general point, consider the two vectors

$$V_1 = \begin{bmatrix} 1 \\ 12 \end{bmatrix}, \quad V_2 = \begin{bmatrix} 2 \\ 2 \end{bmatrix}.$$

Since $1 + 12 > 2 + 2$, it might be tempting to consider that, in some sense, V_1 is longer or more important than V_2. Indeed, many published papers use the "sum of the components" as a measure of size for nonnegative vectors; as justification they confuse it with the well-known L_1 measure, which of course requires that all components have a common unit of measurement. If additional information discloses that the first component is measured in feet and the second in inches, then a weighted L_1 measure $12x_1 + x_2$ reverses the conclusion that V_1 is the larger.

However, the modeler may object that he is thinking of the familiar Euclidean or L_2 measure $\sqrt{x_1^2 + x_2^2}$. We face two separate problems here: (a) what metric to choose, and (b) how to deal with vectors whose individual components have no natural common unit of measurement.

In many operations research contexts, given a nonnegative deficiency vector

$$V = \begin{bmatrix} x_1 \\ \vdots \\ x_n \end{bmatrix},$$

the user's goal is to improve his technology so as to replace V by a "smaller" vector V^*. If he wishes to do this without worsening any component, he requires $V^* \leq V$. Then $Z = V - V^*$ may be regarded as a vector of improvements and the overall improvement can be measured by $G^T Z$, where the positive vector G measures the relative value of unit improvements. That is, for each i and j, the quantity g_i/g_j measures the ratio of the value of a unit change in component i divided by the value of a unit change in component j. The overall value of the improvement Z is proportional to $G^T Z$ in the sense that for improvements Z and Z',

$$G^T Z/G^T Z'$$

measures the ratio of the value of Z to that of Z'.

Clearly, the choice of the vector G represents value judgments which should reflect the goals of the model user and may differ from one user to the next.

The question of valuation of vectors is one that has had much study by scientists, economists, and social scientists and continuing study is warranted.

2.6 Conclusion

These remarks on model building do not pretend to cover the topic, but are offered as viewpoints I have developed from study of published models and from my own attempts at model construction. However important the role of the mathematical scientist may be, he is ordinarily just one member and not the leader of a model development team. It is a challenge to mathematical scientists to play their role effectively and this role can be expected to increase as mathematical model building becomes more complex and comprehensive.

2.7 References

[1] A. Charnes, W. W. Cooper, and E. Rhodes, "Measuring the efficiency of decision making units," *European Journal of Operational Research* **2**, 429–444 (1978).

[2] A. Charnes, W. W. Cooper, and E. Rhodes, "Short communication: measuring the efficiency of decision making units," *European Journal of Operational Research* **3**, 339 (1979).

[3] A. Charnes and W. W. Cooper, "Preface to topics in data envelopment analysis," *Annals of Operations Research* **2**, 59–94 (1985).

[4] A. Charnes, W. W. Cooper, A. Y. Lewin, R. C. Morey, and J. Rousseau, "Sensitivity and stability analysis in DEA," *Annals of Operations Research* **2**, 139–156 (1985).

[5] A. Charnes, W. W. Cooper, and R. M. Thrall, "A structure for classifying and characterizing efficiencies and inefficiencies in data envelopment analysis," *The Journal of Productivity Analysis* **2**, 197–237 (1991).

[6] L. H. Seiford, "A bibliography of data envelopment analysis 1978–1996," *Annals of Operations Research* **73**, 393–438 (1997).

[7] R. G. Thompson, L. N. Langemeier, C.-T. Lee, E. Lee, and R. M. Thrall, "The role of multiplier bounds in efficiency analysis with an application to Kansas farming," *Journal of Econometrics* **46**, 93–108 (1990).

[8] R. G. Thompson and R. M. Thrall, "The need for OR/MS in public policy making," *Applications of Management Science*, Special Issue on Public Policy Usage, Volume 7, E. Rhodes (Ed.), 3–21 (1993).

[9] R. G. Thompson, P. S. Dharmapala, and R. M. Thrall, "DEA sensitivity analysis of efficiency measures, with applications to Kansas farming and Illinois coal mining," in *Data Envelopment Analysis: Theory, Methodology and Applications*, A. Charnes, W. W. Cooper, L. Seiford, A. Lewin (Eds.), Kluwer, Boston, 1994, 393–422.

[10] R. G. Thompson, P. S. Dharmapala, J. Diaz, M. G. Gonzalez-Lima, and R. M. Thrall, "DEA multiplier analytic sensitivity with an illustrative application to independent oil companies," *Annals of Operations Research* **66**, 163–177 (1996).

[11] R. M. Thrall, "Principles of model development with applications to slacks in DEA," in *Operations Research: Methods, Models, and Applications*, J. E. Aronson and S. Zionts (Eds.), Quorum Books, Westport, CN, 1998, 49–56.

[12] R. M. Thrall, "Duality, classification and slacks in DEA," *Annals of Operations Research* **66**, 109–138 (1996).

Chapter 3

Understanding the United States AIDS Epidemic: A Modeler's Odyssey

James R. Thompson

3.1 Introduction

This chapter presents a time-indexed investigation of the author's fifteen years of work in studying the AIDS epidemic and attempting to come up with strategies for its defacilitation. This saga of modeling the AIDS epidemic also demonstrates how one poses and solves a problem in a continually changing data environment. Early on in this odyssey (around 1983), a differential equation model was developed based on currently available data to see what might have facilitated the AIDS endemic becoming an epidemic. This model indicated that a "core" high activity subpopulation within the gay community could be the cause — not as a result of increasing total aggregate contacts, but by virtue of their very high contact rate. Such skewed contact rates have the effect of keeping aggregate contacts constant, yet driving the epidemic the same way as if aggregate contacts had more than doubled.

Time passed, and when such results were presented, they were implicitly accepted by professional audiences. However, attempts to instigate control measures in the city of Houston met with political resistance and lack of action. This points out another important (and sobering) lesson of practical modeling: that a convincing mathematical model need not lead to policy change and action.

In the 1990s, joint work with a graduate student produced a much more complicated model for dealing with the mature epidemic. This model suggested that there was little to be gained in shutting down high contact establishments once the proportion of infectives reached 40%.

In 1998, a really fine data set was obtained from the World Health Organization (WHO). These data showed that there were modest decreases in the incidence of new AIDS cases throughout the First World. A kinetic model based on the data gave the amazing result that the piecewise exponential growth rate was the same for all American and European First World countries. This occurred, in spite of the fact that incidences per 100,000 varied greatly from country to country (by a factor of ten). This led to another model, giving the strong indication that it is cheap travel to and from the U.S.A. which drives the epidemic in Canada and Europe.

3.2   Prelude: The Postwar Polio Epidemic

Effective immunizations against many of the killing diseases of the 19th century, plus antibiotics massively utilized during World War II, gave the promise of the end to life-threatening contagion in the United States. The killers of the future would be those largely associated with the aging process, such as cancer, stroke, and heart attack.

However, in the postwar years, polio, which already had stricken some (including President Roosevelt), became a highly visible scourge in a number of American cities, particularly in the South, particularly among the young. In 1952, over 55,000 cases were reported. Mortality rates in America, due to good care, had by that time dropped to well under 10%. Nonetheless, the spectacle of children confined to wheelchairs or iron lungs was a disturbing one.

This was in the years before the emigration of the middle classes to suburbia and most schools tended to have representation from a wide range of socioeconomic groups. Incidence rates were the highest in the summers, when the schools were closed. But, at the intuitive level, it was clear that polio was a disease predominantly of school age children, and that there was a fair amount of clustering of cases. Although the causative agent had not been isolated, there was little doubt that it was a virus, that it favored young hosts, that the throat was the likely pathway, and that transmission was greatest in the hot weather.

In such a situation, it might appear that a prudent public health policy would be to discourage summer gatherings of children, particularly in confined indoor settings or in swimming pools. Such an inference might well be put down as a prejudice of causation where none existed.

Indeed, this was the era of the kiddie matinee and new municipal swimming facilities given by city governments to their citizens in celebration of a perceived affluence following the War. Some parents did, to the displeasure of their children, attempt to deprive them of matinees and swimming excursions; but such were in the distinct minority. From time to time, city officials would take such steps as shutting down municipal swimming pools, but this was unusual and always temporary. There was a large economic constituency for matinees and swimming pools. The constituency for shutting them down was acting on intuition and without business support. The results were that the movies and pools generally stayed open all summer. The epidemic flourished.

There was a great deal of expectation that "the cavalry will soon ride to the rescue" in the form of an expected vaccine against the disease. In 1955, the Salk vaccine[1] did appear, and new polio cases, for the United States, became a thing of the past. Of course, a residual population of tens of thousands of Americans remained, crippled by polio.

There was very little in the way of a postmortem examination about how effective public health policy had been in managing the American polio epidemic. In fact, there had been essentially no proactive policy at all. But two effective anti-polio vaccines (Salk and then Sabin) seemed to have brought everything right in the end. If there were serious efforts to learn from the mistakes in management of the American polio epidemic, this author has not seen them.

Polio had, apparently, been simply a bump in the road toward a time in which life-threatening contagious diseases in America would be a thing of the past. However, having spent my childhood in Memphis, Tennessee (one of the epicenters of the postwar polio epidemic), that epidemic was something I would never forget. My parents were among the number of those who forbade matinees and swimming pools to their children. But among my childhood friends there were several who died from polio, and many others crippled by it.

3.3 AIDS: A New Epidemic for America

In 1983, I was investigating the common practice of using stochastic models in dealing with various aspects of diseases. When attempting to model the progression of cancer within an individual, a good case could

---

[1] In 1999, evidence started to appear that contamination of the Salk vaccine by a monkey virus, not unrelated to HIV, was causing many of the recipients of the Salk vaccine to develop a variety of cancers, possibly due to a destruction of parts of their immune system.

be made for going stochastic. For example, one matter of concern with solid tumors is whether the primary tumor throws off a metastasis before it has been surgically removed. Whether it has or has not will largely determine whether surgical removal of the primary tumor has cured the patient. Such a phenomenon needs to be modeled stochastically.

On the other hand, when modeling the progression of a contagious disease through a population, the common current practice of using a stochastic model and then finding, for example, the moment generating function of the number $Y(t)$ of infectives seems unnecessarily complicated, particularly if, at the end of the day, one decides simply to extract $E(Y(t))$, the expected number of infectives. Moreover, any sociological data, if available, are likely to be in terms of aggregate information, such as the average number of contacts per day.

I had decided to write a paper giving examples where deterministic modeling would probably be appropriate. I selected the AIDS epidemic because it was current news, with a few hundred cases reported nationally. Although reporting at the time tended to downplay the seriousness of the epidemic (and, of course, the name was pointedly innocuous, the same as an appetite suppressant of the times), there was a palpable undercurrent of horror in the medical community. It looked like a study that might be important.

Even at the very early stage of an observed United States AIDS epidemic, several matters appeared clear to me:

- The disease favored the homosexual male community and outbreaks seemed most noticeable in areas with sociologically identifiable gay communities.

- The disease was also killing (generally rather quickly) people with acute hemophilia.

- Given the virologist's maxim that there are no new diseases, AIDS, in the U.S.A., had been identified starting around 1980 because of some sociological change. A disease endemic under earlier norms, it had blossomed into an epidemic due to a change in society.

At the time, which was before the HIV virus had been isolated and identified, there was a great deal of commentary both in the popular press and in the medical literature (including that of the Centers for Disease Control) to the effect that AIDS was a new disease. Those statements were not only putatively false, but were also potentially harmful. First of all, from a practical virological standpoint, a new disease might have as a practical implication genetic engineering by a hostile foreign power. This was a time of high tension in the Cold War, and such an allegation had the potential for causing serious ramifications at the level of national defense.

Second, treating an unknown disease as a new disease essentially removes the possibility of stopping the epidemic sociologically by simply seeking out and removing (or lessening) the cause(s) that resulted in the endemic being driven over the epidemiological threshold.

For example, if somehow a disease (say, the Lunar Pox) has been introduced from the moon via the return of moon rocks by American astronauts, that is an entirely different matter than, say, a mysterious outbreak of dysentery in St. Louis. For dysentery in St. Louis, we check food and water supplies, and quickly look for "the usual suspects" — unrefrigerated meat, leakage of toxins into the water supply, etc. Given proper resources, eliminating the epidemic should be straightforward.

For the Lunar Pox, there are no usual suspects. We cannot, by reverting to some sociological *status quo ante*, solve our problem. We can only look for a bacterium or virus and try for a cure or vaccine. The age-old way of eliminating an epidemic by sociological means is difficult — perhaps impossible.

In 1982, it was already clear that the United States public health establishment was essentially treating AIDS as though it were the Lunar Pox. The epidemic was at levels hardly worthy of the name in Western Europe, but it was growing. Each of the European countries was following classical sociological protocols for dealing with a venereal disease. These all involved some measure of defacilitating contacts between infectives and susceptibles. The French demanded bright lighting in gay "make-out" areas. Periodic arrests of transvestite prostitutes on the Bois de Bologne were widely publicized. The Swedes took much more draconian steps — mild in comparison with those of the Cubans. The Americans took no significant sociological steps at all.

However, as though following the Lunar Pox strategy, the Americans outdid the rest of the world in money thrown at research related to AIDS. Some of this was spent on isolating the unknown virus. However, it was the French, spending pennies to the Americans' dollars, at the Pasteur Institute (financed largely by a legacy from the late Duke and Duchess of Windsor) who first isolated HIV. In the intervening fifteen years since isolation of the virus, no effective vaccine or cure has been produced.

3.4   Why An AIDS Epidemic in America?

Although the popular press in the early 1980s talked of AIDS as being a new disease, as noted above, prudence and experience indicated that it was not. Just as new species of animals have not been noted during human history, the odds for a sudden appearance (absent genetic engineering) of a new virus are not good. My own discussions with pathologists with some years of experience gave anecdotal cases of young Anglo males who had presented with Kaposi's sarcoma at times going back to early days in the pathologists' careers. This pathology, previously seldom seen in persons of Northern European extraction, now widely associated with AIDS, was at the time simply noted as isolated and unexplained. Indeed, a few years after the discovery of the HIV virus, HIV was discovered in decades old refrigerated human blood samples from both Africa and America.

Although it was clear that AIDS was not a new disease, as an epidemic it had never been recorded. Because some of the early cases were from the Congo, there was an assumption by many that the disease might have its origins there. Clearly, record keeping in the Congo was not and is not very good. But Belgian colonial troops had been located in that region for many years. Any venereal disease acquired in the Congo should have been vectored into Europe in the 19th century. But no AIDS-like disease had been noted. It would appear, then, that AIDS was not contracted easily as is the case, say, with syphilis. Somehow, the appearance of AIDS as an epidemic in the 1980s, and not previously, might be connected with higher rates of promiscuous sexual activity made possible by the relative affluence of the times.

Then there was the matter of the selective appearance of AIDS in the American homosexual community. If the disease required virus in some quantity for effective transmission (the swift progression of the disease in hemophiliacs plus the lack of notice of AIDS in earlier times gave clues that such might be the case), then the profiles in Figures 3.1 and 3.2 give some idea why the epidemic seemed to be centered in the American homosexual community. If passive to active transmission is much less likely than active to passive, then clearly the homosexual transmission patterns facilitate the disease more than the heterosexual ones.

One important consideration that seemed to have escaped attention was the appearance of the epidemic in 1980 instead of ten years earlier. Gay lifestyles had begun to be tolerated by law enforcement authorities in the major urban centers of America by the late 1960s. If homosexuality was the facilitating behavior of the epidemic, then why was there no epidemic before 1980? Of course, believers in the "new disease" theory could simply claim that the causative agent was not present until around 1980. In the popular history of the early American AIDS epidemic, *And the Band Played On*, Randy Shilts points at a gay flight attendant from Quebec as a candidate for "patient zero." But this "Lunar Pox" theory was not a position that any responsible epidemiologist could take (and, indeed, as pointed out earlier, later investigations revealed HIV samples in human blood going back into the 1940s).

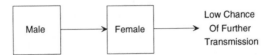

FIGURE 3.1   **Heterosexual transmission of AIDS**
(From J. R. Thompson, *Empirical Model Building*, Wiley, New York, 1989. Reprinted with permission of John Wiley & Sons, Inc.)

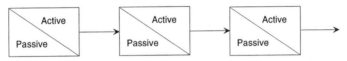

FIGURE 3.2   **Homosexual transmission of AIDS**
(From J. R. Thompson, *Empirical Model Building*, Wiley, New York, 1989. Reprinted with permission of John Wiley & Sons, Inc.)

What accounts for the significant time differential between civil tolerance of homosexual behavior prior to 1970 and the appearance of the AIDS epidemic in the 1980s? Were there some other sociological changes that had taken place in the late 1970s that might have driven the endemic over the epidemiological threshold?

It should be noted that in 1983 data were skimpy and incomplete. As is frequently the case with epidemics, decisions need to be made at the early stages when one needs to work on the basis of skimpy data, analogy with other historical epidemics, and a model constructed on the best information available.

I remember in 1983 thinking back to the earlier American polio epidemic that had produced little in the way of sociological intervention and less in the way of models to explain the progress of the disease. Although polio epidemics had been noted for some years (the first noticed epidemic occurred around the time of World War I in Stockholm), the American public health service had indeed treated it like the "Lunar Pox." That is, they discarded sociological intervention based on past experience of transmission pathways and relied on the appearance of vaccines at any moment. They had been somewhat lucky, since Salk started testing his vaccine in 1952 (certainly they were luckier than the thousands who had died and the tens of thousands who had been permanently crippled). But basing policy on hope and virological research was a dangerous policy (how dangerous we are still learning as we face the prospect of one million American dead from AIDS).

Although some evangelical clergymen inveighed against the epidemic as divine retribution on homosexuals, the function of epidemiologists is to use their God-given wits to stop epidemics. In 1983, virtually nothing was being done except to wait for virological miracles.

One possible candidate was the turning of a blind eye by authorities to the gay bathhouses that started in the late 1970s. These were places where gays could engage in high frequency anonymous sexual contact. By the late 1970s they were allowed to operate without regulation in the major metropolitan centers of America. My initial intuition was that the key was the total average contact rate among the target population. Was the marginal increase in the contact rate facilitated by the bathhouses sufficient to drive the endemic across the epidemiological threshold? It did not seem likely. Reports were that most gays seldom (many, never) frequented the bathhouses.

But perhaps my intuitions were wrong. Perhaps it was not only the total average contact rate that was important, but a skewing of contact rates, with the presence of a high activity subpopulation (the bathhouse customers) somehow driving the epidemic. It was worth a modeling try.

The model developed in [2] considered the situation in which there are two subpopulations: the majority, less sexually active, and a minority with greater activity than that of the majority. We use the subscript "1" to denote the majority portion of the target (gay) population, and the subscript "2" to denote the minority portion. The latter subpopulation, constituting fraction p of the target population, will be taken to have a contact rate τ times the rate k of the majority subpopulation. The following differential equations model the growth of the number of susceptibles X_i and infectives Y_i in subpopulation i $(i = 1, 2)$.

$$\frac{dY_1}{dt} = \frac{k\alpha X_1(Y_1 + \tau Y_2)}{X_1 + Y_1 + \tau(Y_2 + X_2)} - (\gamma + \mu)Y_1$$

$$\frac{dY_2}{dt} = \frac{k\alpha\tau X_2(Y_1 + \tau Y_2)}{X_1 + Y_1 + \tau(Y_2 + X_2)} - (\gamma + \mu)Y_2 \qquad (3.1)$$

$$\frac{dX_1}{dt} = -\frac{k\alpha X_1(Y_1 + \tau Y_2)}{X_1 + Y_1 + \tau(Y_2 + X_2)} + (1 - p)\lambda - \mu X_1$$

$$\frac{dX_2}{dt} = -\frac{k\alpha\tau X_2(Y_1 + \tau Y_2)}{X_1 + Y_1 + \tau(Y_2 + X_2)} + p\lambda - \mu X_2,$$

where

k = number of contacts per month,
α = probability of contact causing AIDS,
λ = immigration rate into the susceptible population,
μ = emigration rate from the population,
γ = marginal emigration rate from the infective population
due to sickness and death.

In [2], it was noted that if we started with 1,000 infectives in a target population with $k\alpha = 0.05$, $\tau = 1$, a susceptible population of 3,000,000 and the best guesses then available ($\mu = 1/(15 \times 12) = 0.00556$, $\gamma = 0.1$, $\lambda = 16{,}666$) for the other parameters, the disease advanced as shown in Table 3.1.

Table 3.1 Extrapolated AIDS cases:
$k\alpha = 0.05$, $\tau = 1$

Year	Cumulative Deaths	Fraction Infective
1	1,751	0.00034
2	2,650	0.00018
3	3,112	0.00009
4	3,349	0.00005
5	3,571	0.00002
10	3,594	0.000001

(From J. R. Thompson, *Empirical Model Building*, Wiley, New York, 1989. Reprinted with permission of John Wiley & Sons, Inc.)

Next, a situation was considered in which the overall contact rate was the same as in Table 3.1, but it was skewed with the more sexually active subpopulation 2 (of size 10%) having contact rates 16 times those of the less active population. Even though the overall average contact rate in Table 3.1 and Table 3.2 is the same $(k\alpha)_{\text{overall}} = 0.05$, the situation is dramatically different in the two cases. Here, it seemed, was a *prima facie* explanation as to how AIDS was pushed over the threshold to a full blown epidemic in the United States: a small but sexually very active subpopulation.

I note that nothing more sophisticated than some numerical quadrature was required to obtain the results in these tables. In the ensuing arguments concerning why AIDS became an epidemic in the United States, everything beyond the rather simple deterministic model (3.1) will be, essentially, frosting on the cake. This was the way things stood in 1984 when I presented the paper at the summer meetings of the Society for

Table 3.2   Extrapolated AIDS cases:
$k\alpha = 0.02$, $\tau = 16$, $p = 0.10$

Year	Cumulative Deaths	Fraction Infective
1	2,184	0.0007
2	6,536	0.0020
3	20,583	0.0067
4	64,157	0.0197
5	170,030	0.0421
10	855,839	0.0229
15	1,056,571	0.0122
20	1,269,362	0.0182

(From J. R. Thompson, *Empirical Model Building*, Wiley, New York, 1989. Reprinted with permission of John Wiley & Sons, Inc.)

Computer Simulation in Vancouver. It hardly created a stir among the mainly pharmacokinetic audience who attended the talk. And, frankly, at the time I did not think too much about it because I supposed that probably even as the paper was being written, the "powers that be" were shutting down the bathhouses. The deaths at the time were numbered in the hundreds, and I did not suppose that things would be allowed to proceed much longer without sociological intervention. Unfortunately, I was mistaken.

In November 1986, the First International Conference on Population Dynamics took place at the University of Mississippi where there were some of the best biomathematical modelers from Europe and the United States. I presented my AIDS results [4], somewhat updated, at a plenary session. By this time, I was already alarmed by the progress of the disease (over 40,000 cases diagnosed and the bathhouses still open). The bottom line of the talk had become more shrill: namely, every month delayed in shutting down the bathhouses in the United States would result in thousands of deaths. The reaction of the audience this time was concern, partly because the prognosis seemed rather chilling, partly because the argument was simple to follow and seemed to lack holes, and partly because it was clear that something was pretty much the matter if things had gone so far off track.

After the talk, the well-known Polish probabilist Robert Bartoszyński, with whom I had carried out a lengthy modeling investigation of breast cancer and melanoma, took me aside and asked whether I did not feel unsafe making such claims. "Who," I asked, "will these claims make unhappy?" "The homosexuals," said Bartoszyński. "No, Robert," I said, "I am trying to save their lives. It will be the public health establishment who will be offended."

And so it has been in the intervening years. I have given AIDS talks before audiences with significant gay attendance in San Francisco, Houston, and other locales without any gay person expressing offense. Indeed, in his 1997 book [1], Gabriel Rotello, one of the leaders of the American gay community, not only acknowledges the validity of my model but also constructs a survival plan for gay society in which the bathhouses have no place.

3.5   A More Detailed Look at the Model

A threshold investigation of the two-activity population model (3.1) is appropriate here. Even today, let alone in the mid-1980s, there was no chance that one would have reliable estimates for all the parameters k, α, γ, μ, λ, p, τ. Happily, one of the techniques sometimes available to the modeler is the opportunity to express the problem in such a form that most of the parameters will cancel. For the present case, we will attempt to determine the $k\alpha$ value necessary to sustain the epidemic when the number of infectives is very small. For this epidemic in its early stages one can manage to get a picture of the bathhouse effect using only a few parameters: namely, the proportion p of the target population which is sexually very active and the activity multiplier τ.

For $Y_1 = Y_2 = 0$, the equilibrium values for X_1 and X_2 are $(1-p)(\lambda/\mu)$ and $p(\lambda/\mu)$, respectively. Expanding the right-hand sides of (3.1) in a Maclaurin series, we have (using lower case symbols for the perturbations from 0)

$$\frac{dy_1}{dt} = \left[\frac{k\alpha(1-p)}{1-p+\tau p} - (\gamma+\mu)\right] y_1 + \frac{k\alpha(1-p)\tau}{1-p+\tau p} y_2$$

$$\frac{dy_2}{dt} = \frac{k\alpha\tau p}{1-p+\tau p} y_1 + \left[\frac{k\alpha\tau^2 p}{1-p+\tau p} - (\gamma+\mu)\right] y_2.$$

Summing then gives

$$\frac{dy_1}{dt} + \frac{dy_2}{dt} = [k\alpha - (\gamma+\mu)] y_1 + [k\alpha\tau - (\gamma+\mu)] y_2. \qquad (3.2)$$

In the early stages of the epidemic,

$$\frac{dy_1/dt}{dy_2/dt} = \frac{(1-p)}{p\tau}.$$

That is to say, the new infectives will be generated proportionately to their relative numerosity in the initial susceptible pool times their relative activity levels. So, assuming a neglible number of initial infectives, we have

$$y_1 = \frac{(1-p)}{p\tau} y_2.$$

Substituting in (3.2) shows that for the epidemic to be sustained, we must have

$$k\alpha > \frac{(1-p+\tau p)}{1-p+\tau^2 p}(\gamma+\mu). \tag{3.3}$$

Accordingly we define the *heterogenous threshold* via

$$k_{\text{het}}\alpha = \frac{(1-p+\tau p)}{1-p+\tau^2 p}(\gamma+\mu).$$

Now, in the homogeneous contact case (i.e., $\tau = 1$), we note that for the epidemic to be sustained, the following condition must hold:

$$k\alpha > \gamma + \mu.$$

Accordingly we define the *homogeneous threshold* by

$$k_{\text{hom}}\alpha = \gamma + \mu.$$

The heterogeneous contact case with k_{het} has the average contact rate

$$k_{\text{ave}}\alpha = p\tau(k_{\text{het}}\alpha) + (1-p)(k_{\text{het}}\alpha) = \frac{(1-p+\tau p)^2}{1-p+\tau^2 p}(\gamma+\mu).$$

Dividing the sustaining value $k_{\text{hom}}\alpha$ by the sustaining value $k_{\text{ave}}\alpha$ for the heterogeneous contact case then produces

$$Q = \frac{1-p+\tau^2 p}{(1-p+\tau p)^2}.$$

Notice that we have been able here to reduce the parameters necessary for consideration from seven to two. This is fairly typical for model-based approaches: the dimensionality of the parameter space may be reducible in answering specific questions. Figure 3.3 shows a plot of this "enhancement factor" Q as a function of τ. Note that the addition of heterogeneity to the transmission picture has roughly the same effect as if all members of the target population had more than doubled their contact rate. Remember that the picture has been corrected to discount any increase in the overall contact rate which occurred as a result of adding heterogeneity. In other words, the enhancement factor is totally due to heterogeneity. It is this heterogeneity effect which I have maintained (since 1984) to be the cause of AIDS getting over the threshold of sustainability in the United States.

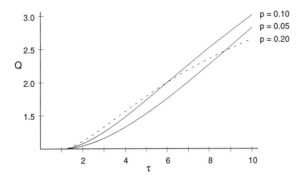

FIGURE 3.3   **Effect of a high activity subpopulation**

If this all still seems counterintuitive, then let us consider the following argument at the level of one individual infective. Suppose, first of all, that the disease is such that one contact changes a susceptible to an infective. Then let us suppose we have an infective who is going to engage in five contacts. What number of susceptibles (assuming equal mixing) will give the highest expected number of conversions of susceptibles to infectives? Note that if the number of susceptibles is small, the expectation will be lessened by the "overkill effect": i.e., there is the danger that some of the contacts will be "wasted" by being applied to an individual already infected by one of the other five contacts. Clearly, here the optimal value for the size N of the susceptible pool is infinity, for then the expected number of conversions from susceptible to infective $E(\mathcal{I} \mid N = \infty)$ is five.

Now let us change the situation to one in which two contacts, rather than one, are required to change a susceptible to an infective. We will still assume a total of five contacts. Clearly, if $N = 1$ then the expected number of conversions is $E = 1$; there has been wastage due to overkill. Next, let us assume the number of susceptibles has grown to $N = 2$. Then the probability of two new infectives is given by

$$P(2 \mid N = 2) = \sum_{j=2}^{3} \binom{5}{j} \left(\frac{1}{2}\right)^5 = \frac{20}{32}.$$

The probability of only one new infective is $1 - P(2 \mid N = 2)$. Thus the expected number of new infectives is

$$E(\mathcal{I} \mid N = 2) = 2 \left(\frac{20}{32}\right) + 1 \left(\frac{12}{32}\right) = 1.625.$$

Now when there are $N = 3$ susceptibles, the contact configurations leading to two new infectives are of the type $(2, 2, 1)$ and $(3, 2, 0)$. All other configurations will produce only one new infective. So the probability of two new infectives is given by

$$P(2 \mid N = 3) = \binom{3}{1} \frac{5!}{2!2!1!} \left(\frac{1}{3}\right)^5 + \binom{3}{1}\binom{2}{1} \frac{5!}{3!2!} \left(\frac{1}{3}\right)^5 = \frac{150}{243}$$

and the expected number of new infectives is

$$E(\mathcal{I} \mid N = 3) = 2 \left(\frac{150}{243}\right) + 1 \left(\frac{93}{243}\right) = 1.617 \ .$$

Further calculations give $E(\mathcal{I} \mid N = 4) = 1.469$, $E(\mathcal{I} \mid N = 5) = 1.314$. For very large N, $E(\mathcal{I})$ is of order $1/N$. Apparently, for the situation where there are a total of five contacts, the value of the number in the susceptible pool that maximizes the total number of new infectives from the one original infective is $N = 2$, not ∞. Obviously, we are oversimplifying, since we stop after only the contacts of the original infective. The situation is much more complicated here, since an epidemic is created by the new infectives infecting others and so on. As well, there is the matter of a distribution of the number of contacts required to give the disease. We have in our main model (3.1) avoided the complexities of branching process modeling by going deterministic. The argument above is given to present an intuitive feel as to the facilitating potential of a high contact core in driving a disease over the threshold of sustainability.

In the case of AIDS, the average number of contacts required to break down the immune system sufficiently to cause the person ultimately to get AIDS is much larger than two. The obvious implication is that a great facilitator for the epidemic being sustained is the presence of a subpopulation of susceptibles whose members have many contacts. In the simple example above, we note that even if the total number of contacts were precisely five, from a standpoint of facilitating the epidemic, it would be best to concentrate the contacts into a small pool of susceptibles. In other words, if the total number of contacts is fixed at some level, it is best to start the epidemic by concentrating the contacts within a small subpopulation. Perhaps the analogy to starting a fire, not by dropping a match onto a pile of logs, but rather onto some kindling beneath the logs, is helpful.

3.6   Forays into the Public Policy Arena

The senior Professor of Pathology at the Baylor College of Medicine in the 1980s was Raymond McBride. McBride had been one of the pioneers in immunosuppression for organ transplantation and was the Chief of Pathology Services for the Harris County (Houston) Medical District.

Distressed to see the ravages of AIDS on autopsied victims, he was quite keen to have municipal authorities act to close down the bathhouses. He and I co-authored a front page op-ed piece for the *Houston Chronicle* entitled "Close Houston's Gay Bathhouses" [7], taking care not to mention the names and addresses of the two major offending establishments lest some vigilante act be taken against them. Hardly a ripple of interest, even though Houston, with less than one-tenth the population of Canada, had more AIDS cases than that entire country. We tried to motivate members of the City Council. When interviewed by a reporter, the office of the Councilman in whose district these two bathhouses were situated shrugged the whole matter off by asking, "What's a bathhouse?" I served on the American Statistical Association's Ad Hoc Committee on AIDS from its inception until its demise, but our mandate was never allowed to extend to modeling. Only the methodology of data analysis was permitted. Nor were we allowed, as a committee, to compare America's AIDS incidence with that from other countries.

The situation was not unlike that of the earlier polio epidemic. There were specific interests for not addressing the bathhouse issue, but there was only a nonspecific general interest for addressing it.

Although I had no experience with the blood-testing issue, it should be noted that early on in the epidemic, long before the discovery of HIV, it was known that over 90% of the persons with AIDS tested positive to antibodies against Hepatitis-B. For many months, the major blood collecting agencies in the United States resisted employing the surrogate Hepatitis test for contaminated blood. The result was rampant death amongst hemophiliacs and significant AIDS infections among persons requiring large amounts of blood products for surgery.

The statistician/economist/sociologist Vilfredo Pareto remarked that Aristotle had made one mistake when he presented to the world the system of logical thinking. The mistake was Aristotle's assumption that once humankind understood logical consistency, actions, including public policy, would be made on the basis of reason. Pareto noted that the historical record showed otherwise. The more important the decision, Pareto noted, the less likely was logical inference based on facts — a significant concern in decision making. So, it has unfortunately been with policy concerning AIDS.

3.7   Modeling the Mature Epidemic

In the United States, the AIDS epidemic crossed the threshold of viability long ago. Consequently, we should investigate the dynamics of the mature epidemic. Unfortunately, we then lose the ability to disregard five of the seven parameters and must content ourselves with picking reasonable values for those parameters. A detailed analysis is given in [6]. In the following, we will make certain ballpark assumptions about some of the underlying parameters. Suppose the contact rate before the possible bathhouse closings is given by

$$(k\alpha)_{\text{overall}} = (1 - p + \tau p)(\gamma + \mu). \tag{3.4}$$

This represents an average contact rate for the two-activity model. We shall take $\mu = 1/(180 \text{ months})$ and $\lambda = 16{,}666$ per month. (We are assuming a target population, absent the epidemic, of roughly 3,000,000.) For a given fraction π of infectives in the target population, we ask what is the ratio of contact rates causing elimination of the epidemic for the closings case divided by that without closings.

Figure 3.4 shows the ratio of contact rates (with closings relative to without closings) as a function of π for $p = 0.1$ and $\gamma = \frac{1}{60}$. It appears that as long as the proportion of infectives π is at most 40% of the target population, there would be a significant benefit from bathhouse closings; the benefit decreases once we get to 40%. However, because there appears to be a continuing influx of new entrants into the susceptible pool, there is good reason to close these establishments. Generally, restoring the sociological *status quo ante* is an effective means of stopping an epidemic; often this is difficult to achieve. Closing the bathhouses continues to be an appropriate action, even though a less effective one than if it had been taken early on in the history of the epidemic.

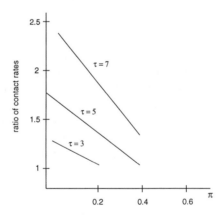

FIGURE 3.4   **Effect of bathhouse closings in a mature epidemic**

Next, we look at the possible effects on the AIDS epidemic of administering a drug, such as AZT, to the entire infective population. Obviously, infectives who die shortly after contracting a contagious disease represent less of an enhancement to the viability of an epidemic than those who live a long time in the infective state. In the case of AIDS, it is probably unreasonable to assume that those who, by the use of medication, increase their T cell count to an extent where apparently normal health has returned, will decide to assume a chaste life style for the rest of their lives. We shall assume that the drug increases life expectancy by two years. Figure 3.5 demonstrates the change in the percent infective if the drug also increases the period of infectivity by two years for various proportions π of infective at the time that the drug is administered. The curves plot the ratio of the proportion infective using AZT to the proportion infective if AZT is not used (with $\gamma = \frac{1}{60}$) and they asymptote to $1.4 = 84/60$, as should be the case. The greater pool of infectives in the target population can, under certain circumstances, create a kind of "Typhoid Mary" effect, where long-lived infectives wander around spreading the disease. Clearly, it should be the policy of health care professionals to help extend the time of quality life for each patient treated. However, it is hardly responsible to fail to realize that, by so doing, in the case of AIDS, there is an obligation of the treated infective to take steps to ensure that he does not transmit the disease to susceptibles. To the extent that this is not the case, the highly laudable use of AZT to improve the length and quality of life for AIDS victims is probably increasing the number of deaths from AIDS.

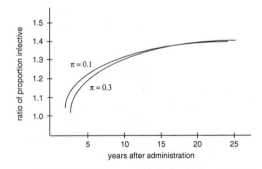

FIGURE 3.5   **AZT effect on sustaining an AIDS epidemic**

3.8   AIDS as a Facilitator of Other Epidemics

In 1994 Webster West [9] completed a doctoral dissertation attempting to see to what extent AIDS could enhance the spread of tuberculosis in America. Since we are primarily concerned here with the spread of

AIDS itself, we shall not dwell very long on the tuberculosis adjuvancy issue. The reader is referred to relevant papers elsewhere [10, 11].

West did discover that if one used stochastic process models and then took the mean trace, one obtained the same results as those obtained simply by using deterministic differential equation models. In the United States, since the Second World War at least, tuberculosis has been a cause of death mainly of the elderly (for example, Mrs. Eleanor Roosevelt died of it). Tuberculosis is carried by the air, and its epidemiological progression is enhanced by infected persons who are well enough to walk around in elevators, offices, etc. When tuberculosis is confined to elderly persons, essentially not moving freely about, it is largely self-contained. But HIV infected persons are generally young and generally employed, at least before the latter stages of full blown AIDS.

West discovered that the result of AIDS facilitating tuberculosis was likely to be only a few hundred additional deaths per year. His model further revealed that modest resources expended in the early treatment of persons infected with tuberculosis could bring even these relatively modest numbers down.

3.9   Comparisons with First World Countries

As noted in Section 3.3, the position of other developed countries toward defacilitating contacts between infectives and susceptibles was quite different from that in the United States. In a very real sense, these other countries can be used as a "control" when examining the epidemic in the United States. Good data for new cases did not become easier and easier to obtain as the epidemic progressed. Whereas in the earlier time span of the epidemic fairly good data for all First World countries could be obtained via "gopher" sites, increasingly it became more and more disconnected as data bases supposedly moved to the Internet. The reality was that the information on the gopher sites stayed in place but was not brought up to date, while data on the Internet appeared temporally disconnected. Great patience was required to follow a group of countries over a period of time and, because of holes in the data, it was not at all clear whether anything but snippet comparisons could be made. I published one of these at a conference in 1995 [5], but the data available to me at the time gave only suggestions of what was happening. There seemed to be something important going on that went to the issue of the United States being a source of infection for other First World countries.

I kept sending out queries to the Centers for Disease Control and the World Health Organization (WHO), but without much success. Finally,

in early 1998, Ms. Rachel Mackenzie of the WHO contacted me and
provided me, not with a URL, but with the data itself, which was in the
hands of the Working Group on Global HIV/AIDS, and STD Surveillance which is a joint Working Group between WHO and UNAIDS. I
wish to acknowledge my gratitude to Ms. Mackenzie and her colleagues
for allowing me to use their data base.

Figure 3.6 shows the staggering differences in cumulative number of
AIDS cases between the United States and France (FR), Denmark (DK),
the Netherlands (NL), Canada (CAN), and the United Kingdom (UK).
The pool of infectives in the U.S.A. dwarfs those of the other First
World countries. Whenever I would bring up the enormous differential
between the AIDS rate in the United States and those in Europe, my
European colleagues would generally attribute all this to a time lag
effect. Somehow the United States had a head start on AIDS, but in time
the European countries would catch up. If other First World countries
were lagging the U.S.A., then one would expect some sort of variation
in new AIDS cases such as that depicted in Figure 3.7. However, Figure
3.8 demonstrates that the time-lagging hypothesis is not supported by
the data. No other First World country is catching up to the U.S.A.
Moreover, a downturn in new case rates is observable in all the countries
shown.

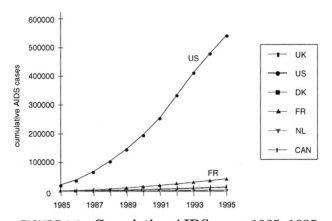

FIGURE 3.6   **Cumulative AIDS cases 1985–1995**

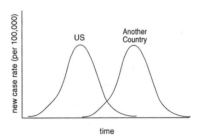

FIGURE 3.7   **A time-lagged scenario**

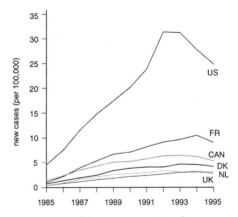

FIGURE 3.8   **New case rates by country**

Further insight is provided by Figure 3.9, in which we divide the annual incidence of AIDS per 100,000 in the U.S.A. by that for various
other First World countries. Note the relative constancy of the new case
ratio across the years for each country when compared to the U.S.A.
Thus, for the United Kingdom, it is around 9, for Denmark 6, etc. It is
a matter of note that this relative constancy of new case rates is maintained over the period examined (eleven years). In a similar comparison,
Figure 3.10 shows that the cumulative cases of AIDS per 100,000 in the
U.S.A. divided by that for other First World countries gives essentially
the same values observed for the new case rates in Figure 3.9.

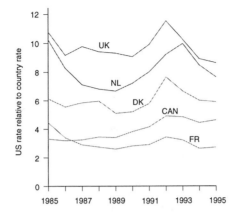

FIGURE 3.9   **Comparative new case rates**

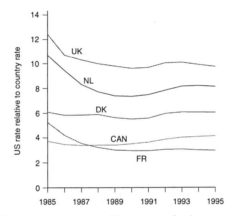

FIGURE 3.10   **Comparative cumulative case rates**

To investigate further, let us consider a piecewise (in time) exponential model for the number of AIDS cases, say in Country A:

$$\frac{dy_A}{dt} = k_A(t)y_A.$$

Figure 3.11 gives estimates for the rates k on a year-by-year basis using

$$k_A(t) \approx \frac{\text{new cases per year}}{\text{cumulative cases}}.$$

Note the apparent near equality of rates for the countries considered. To show this more clearly, Figure 3.12 displays the ratio of the annual estimated piecewise national rates divided by the annual estimated rate of the U.S.A.

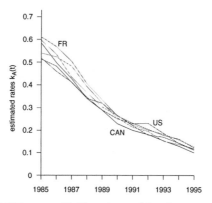

FIGURE 3.11   **Estimates of k_A by country**

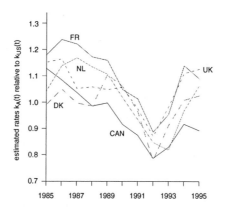

FIGURE 3.12   **Ratios of k_A estimates**

It is a matter of some interest that the k values are essentially the same for each of the countries shown in any given year. How shall we explain a situation where one country has a much higher incidence of new cases, year by year, yet the rate of increase for all countries is the same? For example, by mid-1997, the United Kingdom had a cumulative total of 15,081 cases compared to 612,078 for the United States. This ratio is 40.59 whereas the ratio of populations is only 4.33. This gives us a comparative incidence proportion of 9.37. On the other hand, at the same time, Canada had a cumulative AIDS total of 15,101. The United States population is 9.27 times that of Canada, so the comparative incidence proportion for the U.S.A. versus Canada in mid-1997 was 4.37. The comparative incidence of the U.S.A. vis-a-vis the United Kingdom is over twice that of the U.S.A. vis-a-vis Canada. Yet, in all three countries the rate of growth of AIDS cases is nearly the same. This rate changes from year to year, from around 0.54 in 1985 to roughly 0.12 in 1995. Yet it is very nearly the same for each country in any given year. One could therefore predict the number of new cases in France in a given year, just about as well knowing the case history of the United States instead of that in France. The correlation of new cases for the United States with that for each of the other countries considered is extremely high, generally around 0.96. It is hard to explain this by an appeal to some sort of magical synchronicity — particularly since we have the fact that though the growth rates of AIDS in the countries are roughly the same for any given year, the new case relative incidence per 100,000 for the United States is several times that of any of the other countries.

Recall from Section 3.4 the conjecture made in the mid-1980s that it was the bathhouses which caused the stand-alone epidemic in the United States. But, as we have seen, the bathhouse phenomenon really does not exist in the rest of the First World. How is it, then, that there are stand-alone AIDS epidemics in each of these countries? I do not believe there are stand-alone AIDS epidemics in these countries.

To model this situation, let us suppose there is a country, say Country Zero, in which the sociology favors a stand-alone AIDS epidemic. From other First World countries there is extensive travel to and from Country Zero, as indicated by Figure 3.13. If AIDS, with its very low infectivity rates, breaks out in Country Zero, then naturally the disease will spread to the other countries. But if the infectivity level is sufficiently low, then the maintenance of an apparent epidemic in each of the countries will be dependent on continuing visits to and from Country Zero.

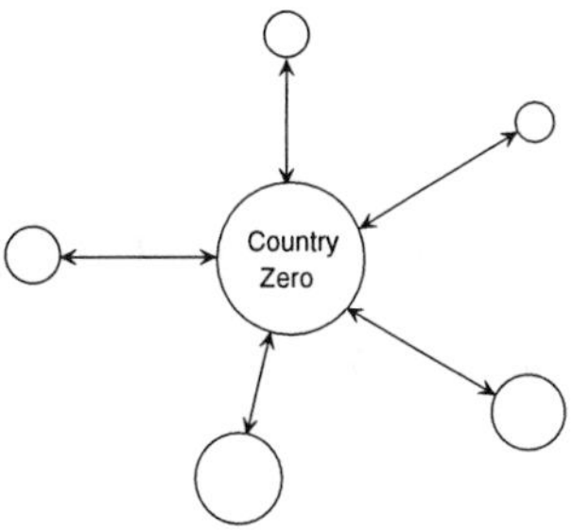

FIGURE 3.13   **Model for spread of disease from Country Zero**

Now let us suppose the fraction of infectives is rather low in country j. Thus, we shall assume that the susceptible pool is roughly constant. Let x_j be the number of infectives in country j and let z be the number of infectives in Country Zero. Let us suppose we have the empirical fact that, both for Country Zero and the other countries, we can use the same β_t in the growth models

$$\frac{dz}{dt} = \beta_t z \tag{3.5}$$

$$\frac{dx_j}{dt} = \beta_t x_j. \tag{3.6}$$

Let the population of country j be given by N_j and that of Country Zero be given by N_Z. Suppose the new case rate in Country Zero divided by that for country j is relatively constant over time, namely

$$\frac{z/N_Z}{x_j/N_j} = c_j. \tag{3.7}$$

Let us suppose that, at any given time, the transmission of the disease in a country is proportional both to the number of infectives in the country and the number of infectives in Country Zero. Then from (3.6)–(3.7)

$$\frac{dx_j}{dt} = \alpha_{jt} x_j + \eta_{jt} z = \left(\alpha_{jt} + \frac{N_Z}{N_j} c_j \eta_{jt}\right) x_j = \beta_t x_j, \tag{3.8}$$

where α_{jt} and η_{jt} are the transmission rates into country j from that country's infectives and Country Zero's infectives, respectively. We are assuming that infectives from other countries will have relatively little effect on the increase of infectives in Country Zero. Thus, for a short time span, (3.5) gives

$$z(t) \approx z(0) e^{\beta_t t}$$

and (3.8) is roughly

$$\frac{dx_j}{dt} = \alpha_{jt} x_j + \eta_{jt} z(0) e^{\beta_t t}.$$

Now, we note that the epidemic in a country can be sustained even if α_{jt} is negative, provided the transmission from the Country Zero infectives is sufficiently high. If we wish to look at the comparative effect of Country Zero transmission on country j vis-a-vis country i, we have

$$\eta_{jt} = \frac{c_i}{c_j} \frac{N_j}{N_i} \eta_{it} + \frac{\alpha_{it} - \alpha_{jt}}{c_j} \frac{N_j}{N_Z}.$$

If for two countries i and j we have $\alpha_{it} = \alpha_{jt}$, then

$$\eta_{jt} = \frac{c_i}{c_j} \frac{N_j}{N_i} \eta_{it}.$$

Using (3.7) this can be expressed as

$$\frac{x_j}{x_i} = \frac{\eta_{jt}}{\eta_{it}}.$$

If η_{jt} doubles, then according to the model, the number of infectives in country j doubles.

Let us see what the situation would be in Canada if, as a stand alone, the epidemic is just at the edge of sustainability: i.e., $\alpha_{CAN,t} = 0$. Then, going back to a universal β_t for all countries including Country Zero (America) and using the c_{CAN} value of 4.14 for 1995, we have from (3.8)

$$
\begin{aligned}
\eta_{CAN,t} &= \frac{N_{CAN}}{N_{USA}} \frac{1}{c_{CAN}} \beta_t \\
&= \frac{26,832,000}{248,709,873} \frac{1}{4.14} \beta_t \\
&= 0.026\beta_t.
\end{aligned}
$$

Thus, according to the model, activity rates from U.S.A. infectives roughly 2.6% of that experienced in the U.S.A. could sustain a Canadian epidemic at a comparative incidence ratio of around 4 to 1, American to Canadian. (If someone would conjecture that it is rather the Canadian infectives who are causing the epidemic in the United States, that would require the activity rate of Canadian infectives with American susceptibles to be $1/0.026 = 38.5$ times that of Canadian infectives with Canadian susceptibles.) If this activity rate would double to 5.2%, then the Canadian total infectives would double, but the rate $(1/x_{CAN})\,dx_{CAN}/dt$ would still grow at rate β_t. Similar calculations show that

$$
\eta_{FR,t} = 0.076\beta_t,\ \eta_{UK,t} = 0.024\beta_t,\ \eta_{DK,t} = 0.0034\beta_t,\ \eta_{NL,t} = 0.0075\beta_t.
$$

In summary, we have observed some surprises and tried to come up with plausible explanations for those surprises. The relative incidence of AIDS for various First World countries when compared to that of the United States appears, for each country, to be relatively constant over time and this incidence appears to be roughly the same for cumulative ratios and for ratios of new cases. The rate of growth β_t for AIDS changes year by year, but it seems to be nearly the same for all the First World countries considered (Figure 3.11), including the U.S.A. The bathhouse phenomenon is generally not present in First World countries other than the United States. Yet AIDS has a continuing small (compared to that of the U.S.A.), though significant, presence in First World countries other than the United States. The new case (piecewise exponential) rate there tracks that of the United States rather closely, country by country. We have shown that a model where a term for "travel" from and to the U.S.A. is dominant does show one way in which these surprises can be explained. Some years ago [2, 3, 4, 8], I pointed out that the American gay community was made unsafe by the presence of a small subpopulation which visited the bathhouses, even though the large majority of gays, as individuals, might not frequent these establishments. The present analysis gives some indication that the high AIDS incidence in the United States should be a matter of concern to other First World countries as long as travel to and from the U.S.A. continues at the brisk rates seen since the early 1980s.

Developing a model requires risk taking. The model, if it is to be useful, will be developed almost always without anything approaching a full data set. We could always find, as the fuller story comes in, that we were wrong. Then, in the case of epidemiology, we might find that by the time we publish our results, the virologists will have come up with a vaccine, perhaps rendering our model interesting but less than relevant. Most perilous of all, however, is to neglect the construction of a model.

3.10   Conclusion: A Modeler's Portfolio

This chapter has given an overview of around fifteen years of my work on the AIDS epidemic. I did not treat this work as an academic exercise. Rather, by public talks, articles in the popular press, service on the ASA AIDS Committee, and meetings with public officials, I tried to change the public policy on the bathhouses, without effect. So it is correct to say that I have not been successful in influencing public policy as I had wished. I well recall, by the late 1980s certainly, that things were not going as I had wished.

I never had the experience of somebody getting up at a professional meeting and poking holes in my AIDS model. I would get comments like, "Well, we see that you have shown a plausible way that the epidemic got started, but that does us little good in providing a plan of action now that the epidemic is well under way." Of course, this statement is not correct, for two reasons. First of all, I have addressed what the effect of closing the bathhouses would be during the mature epidemic. Second, effective restoration of the *status quo ante* will, almost always, reverse the course of an epidemic. In the case of polio, for example, closing of the public swimming pools and the suburban cinemas would have greatly defacilitated the epidemic, even after it was well under way.

To my shock, some colleagues took me aside to say that AIDS might be a very good thing, since it was discouraging a lifestyle of which neither these colleagues nor I approved. I always responded that our obligation in health care was to improve the lives of all persons, whether we liked their lifestyles or not. Moreover, I noted that a continuing entry of young males into the sociologically defined gay communities showed that the discouragement induced by the dreadful deaths generally associated with AIDS was not working the way they supposed. For example, in Houston, most of the leadership of the gay community had died off by the early 1990s. The death toll in Houston was staggering, more than in all Canada which has over ten times Houston's population. And yet, the people who died were replaced by a new wave of infectives.

Perhaps most significant of all, I would hear amazement that my modeling research was receiving any government support since there seemed to be little statistical interest in such public policy consequential modeling. Vast sums had been spent, for example, in support of the design of procedures whereby blood samples could be anonymously dumped into a pool with that of, say, nine other individuals and this exercise repeated many times in such a way to determine the fraction of AIDS infectives in the U.S.A., while ensuring the privacy of those tested. But modeling the progression of the epidemic was not receiving much NIH or PHS support. I was fortunate indeed that the Army Research Office has allowed me to work on modeling problems generally.

The notion of becoming some sort of full-time activist for modification of government policy toward defacilitating the epidemic was tempting. Some hold that, like an entrepreneur with a good idea for a product, the researcher should put all his/her energy into one enterprise at a time. Certainly, to save the hundreds of thousands of lives which have been needlessly lost to AIDS, such single-minded fanaticism would have been more than justified. However, based on the considerable effort that I had expended, it seemed to me that public policy was not going to be changed. If there had been some sort of focused attack on my AIDS model, then I might simply have hoped that a better explanation or a more complete model might win the day. But I had received the worst possible response — "We see your model, find no mistakes in it, and concede that it squares with the data, but it must be flawed because it does not square with policy."

So I continued my general career policy, which is somewhat similar to that of an investment portfolio. The basis of portfolio theory is that putting all of one's assets in one stock, even one with enormous expected return, is generally not a good idea. One is much better advised to use the weak law of large numbers and put one's capital in several enterprises of reasonably good expectation of return, so that the variability of the return of the overall portfolio will be brought down to much better levels than those associated with a single stock. It seems to me that this is a good idea for modeling researchers in allocating their intellectual assets.

During the period since the start of my work on AIDS, I founded the Department of Statistics at Rice, which now has eight faculty, four of them Fellows of the ASA. Again, during this period, I wrote eight books (AIDS figured in only three of these and only as chapters). I produced seven doctoral students during the interval, only one of these writing on AIDS. I managed to obtain United Nations funding to start a Quality Control Task Force in Poland following the fall of Russian domination of that country. I developed computer intensive strategies for simulation based estimation and continuous resampling, largely in connection with modeling work in cancer. I did a modest amount of consulting, saving in the process one or two companies from bankruptcy. I started the development of anti-efficient market theory models which work fine as stochastic simulations, but cannot be handled in closed form. And so on. If AIDS was part of my professional "portfolio," it accounted for only, say, five percent of the investment.

Since I have so far been unable to find political support for closing down bathhouses in America, it could be argued that the AIDS modeling part of the portfolio was not productive. I disagree. Our business as modelers is, first of all, to understand the essentials of the process we are modeling. Only rarely, and generally in relatively simple situations, such as changing the quality control policy of a corporation, should we expect to be able to say, "There; I have fixed it."

The optimism concerning a quick discovery of an AIDS cure has dimmed. No doubt, one will be found at some time in the future. However, after tens of billions of dollars already expended without a cure or vaccine, it is unwise to continue on our present route of muddling through until a miracle occurs. By this time, so many hundreds of thousands of American lives have been wasted by not shutting down high contact facilitating establishments that changing policy could leave open a myriad of litigious possibilities. The families of the dead or dying might have good reason to ask why such policies were not taken fifteen years ago.

Modelers are not generally members of the political/economic power structure, which Pareto termed the "circle of the elites." We cannot ourselves hope to change public policy. But it is certainly our business to develop models that increase understanding of some system or other which appears to need fixing. We should follow the path of Chaucer's poor Clerk of Oxford: "...gladly would he learn and gladly teach."

Following the American polio epidemic of the postwar years, no modeler appears to have attempted to describe what went wrong with its management. Had that been done, perhaps a totally different response might have taken place when AIDS came on the scene. At the very least, I hope that my modeling of AIDS will have some impact on public policy concerning the next plague when it comes, and come it surely will.

Acknowledgments

This research was supported by the Army Research Office (Durham) under DAAH04-95-1-0665 and DAAD19-99-1-0150.

3.11 References

[1] G. Rotello, *Sexual Ecology: AIDS and the Destiny of Gay Men*, Dutton, New York, 1997.

[2] J. R. Thompson, "Deterministic versus stochastic modeling in neoplasia," in *Proceedings of the 1984 Computer Simulation Conference*, Society for Computer Simulation, San Diego, 1984, 822–825.

[3] J. R. Thompson, *Empirical Model Building*, Wiley, New York, 1989.

[4] J. R. Thompson, "AIDS: the mismanagement of an epidemic," *Computers Math. Applic.* **18**, 965–972 (1989).

[5] J. R. Thompson, "The United States AIDS epidemic in first world context," in *Advances in Mathematical Population Dynamics: Molecules, Cells and Man*, O. Arino, D. Axelrod, and M. Kimmel (Eds.), World Scientific Publishing Company, Singapore, 1998, 345–354.

[6] J. R. Thompson and K. Go, "AIDS: modeling the mature epidemic in the American gay community," in *Mathematical Population Dynamics*, O. Arino, D. Axelrod, and M. Kimmel (Eds.), Dekker, New York, 1991, 371–382.

[7] J. R. Thompson and R. A. McBride, "Close Houston's Gay Bathhouses," *Houston Chronicle*, Outlook, April 9, 1989.

[8] J. R. Thompson and R. A. Tapia, *Nonparametric Function Estimation, Modeling and Simulation*, SIAM, Philadelphia, 1990.

[9] R. W. West, *Modeling the Potential Impact of HIV on the Spread of Tuberculosis in the United States*, Ph.D. Dissertation, Rice University, Houston, TX (1994).

[10] R. W. West and J. R. Thompson, "Models for the simple epidemic," *Mathematical Biosciences* **141**, 29–39 (1997).

[11] R. W. West and J. R. Thompson, "Modeling the impact of HIV on the spread of tuberculosis in the United States," *Mathematical Biosciences* **143**, 35–60 (1997).

Chapter 4

A Model for the Spread of Sleeping Sickness

Marc Artzrouni
Jean-Paul Gouteux

Prerequisites: differential equations, linear algebra

4.1 Introduction

Malaria and sleeping sickness are two well-known examples of tropical diseases that are spread through flying insects. Malaria is an ancient, serious, acute, and chronic relapsing infection that is transmitted through mosquito bites. It still affects tens of millions in Central and South America and in North and Central Africa.

Sleeping sickness (also called African Human Trypanosomiasis) is endemic in 36 countries of Sub-Saharan Africa. Approximately 50 million people are exposed to this disease, which is transmitted by the tsetse fly and is characterized by fever and inflammation of the lymph nodes. This illness leads to profound lethargy (hence the name) and to death if left untreated. The World Health Organization spends millions of dollars and has large-scale ongoing programs aimed at fighting both malaria and sleeping sickness.

The second author of this chapter is a *medical entomologist*, i.e., a biologist who studies insects, and particularly their role in the spread of tropical diseases. He has spent 17 years in six different African countries studying the tsetse fly and the population dynamics of sleeping sickness (i.e., the interrelationships between the human and the fly populations). Upon his return to France, he started a collaborative effort with the first author, an applied mathematician who specializes in population dynamics. The goal is to develop a variety of mathematical models to help better understand the dynamics of the disease. It is also hoped that an improved understanding of the dynamics will lead to better control strategies, an important goal of epidemiologic modeling. Indeed, it is our strong belief that the modeling of epidemic diseases must lead to concrete results, and what could be more concrete than the prospect of a theoretical mathematical result leading to a certain degree of control (if not the eradication) of a very serious disease affecting a whole continent?

What do we mean by "control?" Sleeping sickness spreads through a simple cycle of infected flies biting uninfected humans who are in turn bitten by uninfected flies who thus become infected (Figure 4.1). In the present context there are two common-sense methods of control that have been known for a long time: (a) an increase in the mortality of flies (through spraying of insecticides and "trapping") and (b) the screening, treatment, and isolation of infected individuals. It stands to reason that if flies have a shorter life expectancy, they will have less time to infect humans. Also, if every infected person were immediately screened and isolated, the disease could not spread. A modeler's challenge is then to figure out by how much the mortality of flies must be increased and how quickly infected persons should be removed in order to bring about the eventual extinction of the disease.

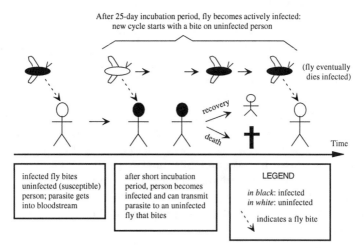

FIGURE 4.1   **Transmission cycle of parasite: humans and flies**

A related and crucial question is that of the cost of such strategies. Indeed, countries affected by sleeping sickness have very limited resources to fight such deadly and complex diseases as sleeping sickness, malaria, or AIDS for that matter. It is thus essential to find control strategies that are also *optimal*: i.e., strategies that cost the least amount of money to obtain the desired result.

4.2   The Compartmental Model

4.2.1   A Brief History

The starting point of our efforts was a very simple differential equations model of sleeping sickness proposed by Professor David Rogers of Cambridge University [6, 7, 8]. His model has only two variables (and hence two "compartments" in epidemiologic parlance): the infected fly and infected human populations. (The susceptible, or uninfected, populations are then automatically known because total fly and human populations are assumed constant.)

In reality the population dynamics of sleeping sickness are not that simple. First there is an incubation period for both humans and flies. This period is about 25 days for flies, whose total life expectancy is approximately 45 days. The incubation period for humans is short, about 12 days. Another important factor is the removal of infected humans; when their infection is diagnosed they are removed to be treated and are then unlikely to be bitten. Such a model with both populations still assumed constant was the object of our earlier work and resulted in a five-compartment differential equations model of sleeping sickness [1, 2].

Another factor that has not been incorporated to date is the fact that the fly population is not closed and constant. Thus, in our ongoing program to improve existing models and make them more realistic, we will describe here a variant of our model that allows for in- and out-migrations of flies. This is an important step toward a more realistic model. (The description below is self-contained and does not require any knowledge of the previous work [1, 2].)

4.2.2   General Description

We consider an "epidemiologic unit" composed of an isolated village surrounded by an open population of tsetse flies. (Flies are sometimes called "vectors" as they are the means through which the parasite is spread; Figure 4.1 illustrates the analogy with an arrow indicating a bite.) There are three vector compartments and three human compartments (Figure 4.2). Vectors are at first "susceptible" — that is, uninfected. When they bite an infected human for their "blood meal" and become infected, they enter an incubation period during which they cannot yet transmit the parasite. At the end of the incubation period they enter a stage of "active infection" during which they can infect a human they bite. Flies eventually die infected — there is no recovery. Thus we define $V_s(t)$, $V_i(t)$, $V_a(t)$ as the susceptible, incubating, and actively infected vectors at time t, and $V(t) = V_s(t) + V_i(t) + V_a(t)$ is the total vector population.

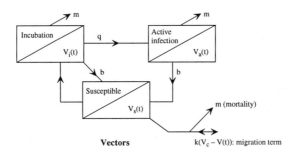

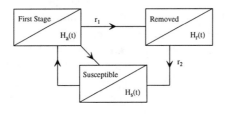

FIGURE 4.2   **Compartments of the model**

The cycle of the disease is as follows for humans. After a short incubation period a person enters a so-called "first stage" of the epidemic during which the person can transmit the parasite when bitten. A second ("removed") stage is reached when symptoms become more severe and the parasite leaves the blood stream and enters the spine. The patient becomes severely incapacited and is removed either to be treated, or stays inside his/her home; the patient is no longer at risk of being bitten (or at least the risk becomes negligible).[1]

---

[1]Parenthetically we contrast here and elsewhere the intellectual *modus operandi* of the biologist and of the mathematician. The former rightly insists that the probability of being bitten becomes very small after removal but is never zero — things that are rarely black and white. The latter insists that the art of modeling is to know what simplifying assumptions must be made without compromising the usefulness of the exercise; so we agreed here that the presence of a removed compartment with a zero probability of being bitten was a reasonable approximation of reality.

We assume that the person eventually recovers and re-enters the susceptible compartment. In order to reduce the dimension of this model, made more complex with a variable total fly population subjected to migrations, we will ignore the 12-day incubation period for humans which is short compared to the duration of the first stage (from several months to several years). Hence the three human populations are $H_s(t)$, $H_a(t)$, $H_r(t)$: susceptible, first stage carriers, and removed humans at time t. The total human population H is assumed constant.

4.2.3 The Fly-Bite Model

Because flies bite a human on average once every three days we will find it convenient to consider a continuous-time model where the unit of time is three days. Tsetse flies have their blood meals either on humans or on animals (e.g., pigs). Thus each fly has a blood meal every three days, but not necessarily on a human.

The standard approach to the fly-bite model so far has been to simply assume that there is a probability τ_1 that a meal will be on a human. (For example, a village with many pigs will have a lower τ_1 because flies prefer to bite pigs than humans.) However, in our effort to improve the realism of such a model, we wish to incorporate the fact that flies are opportunistic in their feeding habits. If, for example, many sick people are removed then there are few humans available and flies are more likely to feed on whatever animal may be available (pigs, lizards, etc.).

We can model in a simple way the opportunistic feeding habits of flies by assuming that there is a pool of A_0 animals that a fly may bite. At time t there are also $H_s(t) + H_a(t) = H - H_r(t)$ humans that a fly may bite. We assume that a fly's preference for animals is measured by a weight factor w that is between 0 and 1, where $w = 0.5$ indicates indifference between the two groups. The probability τ_1 of biting a human during one time unit (i.e., one three-day interval) is then a function of the number of removed individuals $(H_r(t))$ that is equal to the number of available humans $(H - H_r(t))$ divided by the sum of the available humans and of the number A_0 of animals (with weights $(1 - w)$ and w for each one of the two groups): that is,

$$\tau_1(H_r(t)) = \frac{(H - H_r(t))(1 - w)}{(H - H_r(t))(1 - w) + wA_0}$$
$$= \frac{H - H_r(t)}{H - H_r(t) + wA_0/(1 - w)}.$$

We can thus consider that each fly bites with equal probability either one of $H - H_r(t)$ humans or one animal out of a fictitious population of $A \equiv wA_0/(1 - w)$ animals. The probability that a fly will bite an infected individual is then $H_a(t)/(H - H_r(t) + A)$ and that it will bite a susceptible one is $(H - H_a(t) - H_r(t))/(H - H_r(t) + A)$.

4.2.4   The Population Dynamics of Vectors

We assume that flies have constant birth and mortality rates b and m. In addition we assume a simple regulating migratory mechanism: during a time interval dt the net growth of the susceptible vector population due to migration is of the form $k(V_c - V(t))dt$; $k \geq 0$ measures the strength of the feedback (i.e., the magnitude of the migratory flows) and V_c is a critical value of the total population below which there is in-migration, and above which there is out-migration. We assume for simplicity that migrations occur only in the susceptible compartment: healthy flies alone migrate. (This assumption is reasonable as a first approximation because the overwhelming majority of flies are uninfected. Indeed, severe sleeping sickness epidemics occur with less than one percent of flies being infected.)

Flies can become infected primarily during their first blood meal (i.e., while in the first three-day age group). Given that the number of flies in this first age group is approximately $V(t)b$, the number of flies at risk of infection is also $V(t)b$. The equation for the incubating fly population is then

$$\frac{dV_i(t)}{dt} = \frac{V(t)b\tau_2 H_a(t)}{H - H_r(t) + A} - V_i(t)(q + m), \tag{4.1}$$

where the first term on the right-hand side represents the newly infected vectors. Indeed, $H_a(t)/(H - H_r(t) + A)$ is the probability of biting an infected individual, and τ_2 is the *vector susceptibility*, the probability that a susceptible vector eventually becomes infected after biting an infected human. (Because of the complexities of the biological processes involved, a susceptible vector biting a human does not necessarily become infected.) Finally, q is the rate at which vectors leave the incubating stage and $1/q$ is therefore the average duration of the incubation period. The term $V_i(t)(q + m)$ thus reflects losses due to new infections and deaths. Changes in $V_a(t)$ occur only through new active infections and deaths, so

$$\frac{dV_a(t)}{dt} = V_i(t)q - V_a(t)m = (V(t) - V_s(t) - V_a(t))q - V_a(t)m. \tag{4.2}$$

The equation for the susceptible compartment is

$$\frac{dV_s(t)}{dt} = V(t)b\left(1 - \frac{\tau_2 H_a(t)}{H - H_r(t) + A}\right) + k(V_c - V(t)) - mV_s(t), \tag{4.3}$$

which captures gains due to births $(V(t)b)$ and losses due to new infections as well as losses due to mortality within the compartment $(mV_s(t))$; $k(V_c - V(t))$ is the migration term. Adding (4.1)–(4.3) yields

$$\frac{dV(t)}{dt} = V(t)(b - m) + k(V_c - V(t)). \tag{4.4}$$

Not surprisingly this is a differential equation in the single variable $V(t)$ that expresses the fact that the total population simply changes through migration and through the balance between mortality and natality. When $k + m - b = 0$ then $V(t)$ increases linearly if $k > 0$ and remains constant if $k = 0$. When $k + m - b \neq 0$ the solution to the differential equation (4.4) is

$$V(t) = V^* - (V^* - V(0))e^{-(k+m-b)t}, \tag{4.5}$$

where $V^* = kV_c/(k + m - b)$. To simplify matters we assume that the feedback term k is large enough to insure that $k + m - b > 0$, which means that the vector population $V(t)$ does converge to an equilibrium value V^*. (V^* is called the *carrying capacity*, the maximum population the environment can "carry.")

Equation (4.5) is a classical growth equation of population dynamics that has been used in sleeping sickness modeling [3, 4]. The total population converges monotonically to its carrying capacity V^*, whether the initial population $V(0)$ is greater or smaller than V^*. (The reader is invited to study in more detail the population $V(t)$ — in particular as a function of the initial population $V(0)$ — and to also study what happens if $k + m - b < 0$.)

4.2.5   The Population Dynamics of Humans

The two equations for the human populations are

$$\frac{dH_a(t)}{dt} = V_a(t)\tau_3 \frac{H - H_a(t) - H_r(t)}{H - H_r(t) + A} - H_a(t)r_1 \tag{4.6}$$

$$\frac{dH_r(t)}{dt} = H_a(t)r_1 - H_r(t)r_2, \tag{4.7}$$

where the first term on the right-hand side of (4.6) represents new human infections. Each one of the $V_a(t)$ actively infected flies has a probability $(H - H_a(t) - H_r(t))/(H - H_r(t) + A)$ of infecting a susceptible person and τ_3 is the probability that a susceptible human bitten by an infected fly will become sick. (Here τ_3 is the *human susceptibility* analogous to to the τ_2 for flies: a susceptible person bitten by an infected fly does not necessarily become infected.)

The parameter r_1 is the transition rate between the first-stage and the removed compartments. This transition reflects the natural history of the disease for infected individuals who move from the first to the second (removed) stage when the risk of transmitting the parasite becomes negligible. This transition also reflects the detection and removal of first-stage infected individuals. Indeed, the removal rate is increased when mobile medical units are sent in the field to detect infected individuals who are then removed for treatment.

Equation (4.7) is a routine balance equation for the removed compartment, where r_2 reflects the recovery of treated individuals. (The susceptible population $H_s(t)$ is known through $H_s(t) = H - H_a(t) - H_r(t)$.)

4.3   Mathematical Results

4.3.1   Basic Reproduction Number

The *basic reproduction number* R_0 at the beginning of an epidemic is the average number of persons a single sick individual ("patient zero") will infect. If $R_0 < 1$ then the disease will eventually die out since each infected person is replaced by less than one other infected person. Conversely, the disease spreads if $R_0 > 1$.

Here R_0 will be the number of new human infections generated (via the tsetse flies since this is the only mode of transmission) by a single infected person; R_0 can be found as the product $R_0 = R_1 R_2$, where R_1 is the number of flies infected by patient zero and R_2 is the number of humans each one of the R_1 flies will infect in turn.

PROPOSITION 4.1

When the total population is at an equilibrium value V^* and there are no infected vectors and no removed population, then the basic reproduction number R_0 is

$$R_0 = \frac{V^* b \tau_2 q \tau_1(0)^2 \tau_3}{m(q+m)r_1 H} . \tag{4.8}$$

PROOF   We note that $\tau_1 \equiv \tau_1(0) = H/(H+A)$ is the fly bite rate when the removed compartment is empty. The quantity $R_1 \equiv V^* b \tau_1 \tau_2 /(r_1 H)$ is the number of flies infected by patient zero. Indeed, during one unit of time (three days), each one of the V^*b young flies has a probability $\tau_1\tau_2/H$ of biting patient zero and becoming infected. Furthermore, the duration of this exposure is $1/r_1$ (the time patient zero spends in the first stage). Also $R_2 \equiv \tau_1\tau_3(q/(q+m))/m$ is the average number of new human infections generated by each new infected fly. This follows because $q/(q+m)$ is the probability of reaching the active infective stage, $1/m$ is the life expectancy once a fly has reached that stage, and the probability of biting a susceptible human who will become infected is $\tau_1\tau_3$. Since $R_0 = R_1R_2$, equation (4.8) results.   ∎

4.3.2   Stability Theorem

We define the system under study through the vector $P(t) = [V(t), V_a(t), V_s(t), H_a(t), H_r(t)]^T$ and the corresponding five differential equations (4.4), (4.2), (4.3), (4.6), and (4.7). (The variable $V_i(t)$ does not appear in this system but is known through $V_i(t) = V(t) - V_s(t) - V_a(t)$.) The system can then be written compactly as $dP(t)/dt = F(P(t))$, where $F(P(t))$ represents the right-hand sides of the five equations. We note that the vector $W_0 \equiv [V^*, 0, V^*, 0, 0]^T$ (which we call the "origin") is an *equilibrium point* of the system corresponding to the situation with no epidemic: that is, $F(W_0) = 0$.

Although we indicated above that $R_0 < 1$ was equivalent to extinction, this needs to be proved. We will show that the origin W_0 is *unstable* if $R_0 > 1$ and asymptotically *stable* if $R_0 < 1$. This means that if an initial point $P(0)$ is close to the origin, then for $R_0 < 1$ the function $P(t)$ will converge to W_0 (extinction) and will move away from W_0 if $R_0 > 1$. (If $R_0 = 1$, the analysis becomes quite complex and the behavior of the system hinges on second-order considerations.)

When $R_0 < 1$ we will also give detailed results on the extinction rate of the disease, i.e., on the asymptotic rate at which each component of the vector $W(t) \equiv P(t) - W_0$ approaches 0. For example, if asymptotically $V_a(t) \approx e^{ct}$ for some $c < 0$, then we call c the *extinction rate* of the population $V_a(t)$. In an analogy with the doubling time of a population, we can then calculate the asymptotic *halving time* of the actively infected vector population, defined by $HT = \ln(0.5)/c$.

In the course of proving our main stability theorem we will have to express the roots of a cubic equation, which will involve the following quantities:

$$r = 2m + q + r_1 \tag{4.9}$$

$$s = m(m + q) + r_1(2m + q) \tag{4.10}$$

$$t = m(m + q)r_1(1 - R_0) \tag{4.11}$$

$$p^* = s - r^2/3 \tag{4.12}$$

$$q^* = 2r^3/27 - rs/3 + t \tag{4.13}$$

In the theorem we will make use of the fact that $p^* < 0$ (indeed p^* is a quadratic polynomial in the unknown $m - r_1$ and this quadratic is always negative).

THEOREM 4.1

The origin $W_0 = [V^*, 0, V^*, 0, 0]$ is asymptotically stable if $R_0 < 1$ and unstable if $R_0 > 1$. Moreover, define

$$\lambda = 2\left(-\frac{p^{*3}}{27}\right)^{1/6} \cos\left(\frac{1}{3}\arccos\left[-\frac{q^*}{2}\left(-\frac{p^{*3}}{27}\right)^{-1/2}\right]\right) - \frac{r}{3}. \tag{4.14}$$

If $R_0 > 1$ then $\lambda > 0$; if $R_0 < 1$ then $\lambda < 0$ and the asymptotic growth rates (extinction rates) $\rho(V(t) - V^*)$, $\rho(V_a(t))$, $\rho(V_s(t) - V^*)$, $\rho(H_a(t))$, and $\rho(H_r(t))$ of the five components of $W(t)$ are given by

$$\rho(V(t) - V^*) = -k - m + b \tag{4.15}$$

$$\rho(V_a(t)) = \lambda \tag{4.16}$$

$$\rho(V_s(t) - V^*) = \max\{-k - m + b, \lambda\} \tag{4.17}$$

$$\rho(H_a(t)) = \lambda \tag{4.18}$$

$$\rho(H_r(t)) = \max\{-r_2, \lambda\} \tag{4.19}$$

PROOF   To prove stability results at W_0 for the vector $P(t) = [V(t), V_a(t), V_s(t), H_a(t), H_r(t)]^T$ we calculate the *Jacobian matrix* J at the origin W_0. Recall that the Jacobian matrix is a matrix of partial derivatives of the right-hand sides of the five differential equations (4.4), (4.2), (4.3), (4.6), and (4.7), taken in that order. For (4.4) we see that $\partial[V(b - m) + k(V_c - V)]/\partial V = (-k - m + b)$, with other partial derivatives being zero — hence the first row of J below. As another example, in the third row we have the partial derivatives of the right-hand side of (4.3). This partial derivative with respect to H_a (at W_0) is found to be $-V^* b\tau_2/(H + A)$, giving the element in the fourth column of the third row of the Jacobian matrix

$$J = \begin{bmatrix} -k - m + b & 0 & 0 & 0 & 0 \\ q & -q - m & -q & 0 & 0 \\ b - k & 0 & -m & \frac{-V^* b\tau_2}{H+A} & 0 \\ 0 & \frac{\tau_3 H}{H+A} & 0 & -r_1 & 0 \\ 0 & 0 & 0 & r_1 & -r_2 \end{bmatrix}. \tag{4.20}$$

The theory of systems of differential equations insures that stability at the origin will obtain if all the eigenvalues of the Jacobian matrix J have negative real parts. The origin is unstable if at least one eigenvalue has a positive real part.

Two obvious eigenvalues are $ev_1 = -k - m + b$ and $ev_2 = -r_2$. The reader may want to use algebraic manipulation software such as Mathematica [9] or Mathcad [5] to check that the three other eigenvalues $(ev_3, ev_4,$ and $ev_5)$ are the roots of the following cubic equation in the unknown x:

$$H(x) \equiv (m + x)(m + q + x)(r_1 + x) = R_0 m(m + q)r_1$$
$$= \frac{kV_c b\tau_2 q\tau_1^2 \tau_3}{(k + m - b)H}. \quad (4.21)$$

By expanding $H(x)$ we obtain

$$x^3 + rx^2 + sx + t = 0,$$

where r, s, and t are given in (4.9)–(4.11). Cubic equations are nasty but do have roots with closed-form expressions that are not much more complicated that those of a quadratic equation. These roots can now be handled easily with Mathematica or Mathcad. The closed-form expressions for the roots ev_3, ev_4, and ev_5 are then

$$ev_3 = 2\left(-\frac{p^{*3}}{27}\right)^{1/6} \cos\left(\frac{1}{3}\arccos\left[-\frac{q^*}{2}\left(-\frac{p^{*3}}{27}\right)^{-1/2}\right] + \frac{4\pi}{3}\right) - \frac{r}{3}$$

$$ev_4 = 2\left(-\frac{p^{*3}}{27}\right)^{1/6} \cos\left(\frac{1}{3}\arccos\left[-\frac{q^*}{2}\left(-\frac{p^{*3}}{27}\right)^{-1/2}\right] + \frac{2\pi}{3}\right) - \frac{r}{3}$$

$$ev_5 = 2\left(-\frac{p^{*3}}{27}\right)^{1/6} \cos\left(\frac{1}{3}\arccos\left[-\frac{q^*}{2}\left(-\frac{p^{*3}}{27}\right)^{-1/2}\right]\right) - \frac{r}{3}$$

where p^*, q^* are given in (4.12) and (4.13). Notice that λ of (4.14) is ev_5. The cosine and arccosine function are defined in the complex domain when the real argument

$$W \equiv -\frac{q^*}{2}\left(-\frac{p^{*3}}{27}\right)^{-1/2}$$

of the arccosine function has an absolute value larger than 1. The proof will now proceed in three steps. We will first show that ev_5 has the largest real part among the three roots and is real (Step 1). Then we will show that if $R_0 > 1$ then $ev_5 > 0$, and if $R_0 < 1$ then $ev_5 < 0$ (Step 2). Finally, we will obtain the results on the convergence rates by considering the basis in which the matrix J is in diagonal form (Step 3).

Step 1

We define for any real Z the following three functions, whose real parts are plotted in Figure 4.3:

$$u_3(Z) \equiv \cos\left(\frac{\arccos(Z)}{3} + \frac{4\pi}{3}\right)$$

$$u_4(Z) \equiv \cos\left(\frac{\arccos(Z)}{3} + \frac{2\pi}{3}\right)$$

$$u_5(Z) \equiv \cos\left(\frac{\arccos(Z)}{3}\right)$$

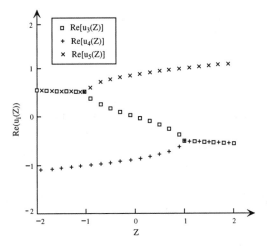

FIGURE 4.3   **The functions $\mathrm{Re}(u_i(Z))$**

The fact that $\mathrm{Re}(u_4(Z)) \leq \mathrm{Re}(u_3(Z)) \leq \mathrm{Re}(u_5(Z))$ and $p^* < 0$ implies

$$\mathrm{Re}(ev_4) \leq \mathrm{Re}(ev_3) \leq \mathrm{Re}(ev_5). \tag{4.22}$$

For $Z \geq -1$ then $u_5(Z) = \mathrm{Re}(u_5(Z))$ (i.e., $u_5(Z)$ is real) and $u_5(Z)$ is then a real increasing function of Z for $Z \geq -1$. In addition, if $Z < -1$ then $u_5(Z)$ is pure imaginary. Given that

$$ev_5 = 2\left(-\frac{p^{*3}}{27}\right)^{1/6} u_5(W) - \frac{r}{3},$$

we see that ev_5 is a real increasing function of W if $W \geq -1$. We will now prove that $W \geq -1$ by considering W as a function $W(r,s,t)$ of the three parameters r, s, and t. For any fixed r, s (i.e., any fixed m, q, r_1) $W(r,s,t)$ is a decreasing affine function of t: $W(r,s,t) = \alpha t + \beta$,

$\alpha < 0$. Also $W(r, s, t)$ reaches a minimum $W_m = W(r, s, m(m + q)r_1)$ for $t = m(m+q)r_1$ (when $R_0 = 0$). However, in this case the three roots of (4.21) are $-m$, $-m - q$, and $-r_1$. Therefore, $W_m \geq -1$ necessarily holds since otherwise ev_5 would be imaginary. This proves that $W \geq -1$ which implies that $u_5(W)$ and ev_5 are real.

Step 2

We now define the right-hand side of (4.21) as

$$R^* = \frac{kV_cb\tau_2q\tau_1^2\tau_3}{(k + m - b)H}$$

and assume that $R_0 > 1$, which is equivalent to $R^* > m(m+q)r_1$. Figure 4.4 depicts a typical example[2] of the cubic $H(x)$. The function crosses the x-axis at $-m$, $-m - q$, $-r_1$, and the equation $H(x) = R^*$ has one positive root (ev_5) as well as two complex roots since the horizontal line at R^* intersects the function $H(x)$ only once. (There would be three real roots for R^* small enough.) The fact that there is a positive root can also be seen from the fact that for $x > 0$, $H(x)$ is an increasing function of x that is equal to $m(m + q)r_1$ at $x = 0$. The equation $H(x) = R^*$ then necessarily has a single positive root that must be ev_5 since ev_5 is a real root that is larger than the real parts of ev_3 and ev_4; see (4.22).

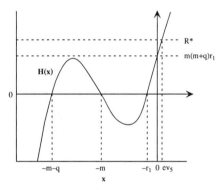

FIGURE 4.4   **Cubic $H(x)$ of (4.21), with $R_0 > 1$**

We next assume that $R_0 < 1$: that is, $R^* < m(m + q)r_1$. Figure 4.5 depicts in this case a typical function $H(x)$. If R_c is the value at which

---

[2]In this figure we have $r_1 < m$ which is usually the case without intervention; the life expectancy $(1/m)$ of the tsetse fly is on the order of one or two months, whereas the time spent in the asymptomatic stage $(1/r_1)$ is several months to several years. The detection of infected individuals can, however, bring r_1 above m.

a horizontal line is tangent to $H(x)$, then there are three negative roots (ev_4, ev_3, and ev_5) if $R^* < R_c$ (e.g., at $R^* = R^*(1)$) and two complex and one negative root if $m(m + q)r_1 > R^* > R_c$ (e.g., at $R^* = R^*(2)$).

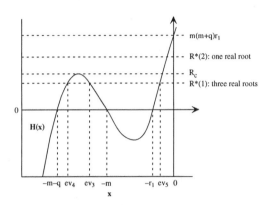

FIGURE 4.5   **Cubic $H(x)$ of (4.21), with $R_0 < 1$**

Step 3

We recall that the system is written in the form $dP(t)/dt = F(P(t))$. If we change variables and define $W(t) = P(t) - W_0 = [V(t) - V^*, V_a(t), V_s(t) - V^*, H_a(t), H_r(t)]^T$, then the system can be written as

$$\frac{dP(t)}{dt} = F(W(t) + W_0) = \frac{dW(t)}{dt}.$$

Because $F(W_0) = 0$, the linear approximation of $F(W(t) + W_0)$ at W_0 yields

$$\frac{dW(t)}{dt} = F(W(t) + W_0) \approx JW(t),$$

where J is the Jacobian matrix of (4.20). The system at W_0 is therefore $dW(t)/dt \approx JW(t)$. Given that the eigenvalues are distinct, J can be diagonalized and written as $J = PDP^{-1}$, where D is a diagonal matrix with the eigenvalues ev_1, ev_2, ev_3, ev_4, and ev_5 on the diagonal, and P is a matrix with the corresponding eigenvectors in its columns. If we define the functions

$$E_1(d) = \frac{(r_1 + d)(r_2 + d)(H + A)}{r_3 H r_1}$$

$$E_2(d) = \frac{-V^* b \tau_2 (r_2 + d)}{(m + d)(H + A) r_1}$$

$$E_3(d) = \frac{r_2 + d}{r_1},$$

then the matrix P is

$$P = \begin{bmatrix} 1 & 0 & 0 & 0 & 0 \\ 0 & 0 & E_1(ev_3) & E_1(ev_4) & E_1(ev_5) \\ 1 & 0 & E_2(ev_3) & E_2(ev_4) & E_2(ev_5) \\ 0 & 0 & E_3(ev_3) & E_3(ev_4) & E_3(ev_5) \\ 0 & 1 & 1 & 1 & 1 \end{bmatrix}. \qquad (4.23)$$

We next define $Z(t) = P^{-1}W(t)$ and note that $dZ/dt = DZ(t)$. This is now a very simple system of differential equations with solution $Z(t) = [c_1 e^{ev_1 t}, c_2 e^{ev_2 t}, \ldots, c_5 e^{ev_5 t}]^T$, where each c_i is real when ev_i is real, and the c_i are in conjugate pairs when they correspond to conjugate pairs of complex eigenvalues. Given that $W(t) = PZ(t)$ and letting P_i be the ith column of P $(i = 1, \ldots, 5)$, we have

$$W(t) = \sum_{k=1}^{5} c_k P_k e^{ev_k t}, \qquad (4.24)$$

where the five scalars c_k $(k = 1, \ldots, 5)$ depend on initial conditions (and are assumed to be nonzero). We now recall that if $R_0 < 1$ then all the eigenvalues ev_k have negative real parts, which shows that $W(t) \to 0$ when $t \to \infty$. This proves the stability of W_0.

Equation (4.24) shows that $V(t) - V^*$ (the first component of $W(t)$) approaches 0 at a rate $ev_1 = -k - m + b$, which is not surprising in view of (4.5). Because only the last three entries of the second row of P are nonzero, the second component $V_a(t)$ of $W(t)$ converges to 0 at a rate equal to $\max\{\mathrm{Re}(ev_3), \mathrm{Re}(ev_4), ev_5\} = ev_5$. Similarly, $V_s(t) - V^*$ converges to 0 at a rate equal to $\max\{-k-m+b, \mathrm{Re}(ev_3), \mathrm{Re}(ev_4), ev_5\} = \max\{-k - m + b, ev_5\}$. This verifies the equations (4.15)–(4.17). The results of (4.18)–(4.19) follow in a similar fashion. ∎

4.4   Discussion

4.4.1   Basic Reproduction Number or Rate of Extinction?

The quantity λ defined in (4.14) plays a central role in the convergence rates of the five variables $V(t)$, $V_a(t)$, $V_s(t)$, $H_a(t)$, and $H_r(t)$. In particular, when $R_0 < 1$ then λ is the extinction rate of the actively infected populations $V_a(t)$ and $H_a(t)$; see (4.16) and (4.18).

It is standard in epidemiologic modeling to focus on R_0 as a stability criterion. Thus, to investigate control strategies, the approach adopted in [1, 2] consists of studying parameters values for which R_0 was either

Table 4.1   Baseline values for parameters

Parameter	Description	Value
V_c	critical value of vector population	5,000
H	total human population	300
τ_1	fly bite – no removed population	0.1
τ_2	vector susceptibility	0.1
τ_3	human susceptibility	0.62
q	incubation rate	0.12 (25 days)
r_1	removal rate	0.0075 (13.33 mos)
b	fly birth rate	1/15
m	fly death rate	1/15 (life expect = 1.5 mos)
k	migratory feedback parameter	0.05

above 1 or below 1. However, bringing R_0 just below 1 may be of little value since the halving time of the epidemic can still be quite long. To illustrate this, consider the set of realistic "baseline" parameter values given in Table 4.1. (Recall that the time unit is 3 days; if we assume that every month has 30 days then, for example, a 25 day incubation period translates into a rate $q = 3/25 = 0.12$.)

With the parameter values in Table 4.1, $V^* = 5{,}000$, $R_0 = 0.886$, $\lambda = -7.539 \times 10^{-4}$, and the halving time expressed in months is $HT = (0.1 \ln(0.5))/\lambda \approx 92$ months. Even though R_0 is well below 1, it takes more than seven years for the infected population to be cut in half. If however the vector death rate m could be increased to $1/10$ (through trapping or any other control method that increases the vector mortality) then R_0 drops to 0.301; V^*, the equilibrium vector population, drops from 5,000 to 3,000, and the corresponding halving time HT is (still) over one year (13.7 months).

If τ_2 and r_1 of Table 4.1 are taken equal to 0.068 and 0.00525 (with other parameters left unchanged) then $R_0 = 0.860$ and $\lambda = -6.705 \times 10^{-4}$, with a corresponding halving time of 103 months. This example shows that, depending on how the parameters are modified, a decrease in the basic reproduction number — which may seem a desirable result — can actually increase the halving time (by almost a year in the present example). The paradox of a decrease in R_0 bringing about a more protracted epidemic confirms that simply lowering R_0 may not be the desirable outcome of a control strategy.

Although R_0 is biologically meaningful, the parameter λ (or the halving time HT) is an epidemiologically more useful criterion for extinction since it incorporates the rate at which the disease dies out. This is particularly significant in the study of optimal control strategies since an important factor is *how long* a given strategy must be used to obtain a given result (for example, a 90% decrease of the infected population).

4.4.2   Control Strategies

We will now briefly explore the implications of these results for the study of optimal control strategies. These strategies involve vector control (which increases the vector mortality m) and the detection of infected individuals (which increases r_1). All other parameters are kept constant. Although the expression (4.14) for λ is a complicated function of r_1 and m, the relationship between λ and these two parameters appears clearly when one considers λ (which is ev_5) as the largest root of $H(x) = R^*$; see Figures 4.4 and 4.5. Indeed, if m increases with fixed r_1, then the intersection of $H(x)$ with the vertical axis moves upward, and ev_5 decreases without ever going below $-r_1$: no matter how much vector mortality is increased, the rate of extinction ev_5 cannot be brought under $-r_1$. Hence, in endemic situations in which infected persons may stay several years in the first stage, the halving time will also be several years, regardless of any attempts at vector control. Similarly, if r_1 is made to increase through aggressive screening, then ev_5 cannot drop below $-m$.

These relationships are explored more precisely in Figure 4.6, which represents λ as a function of r_1 and m, for $0.0075 < r_1 < 0.2$ and $1/15 < m < 0.4$. (The lower bounds thus correspond to the baseline values of Table 4.1.) The figure clearly shows how the "steepest descent" on the surface will depend on the initial ("pre-intervention") level of r_1 and m. If the initial values are those of Table 4.1, we are at the top right corner of the graph (point A) and the steepest decrease in λ is obtained by increasing both r_1 and m in order to stay in the "valley of steepest descent." If $m = 0.23$ (about the middle of its interval, with the same value of r_1, corresponding to point B) then it is apparent that further increasing m leaves λ virtually unchanged. The steepest descent is obtained by increasing r_1 alone. (This increase could go all the way to the valley in which both parameters should then increase in tandem.) Similarly, if r_1 is about 0.10 (point C), then m should be increased first and then both variables should increase together.

4.5   Alternative Models

There is evidence to suggest that vector migrations play an important role in the population dynamics of tsetse flies. Also, the well-documented persistence of sleeping-sickness foci in various areas of Africa suggest that the disease's transmission dynamics are quite slow. For these reasons,

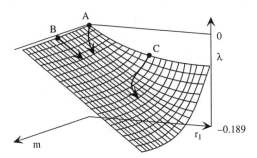

FIGURE 4.6   **Value of λ as a function of r_1 and m**

it is important to have epidemiologic models of sleeping sickness that
include migrating flies and that provide a measure of the speed at which
the disease evolves. We hope such models can contribute to a better
understanding of a serious disease that still affects large areas of Africa.
In this chapter we have discussed the construction and analysis of one
particular model that incorporates migration. Alternative models are
outlined in the present section.

4.5.1   *More Realistic Migration Model*

A more realistic model would allow for in- and out-migration of incubating and infected flies. One possibility is to postulate that the net
flow $k(V_c - V(t))dt$ of flies entering the system (or exiting, depending
on the sign of $V_c - V(t)$) can be divided according to three proportions
π_s, π_i, and π_a ($\pi_s + \pi_i + \pi_a = 1$) among the susceptible, incubating, and
actively infected vector populations. Hence, in the interval $(t, t + dt)$,
the net flow into each one of the three compartments is $\pi_s k(V_c - V(t))dt$,
$\pi_i k(V_c - V(t))dt$, and $\pi_a k(V_c - V(t))dt$. This is fine for in-migrations
when $V_c - V(t) > 0$. If however $V_c - V(t)$ is negative, then for example
the outflow $\pi_i k(V_c - V(t))dt$ from the compartment of incubating flies
could be larger than the number in the compartment and thus lead to
negative populations. For this reason a possible alternative is to have
the above mentioned three different flows *only for* $V_c - V(t) > 0$. In addition to this compensatory in-migration, each compartment can have,
regardless of $V_c - V(t)$, a constant emigration rate e representing the
natural tendency of vectors to migrate from one site to another. The
out-migration from each one of the three compartments in the time interval $(t, t + dt)$ would then simply be $eV_s(t)dt$, $eV_i(t)dt$, and $eV_a(t)dt$,
which avoids the problem of a population possibly becoming negative.

4.5.2 Markov Chain Model

We end by coming full circle and recalling an innocuous-sounding word mentioned in the introduction. We indicated our interest in developing a *variety* of mathematical models of sleeping sickness. Indeed, we feel that good modeling must use various mathematical tools, and that one must avoid simply sticking to a limited toolbox of techniques one is familiar with.

In the present case, it is common practice to use differential equations in population dynamics. However, this toolbox is only appropriate when populations are large enough to warrant considering them as continuous quantities (as opposed to the discrete, integer quantities they really are). In our example, the total human and vector populations are a few hundred and a few thousand, respectively. The infection prevalence can be 5–10% for humans and always less than one percent for flies — we thus typically have a few dozen infected humans or infected flies. But when there is extinction, these infected populations approach zero continuously, which is not realistic. Rather, the reality of extinction is a very discrete and random one. Typically, the disease dies out when a single remaining infected fly manages to infect only one or two individuals who are removed before they have the chance of infecting more flies.

The fact that populations are small, with a mechanism that is both discrete and random, suggests the use of *Markov chain models* (see Chapter 13). We now outline what such a model might be.

The compartments of Figure 4.2 would remain unchanged, with among susceptible vectors one extra sub-compartment of *teneral flies* $V_t(t)$, those young flies in the first age group who are the only ones really susceptible to infection. Given the discrete regular time interval of three days between blood meals, the states of the Markov chain could be defined at discrete points in times by the six integer variables $V_s(t)$, $V_t(t)$, $V_i(t)$, $V_a(t)$, $H_a(t)$, and $H_r(t)$. Each teneral fly bites a human with a certain probability, and this human is either infected or susceptible, so that each teneral fly has a certain probability of moving into the incubation compartment. (The bites of uninfected non-teneral flies or merely incubating flies are of no consequence.) Each actively infected fly also has a blood meal that may infect a susceptible human who then moves into the first stage.

There are other transitions to consider: a certain proportion of incubating flies will become actively infected, some humans will recover, others are removed, etc. Therefore, at time $t + 1$ the state of the system $[V_s(t+1), V_t(t+1), V_i(t+1), V_a(t+1), H_a(t+1), H_r(t+1)]$ can be probabilistically determined by the state of the system at time t, which is in accord with the definition of a Markov chain. In- and out-migrations are easily and realistically modeled in this framework with random numbers of flies leaving or entering each compartment.

4.6   Exercises and Projects

1. Examine the growth model (4.5) for the total population $V(t)$ by varying the initial population $V(0)$ relative to V^*. Also study what happens if $k + m - b < 0$.

2. Verify that p^* defined by (4.12) is a quadratic polynomial in the unknown $m - r_1$ and that this quadratic is always negative.

3. Verify the entries of the Jacobian matrix given in (4.20) by direct computation.

4. Use a symbolic algebra software package, such as Mathematica [9] or Mathcad [5], to check that the eigenvalues ev_3, ev_4, ev_5 of the Jacobian matrix (4.20) are the roots of the cubic equation (4.21).

5. Work out the details needed to verify the expression (4.23) given for P. Also check (numerically on an example, or analytically) that when there is a pair of complex conjugate eigenvalues, the corresponding eigenvectors in P also come in complex conjugate pairs.

6. Establish the results stated in (4.18)–(4.19).

7. (Project) Consider the more realistic migration model described in Section 4.5.1.

   a. Modify equations (4.1)–(4.3) to reflect the new migration model. Note that the vector mortality rate m simply becomes $m + e$ since emigration is just another form of exit.

   b. As before, add the three vector populations to obtain $V(t)$. Your expression for $V(t)$ should not be very different from that found in (4.5).

   c. Prove a stability theorem similar to Theorem 4.1 for this more realistic migration model.

8. (Project) Formulate in detail the Markov chain model outlined in Section 4.5.2 and carry out an analysis of this Markov model.

The focus in this chapter has been on *extinction*. However, depending on parameter values, the system described by the vector $P(t)$ may have nonzero equilibria. That is, for $R_0 > 1$ there will be one or more equilibrium values $P^* = [V^*, V_a^*, V_s^*, H_a^*, H_r^*]$ of $P(t)$. These values are obtained by setting $dV(t)/dt$, $dV_a(t)/dt$, $dV_s(t)/dt$, $dH_a(t)/dt$, and $dH_r(t)/dt$ equal to 0 in (4.4), (4.2), (4.3), (4.6), and (4.7). The following exercises and projects address possible equilibrium values other than the origin W_0.

9. Find and discuss these equilibrium points as a function of R_0 and the other parameters. **Hint:** These points can be found through a quadratic equation that has complex or real roots, depending on parameter values.

10. (Project) Simulate numerically the trajectories of the epidemic, for different values of R_0. (Remember that an equilibrium point can be either stable or unstable.) Show that, depending on the parameter values, we can have the following situations.

a. The origin is the only equilibrium point and it is stable (the result proved in this chapter).

b. The origin is unstable (the result proved here when $R_0 > 1$) and there is one other equilibrium point that appears to be locally stable: trajectories that start close to this equilibrium point converge to it, and it can be checked *numerically* that the eigenvalues of the Jacobian matrix at this point all have negative real parts. (It seems difficult to prove this analytically because the Jacobian matrix is of dimension five and, unlike the case for the origin, the characteristic equation does not reduce to a cubic.)

c. The origin is locally stable and there are two other equilibrium points: one that appears to be locally stable, the other unstable. Again, check these conjectures numerically using the Jacobian matrix. This is the most interesting and intriguing scenario. Try to interpret biologically its implications: namely, that the epidemic will either go to extinction or flare up, depending on the initial number of infected flies and humans. This behavior raises biologically challenging and mathematically difficult questions. For example, one would like to determine the different *domains of attraction*, the set of initial values that will drive the system to extinction and those initial values that will lead to a positive equilibrium. What is the frontier between these domains, and how does the system behave for an initial value on the frontier?

4.7   References

[1] M. Artzrouni and J.-P. Gouteux, "A compartmental model of vector-borne diseases: application to sleeping sickness in central Africa," *Journal of Biological Systems* **4**, 459–477 (1996).

[2] M. Artzrouni and J.-P. Gouteux, "Control strategies for sleeping sickness in central Africa: a model-based approach," *Tropical Medicine and International Health* **1**, 753–764 (1996).

[3] J. W. Hargrove, "Discrepancies between estimates of tsetse fly populations using mark-recapture and removal trapping techniques," *J. Applied Ecology* **18**, 737–748 (1981).

[4] M. J. D. Lebreton and C. Millier, "Modèles dynamiques déterministes définis par des équations différentielles," *Modèles Dynamiques Déterministes en Biologie*, Masson, Paris, 1982, 13–57.

[5] MathSoft, Inc., *Mathcad 7 Users Guide*, MathSoft, Inc., Cambridge, MA, 1997. <http://www.mathcad.com/>

[6] D. J. Rogers, "A general model for the African trypanosomiases," *Parasitology* **97**, 193–212 (1988).

[7] D. J. Rogers, "The dynamics of vector-transmited diseases in human communities," *Philosophical Transactions of the Royal Society of London, Series B* **321**, 513–539 (1988).

[8] D. J. Rogers, "The development of analytical models for human trypanosomiasis," *Annales de la Société Belge de Médecine Tropicale* **69** (suppl. 1), 73–88 (1989).

[9] S. Wolfram, *The Mathematica Book*, 3rd ed., Cambridge University Press, Cambridge, 1996. <http://www.wri.com/>

Chapter 5

Mathematical Models in Classical Cryptology

Joel Brawley
Shuhong Gao

Prerequisites: linear and modern algebra

5.1 Introduction

At the beginning of his 1941 invited address on cryptography to the American Mathematical Society in Manhattan, Kansas, the great University of Chicago algebraist A. A. Albert said "We shall see that cryptography is more than a subject permitting mathematical formulation, for indeed it would not be an exaggeration to state that abstract cryptography is identical with abstract mathematics" [4].

It has been over a half century since Albert made this statement and during this time there have been dramatic new developments in cryptography including the entire concept of public-key cryptography, originated by Diffie and Hellman in 1976 [1]. Consequently, if Albert's claim were true in 1941, it is equally true today because many of the developments in these two areas, cryptology and mathematics, have gone hand-in-hand. The main purposes of the present chapter are to introduce to students of the mathematical sciences several of the many mathematical models used in classical cryptology and to show some of the interplay between the two subjects. In the companion Chapter 6, we will explore mathematical models used in public-key cryptology. It is hoped that these expositions will give some insights into what Albert meant by his statement and why it is as valid today as it was when he made it.

This chapter is intended for advanced undergraduates or beginning graduate students who have had little or no experience in cryptography.

The prerequisites for reading it are, at most, undergraduate study in modern algebra and linear algebra. Chapter 6, while slightly more sophisticated, can also be read within this framework, perhaps with the addition of elementary number theory. We have included a number of student exercises (of varying degrees of difficulty) and most of these can be completed within the above mentioned prerequisites. Additionally, we have tried to include enough references to aid students who desire to study deeper into the subject. In particular, we suggest the recent book [11] by D. R. Stinson which contains treatments of both classical and public-key cryptology.

5.2   Some Terminology of Cryptology

The term *cryptography* refers to the activity of devising secure methods for secret communications between two or more parties, and it is the task of the *cryptographer* to produce such methods. While cryptography includes methods for concealing the fact that communication is happening (e.g., invisible inks and microdots), we shall deal only with the kind of cryptography where the message to be communicated, the so-called *plaintext*, is transformed into a new form called the *ciphertext* (or *cryptogram*), which is presumably unintelligible to an unauthorized reader.

When two (authorized) parties are communicating using such a cryptographic transformation, the process of converting the plaintext to the ciphertext is called the *enciphering* process. Of course, the enciphering transformation must be invertible so that the plaintext can be uniquely recovered without ambiguity from the ciphertext. The process of transforming the ciphertext back to plaintext is called the *deciphering* process.

In contrast, *cryptanalysis* is that activity undertaken by an unauthorized party in an attempt to "read" or make intelligible an intercepted ciphertext, and it is the *cryptanalyst* who engages in this activity. When the cryptanalyst is able to decipher a message, he or she has *decrypted* it.

Of course, one does not generally carry out cryptanalysis without an understanding of cryptography and one does not generally carry out cryptography without an understanding of cryptanalysis; i.e., the cryptographer and the cryptanalyst are often the same person wearing different hats. For example, the designer of a good system must necessarily test the system against onslaughts of cryptanalysis and this often means that the designer becomes a cryptanalyst trying to break his or her own system. The word *cryptology* is a more general term which is used to encompass both cryptography and cryptanalysis, as indicated schematically in Figure 5.1.

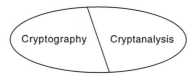

FIGURE 5.1 **Components of cryptology**

5.3 *Simple Substitution Systems within a General Cryptographic Framework*

The present section outlines a general mathematical framework which will serve as a model for several of the classical cryptographic systems described later, and we indicate here how this general framework specializes to the familiar simple substitution system.

Let $\mathcal{L}$ and $\mathcal{K}$ be nonempty finite sets and let $\mathcal{P}(\mathcal{L})$ denote the group of all permutations of $\mathcal{L}$ under the operation of composition. A *cryptographic system* based on $\mathcal{L}$ and $\mathcal{K}$ is a triple $(\mathcal{L}, \mathcal{K}, \phi)$ consisting of the sets $\mathcal{L}$ and $\mathcal{K}$ together with a function $\phi : \mathcal{K} \to \mathcal{P}(\mathcal{L})$ from $\mathcal{K}$ into the group of permutations of $\mathcal{L}$. The sets $\mathcal{L}$ and $\mathcal{K}$ will be called, respectively, the set of *letters* and the set of *keys*. Each permutation $f \in P(\mathcal{L})$ is called a *cipher*. The word "cipher" literally means a method of transforming text to conceal its meaning and, as we shall see, this is the purpose of f.

The set of letters $\mathcal{L}$ in this definition is an abstract set; however, we are thinking of $\mathcal{L}$ as containing the alphabetic characters or perhaps sequences of characters from the alphabet in which we are communicating. For example, $\mathcal{L}$ might be the set of letters in some standard alphabet, like $\{A, B, \ldots, Z\}$ or $\{0, 1\}$, or it might be the set of letters of a standard alphabet together with several punctuation marks, say $\{A, B, \ldots, Z, ?, ., \varnothing\}$ where $\varnothing$ denotes a blank space. It might also be the collection of symbols of some coded alphabet like Morse or ASCII code, or it could be Cartesian products of some given set: e.g., $\{0, 1\}^4 = \{0, 1\} \times \{0, 1\} \times \{0, 1\} \times \{0, 1\} = \{0000, 0001, 0010, 0011, \ldots, 1111\}$. We are thus thinking of the members of $\mathcal{L}$ as being "generalized" alphabetic letters or characters. By a *message* we mean a sequence $p_1, p_2, p_3, \ldots$ of letters from $\mathcal{L}$.

The reader should note that within these definitions it is possible for an entire phrase like

"COMEØATØMIDNIGHT"

to be viewed as a single "generalized" letter; e.g., $\mathcal{L}$ could be the Cartesian product of 16 copies of the set $\{A, B, \dots, Z, ?, ., \varnothing\}$, in which case
the entire message above could be viewed as a single letter (or mem-
ber) of this Cartesian product set. The idea behind the above definition
is explained as follows. The two communicating parties (and perhaps
an unauthorized party as well) all know the ingredients of the crypto-
graphic system $(\mathcal{L}, \mathcal{K}, \phi)$. Privately, however, the communicating parties
have agreed upon a specific key k from $\mathcal{K}$ (which they may change from
time to time), and this key is known only to authorized parties. The
private key k thus identifies via ϕ a single permutation f of $\mathcal{L}$. The
encipherment of a plaintext message, say $p_1, p_2, p_3, \dots$ is the ciphertext
message $c_1, c_2, c_3, \dots$, where c_i is given by

$$c_i = f(p_i), \qquad i = 1, 2, 3, \dots$$

It is assumed in classical cryptography that a knowledge of the specific
key k allows one to easily obtain the inverse function f^{-1}, in which case
the decipherment of $c_1, c_2, c_3, \dots$ is given by

$$p_i = f^{-1}(c_i), \qquad i = 1, 2, 3, \dots$$

Example 5.1

Consider the case where $\mathcal{L}$ is the following 31-letter alphabet consisting of the standard 26 letters in the English alphabet together with
5 punctuation marks $\{ \, . \, ? \, , \, ' \, \varnothing \}$, where again $\varnothing$ denotes a blank
space:
$$\mathcal{L} = \{A\ B\ C\ D \dots Z \, . \, ? \, , \, ' \, \varnothing\} \qquad (5.1)$$
Further, let $\mathcal{K}$ denote the group $\mathcal{P}(\mathcal{L})$ of all 31! permutations of $\mathcal{L}$,
and let $\phi : \mathcal{P}(\mathcal{L}) \to \mathcal{P}(\mathcal{L})$ be the identity mapping. Then clearly the
cryptographic system $(\mathcal{L}, \mathcal{K}, \phi)$ is just the familiar *simple substitution
system*. The key k that both parties agree on in advance is the
permutation of $\mathcal{L}$ used in the encipherment.

Simple substitution systems are the most familiar of all the cryptosystems; in fact, versions of the above example appear in many daily
newspapers across the United States. The keys $\mathcal{K}$ (or permutations of
$\mathcal{L}$) which are used in such newspaper ciphers usually have the following
two additional properties:

- They "fix" the five symbols $\{ \, . \, ? \, , \, ' \, \varnothing \}$ and any other punctuation
  marks.

- They "move" every one of the symbols {A B C D ... Z}; in other words, none of the standard letters are mapped to themselves.

It is also true of these newspaper ciphers that while there are a very large number of possible keys or permutations (more than 1.4×10^{26}), the existence of such a large number of keys does not mean newspaper ciphers are hard to decrypt; indeed, their appeal to the amateur cryptanalyst is that they are relatively easy to read using such things as (a) frequencies of letters (e.g., E is the most frequently occurring letter in standard English), (b) probable words (e.g., THE is a common word in English whereas A and I are the only two single-letter words), and (c) the structure of punctuation (e.g., if one of the conjunctions AND or BUT is used to connect two sentences, the conjunction is preceded by a comma).

The so-called *Caesar ciphers* represent a special collection of 26 simple substitution systems in which one first orders the alphabet, say the natural ordering

$$\pi = (\text{A B C D} \ldots \text{Z}), \qquad (5.2)$$

and thinks of it as a circular ordering (so B follows A, C follows B, ..., and A follows Z). Once a circular ordering is specified, the key set $\mathcal{K}$ is the set of integers $\{k : 0 \leq k \leq 25\}$ and the permutation associated with a given key k is the mapping of $\mathcal{L}$ to $\mathcal{L}$ which transforms each letter into the letter appearing k places down the alphabet. For example, with a key of $k = 3$, the associated permutation takes A to D, B to E, C to F, and so on, so that X goes to A, Y to B, and Z to C. Of course, if we think of the π of (5.2) as a permutation of the letters $\mathcal{L}$ written in its cycle form, then the element in $\mathcal{P}(\mathcal{L})$ to which the key element $k = 3$ corresponds under ϕ is just the mapping π^3, and the image under ϕ of the entire key set $\mathcal{K} = \{k : 0 \leq k \leq 25\}$ is just the cyclic subgroup of $\mathcal{P}(\mathcal{L})$ generated by π.

We shall give further examples in the next section, but we now want to introduce an additional algebraic structure, namely that of a ring, into simple substitution cryptography. The value of introducing a ring is that the ring operations of addition, multiplication, subtraction, and division (when defined), can be used as enciphering and deciphering aids. In other words, a computer which knows (or can be programmed to know) the ring arithmetic can facilitate the processes of enciphering and deciphering.

Let $\mathcal{L}$ denote the set of letters to be used in the secret communications, let $\mathbf{R}$ be a ring with an identity whose cardinality is the same as that of $\mathcal{L}$, and let $\alpha : \mathcal{L} \to \mathbf{R}$ denote a one-to-one correspondence between $\mathcal{L}$ and $\mathbf{R}$. For example, assuming that

$$\mathcal{L} = \{\text{A B C D} \ldots \text{Z}\},$$

we may choose $\mathbf{R}$ to be the ring of integers modulo 26, and we may let α denote the letter-ring correspondence given by

$$\alpha = \begin{array}{|ccccccccccccc|} \hline A & B & C & D & E & F & G & H & I & J & K & L & M \\ 0 & 1 & 2 & 3 & 4 & 5 & 6 & 7 & 8 & 9 & 10 & 11 & 12 \\ \hline N & O & P & Q & R & S & T & U & V & W & X & Y & Z \\ 13 & 14 & 15 & 16 & 17 & 18 & 19 & 20 & 21 & 22 & 23 & 24 & 25 \\ \hline \end{array} \tag{5.3}$$

where each letter is associated with the ring element under it.

Next let $f : \mathbf{R} \to \mathbf{R}$ denote a permutation of $\mathbf{R}$, and note that such a permutation could be selected, if desired, so as to have a "nice" rule involving the ring operations for computing the image of x under f. For example, if $\mathbf{R}$ is the ring of integers modulo 26 and a and b are integers in the range $0 \le a, b \le 25$, where a is relatively prime to 26, then the function defined by

$$y = f(x) = ax + b \tag{5.4}$$

is a permutation of $\mathbf{R}$ where the addition and multiplication operations assumed in (5.4) are the modulo 26 ring operations.

Since a composition of bijective functions is bijective, it is clear that the composition $\alpha^{-1} \circ f \circ \alpha$ (read right to left) defines a simple substitution system from $\mathcal{L}$ to $\mathcal{L}$.

Example 5.2

Let $\mathbf{R}$ denote the ring of integers modulo 26, let $f(x)$ be given by

$$y = f(x) = 9x + 14, \tag{5.5}$$

and let α be given by (5.3). To encipher letters of the message COME AT MIDNIGHT with the composition $\alpha^{-1} \circ f \circ \alpha$, we first apply α; i.e., we first convert the letters to their ring equivalents using (5.3) to obtain

$$2, 14, 12, 4, \varnothing, 0, 19, \varnothing, 12, 8, 3, 13, 8, 6, 7, 19.$$

Next, we apply (5.5) to get

$$6, 10, 18, 24, \varnothing, 14, 3, \varnothing, 18, 8, 15, 1, 8, 16, 25, 3,$$

and finally we apply α^{-1} to get the ciphertext

$$\text{G,K,S,Y,}\varnothing\text{,O,D,}\varnothing\text{,S,I,P,B,I,Q,Z,D,} \tag{5.6}$$

which is written "GKSY OD SIPBIQZD". Here, we have used the newspaper cipher convention that the space is fixed under the transformation.

The decipherment is obtained in a similar manner except that the inverse of $\alpha^{-1} \circ f \circ \alpha$, which is $\alpha^{-1} \circ f^{-1} \circ \alpha$, must be applied. In this case, because 3 is the inverse of 9 (mod 26), it is easily seen that the inverse of (5.5) is

$$x = f^{-1}(y) = 3y + 10.$$

Thus, for instance, to decipher the letter G, we have

$$\alpha^{-1} \circ f^{-1} \circ \alpha(\mathsf{G}) = \alpha^{-1}(f^{-1}(6)) = \alpha^{-1}(2) = \mathsf{C}.$$

Perhaps the reader can begin to see from this very elementary example that many generalizations are possible. For example, one could use two different one-to-one correspondences, $\alpha : \mathcal{L} \to \mathbf{R}$ and $\beta : \mathcal{L} \to \mathbf{R}$ in which case the composition $\beta^{-1} \circ f \circ \alpha$ is a simple substitution system. Other generalizations will be given in the next section.

5.4   *The Vigenére Cipher and One-Time Pads*

In the present section we describe the Vigenére cipher, which is one of many systems using what is called *polyalphabetic substitution*. This word simply means that several (or many) different simple substitution permutations are used in the encipherment process. We explain the Vigenére system with an example.

Consider the square of letters shown in Figure 5.2; this square is called the *standard Vigenére square*. The first row of the square consists of the letters in the alphabet in their standard order, and each row after the first represents a single circular left shift of the row just above it. In the general Vigenére square, the first row can be any arrangement (permutation) of the letters, but just as in the standard square, each of the remaining rows is a single circular left shift of the row immediately above it. There are 26! different Vigenére squares depending on the first row.

Suppose that Bob wants to send the message COME AT MIDNIGHT to Alice using the Vigenére square of Figure 5.2. Suppose further that in advance Bob and Alice have shared a secret sequence of letters; say the sequence ROCKY. This sequence is called the *key* sequence and is used along with the square to encipher a message as follows. First, Bob repeatedly copies the key sequence under the message he wants to send. That is, he writes

COMEATMIDNIGHT
ROCKYROCKYROCK

```
A B C D E F G H I J K L M N O P Q R S T U V W X Y Z
B C D E F G H I J K L M N O P Q R S T U V W X Y Z A
C D E F G H I J K L M N O P Q R S T U V W X Y Z A B
D E F G H I J K L M N O P Q R S T U V W X Y Z A B C
E F G H I J K L M N O P Q R S T U V W X Y Z A B C D
F G H I J K L M N O P Q R S T U V W X Y Z A B C D E
G H I J K L M N O P Q R S T U V W X Y Z A B C D E F
H I J K L M N O P Q R S T U V W X Y Z A B C D E F G
I J K L M N O P Q R S T U V W X Y Z A B C D E F G H
J K L M N O P Q R S T U V W X Y Z A B C D E F G H I
K L M N O P Q R S T U V W X Y Z A B C D E F G H I J
L M N O P Q R S T U V W X Y Z A B C D E F G H I J K
M N O P Q R S T U V W X Y Z A B C D E F G H I J K L
N O P Q R S T U V W X Y Z A B C D E F G H I J K L M
O P Q R S T U V W X Y Z A B C D E F G H I J K L M N
P Q R S T U V W X Y Z A B C D E F G H I J K L M N O
Q R S T U V W X Y Z A B C D E F G H I J K L M N O P
R S T U V W X Y Z A B C D E F G H I J K L M N O P Q
S T U V W X Y Z A B C D E F G H I J K L M N O P Q R
T U V W X Y Z A B C D E F G H I J K L M N O P Q R S
U V W X Y Z A B C D E F G H I J K L M N O P Q R S T
V W X Y Z A B C D E F G H I J K L M N O P Q R S T U
W X Y Z A B C D E F G H I J K L M N O P Q R S T U V
X Y Z A B C D E F G H I J K L M N O P Q R S T U V W
Y Z A B C D E F G H I J K L M N O P Q R S T U V W X
Z A B C D E F G H I J K L M N O P Q R S T U V W X Y
```

FIGURE 5.2   The standard Vigenére square

Table 5.1   Vigenére encipherment of a message

message	C	O	M	E	A	T	M	I	D	N	I	G	H	T
ring equiv	2	14	12	4	0	19	12	8	3	13	8	6	7	19
num key	17	14	2	10	24	17	14	2	10	24	17	14	2	10
+ mod 26	19	2	14	14	24	10	0	10	13	11	25	20	9	3
ciphertext	T	C	O	O	Y	K	A	K	N	L	Z	U	J	D

Now to encipher the first letter C, Bob notes that letter R appears under C. He then goes to the Vigenére square and finds that letter at the intersection of the column headed by C and the row beginning with R; i.e., he locates a T. This means C enciphers as T. Similarly, the encipherment of O is C since C is at the intersection of the column headed by O and the row beginning with O, and the encipherment of M is O because O is at the intersection of the column headed by M and the row headed by C. Continuing in this way, the complete encipherment of the message is

T C O O Y K A K N L Z U J D.

To decipher this message, Alice reverses the process: e.g., to decipher T she locates letter T in the row beginning with R and reads C at the top of the column of T.

We shall now show how to use the modulo 26 ring and the correspondence α of (5.3) to do the above Vigenére encipherment. In the first place, note that the key sequence ROCKY corresponds via (5.3) to the number sequence 17, 14, 2, 10, 24. This means that we shall use the following sequence of mappings f_i to encipher the letters:

$$f_1(x) = x + 17, \ f_2(x) = x + 14, \ f_3(x) = x + 2,$$

$$f_4(x) = x + 10, \ f_5(x) = x + 24.$$

That is, to encipher the first letter we use the mapping $\alpha^{-1} \circ f_1 \circ \alpha$, to encipher the second letter we use $\alpha^{-1} \circ f_2 \circ \alpha$, and so on, with the sixth letter being enciphered using $\alpha^{-1} \circ f_1 \circ \alpha$ for a second time. The entire enciphering process is illustrated concisely in Table 5.1.

Decipherment is handled using the inverse functions, which in this case are

$$f_1^{-1}(x) = x + 9, \ f_2^{-1}(x) = x + 12, \ f_3^{-1}(x) = x + 24,$$

$$f_4^{-1}(x) = x + 16, \ f_5^{-1}(x) = x + 2.$$

(Of course, the inverse of $f(x) = x + a$ is $f^{-1}(x) = x + (-a)$ where $-a$ is the additive inverse of a in the ring **R**.) The deciphering process is illustrated concisely in Table 5.2.

Table 5.2   Vigenére decipherment of a message

ciphertext	T	C	O	O	Y	K	A	K	N	L	Z	U	J	D
ring equiv	19	2	14	14	24	10	0	10	13	11	25	20	9	3
inv key	9	12	24	16	2	9	12	24	16	2	9	12	24	16
+ mod 26	2	14	12	4	0	19	12	8	3	13	8	6	7	19
message	C	O	M	E	A	T	M	I	D	N	I	G	H	T

In the case of the general Vigenére square, if the letter arrangement of the first row of the square is denoted by $(L_0, L_1, \ldots, L_{25})$, then the encipherment process can be modeled mathematically by using the correspondence α defined by $\alpha(L_k) = k$ in place of (5.3) and by changing the key sequence of letters into a numerical sequence $(k_1, k_2, \ldots, k_t)$ where k_i is the image of the ith letter of the sequence under α.

To put the collection of Vigenére ciphers of key length t into the general mathematical framework described in Section 5.3, let L denote a finite set of m alphabetic characters, let $\mathbf{R}$ denote the additive group of integers modulo m, and let α denote a fixed one-to-one correspondence from L to $\mathbf{R}$. Further, let $\mathcal{L} = L^t$ denote the set of all t-tuples of letters from L, and let $\mathcal{K} = \mathbf{R}^t$ denote the set of all t-tuples $k = (k_1, k_2, \ldots, k_t)$ of ring elements from $\mathbf{R}$. With each key element $k \in \mathcal{K}$, we associate the permutation $\phi(k)$ of $\mathcal{L}$ defined as follows: each t-component vector of letters $(L_1, L_2, \ldots, L_t) \in \mathcal{L}$ is mapped under $\phi(k)$ to the t-component vector of letters $(L'_1, L'_2, \ldots, L'_t) \in \mathcal{L}$, where

$$L'_j = \alpha^{-1}(\alpha(L_j) + k_j). \tag{5.7}$$

Then the triple $(\mathcal{L}, \mathcal{K}, \phi)$ describes all Vigenére systems of key length t based on a fixed α.

We can incorporate the correspondence α as part of the key for the system by simply putting $\mathcal{K} = \mathcal{A} \times \mathbf{R}^t$, where $\mathcal{A}$ denotes the set of all possible correspondences α between L and $\mathbf{R}$. In this case, the key necessary for using (5.7) as an encipherment process consists of a pair (α, k) where $\alpha \in \mathcal{A}$ and $k = (k_1, k_2, \ldots, k_t) \in \mathbf{R}^t$.

As demonstrated in the example above, the key sequence in Vigenére encipherment is usually considered as finite, but it need not be. For example, one could employ what is called *ciphertext autokey* by agreeing, say, to use $f(x) = x + 3$ for enciphering the first letter of plaintext, and for the ith letter of plaintext, $i \geq 2$, to use the equation $f(x) = x + c_{i-1}$ where c_{i-1} is the $(i - 1)$st letter of ciphertext. To illustrate, COME AT MIDNIGHT would be enciphered as follows: C = 2 would encipher to $2 + 3 = 5 = $ F (since 3 was initially agreed on), O = 14 would encipher to $14 + 5 = 19 = $ T (since 5 was the last value of ciphertext), M = 12 would encipher to $12 + 19 = 5 = $ F (since 19 was last value of ciphertext), and so on. Of course, one could also use a *plaintext autokey* scheme where the next additive key value depends on the previous plaintext value.

All of the methods described thus far in this section are vulnerable to various kinds of cryptanalysis attacks, usually because of letter frequency considerations. There is, however, one generalization of the Vigenére system called the *one-time pad* which, in its purest form, is not vulnerable to any frequency attack and indeed is truly an *unbreakable* cipher. To describe it, consider an arbitrarily long sequence $k_1, k_2, k_3, \ldots$ of integers, $0 \le k_i \le 25$, which has been generated (let us assume) as values of independent, identically distributed random variables from a uniform probability distribution on $\{0, 1, 2, \ldots, 25\}$: that is,

$$P(k_i = j) = \frac{1}{26}, \qquad 0 \le j \le 25.$$

Further, suppose that once the values $k_1, k_2, k_3, \ldots$ are generated, they are printed sequentially in a booklet (or pad) and that the booklet is duplicated exactly once with a copy given to each of the two parties, Bob and Alice. Then, when Alice wants to send a message to Bob, she simply lets him know exactly that position on the pad from which she will select her first bit of additive key. She then sequentially goes through the pad selecting values for the additive key and, since Bob has a copy of the pad, he knows the various key numbers Alice is using and thus can decipher her message. They both tear out pages from the pad as they are used, never using the same key twice (whence the name "one-time pad"). Then, so long as they follow this one-time procedure, there is no way to break the system without knowledge of the pad.

The Vigenére system with a finite key sequence, the ciphertext autokey, and the one-time pad system described above are special cases of the so-called *stream ciphers* [9] in which a key stream $k_1, k_2, k_3, \ldots$ is used to vary the encipherment from letter to letter.

5.5   *The Basic Hill System and Variations*

The Hill system is named after the mathematician Lester S. Hill (1891–1961) who, while a professor at Hunter College in New York, described his basic ideas in two papers [2, 3] published in 1929 and 1931. Although the system itself has seen very little practical use, it had a real impact on cryptography [4] and was the first system which truly depended on algebra and abstract mathematics in its formulation. We include a description of the Hill system here because it demonstrates so clearly the interplay between interesting theoretical questions in algebra and practical cryptology questions. Also, we believe it provides some valuable insights as to what Albert might have meant in his quote cited in the introduction.

The ingredients of the basic Hill system are: (a) a finite set $\mathcal{L}$ called the set of letters, (b) a finite ring $\mathbf{R}$ with identity (usually taken to be commutative) whose order is that of $\mathcal{L}$, (c) a one-to-one correspondence α from $\mathcal{L}$ to $\mathbf{R}$, and (d) an $n \times n$ invertible matrix A over $\mathbf{R}$ where n is an arbitrary fixed positive integer. In this case, the correspondence α induces, componentwise, a one-to-one function from $\mathcal{L}^n$ to $\mathbf{R}^n$; that is, if $L = (L_1, L_2, \ldots, L_n) \in \mathcal{L}^n$, then

$$\alpha(L) = (\alpha(L_1), \alpha(L_2), \ldots, \alpha(L_n)), \tag{5.8}$$

where we have also called the induced mapping α. It is conventional to think of the members of both $\mathcal{L}^n$ and $\mathbf{R}^n$ as columns, and we do so here. Note that α^{-1}, the inverse of the mapping α, induces a mapping from $\mathbf{R}^n$ to $\mathcal{L}^n$ which is the inverse of (5.8). Again we somewhat abuse the notation and refer to this inverse mapping from $\mathbf{R}^n$ to $\mathcal{L}^n$ as α^{-1}.

Now consider a message consisting of a finite sequence of letters from $\mathcal{L}$. The basic encipherment process outlined by Hill is as follows. Group the first n letters of the message, let us call them $L_{11}, L_{21}, \ldots, L_{n1}$, to form the column vector $P_1 = (L_{11}, L_{21}, \ldots, L_{n1})$. Similarly, group the next n letters $L_{12}, L_{22}, \ldots, L_{n2}$ to form the second column vector $P_2 = (L_{12}, L_{22}, \ldots, L_{n2})$, and so on, until the entire message is partitioned into n-component vectors. In case the number of letters in the message is not divisible by n, the last group is padded by so-called nulls (meaningless letters) to make an n-component vector; hence, the entire message becomes a sequence of n-component column vectors $P_1, P_2, \ldots, P_t$. The encipherment of vector P_i is the n-component vector C_i given by

$$C_i = \alpha^{-1}(A(\alpha(P_i))), \qquad i = 1, 2, \ldots, t, \tag{5.9}$$

where both α and α^{-1} denote the induced componentwise mappings. In other words, the ciphertext is $C_1, C_2, \ldots, C_t$ which is transmitted as a letter sequence $C_{11}C_{21} \ldots C_{n1}C_{12}C_{22} \ldots C_{n2} \ldots C_{1t}C_{2t} \ldots C_{nt}$. The decipherment of (5.9) is, of course, given by

$$P_i = \alpha^{-1}(A^{-1}(\alpha(C_i))), \qquad i = 1, 2, \ldots, t, \tag{5.10}$$

where A^{-1} is the inverse of matrix A.

Example 5.3

Consider the 31-letter alphabet

$$\mathcal{L} = \{\text{A B C D} \ldots \text{Z . ? , ' } \emptyset\}, \tag{5.11}$$

let $\mathbf{R}$ denote the finite field $\mathbf{R} = \{0, 1, 2, \ldots, 30\}$ modulo 31, let α denote the letter-field correspondence

$$
\alpha =
\begin{array}{|ccccccccccccccccc|}
\hline
A & B & C & D & E & F & G & H & I & J & K & L & M & N & O & P \\
0 & 1 & 2 & 3 & 4 & 5 & 6 & 7 & 8 & 9 & 10 & 11 & 12 & 13 & 14 & 15 \\
\hline
Q & R & S & T & U & V & W & X & Y & Z & . & ? & , & ' & \emptyset & \\
16 & 17 & 18 & 19 & 20 & 21 & 22 & 23 & 24 & 25 & 26 & 27 & 28 & 29 & 30 & \\
\hline
\end{array}
\tag{5.12}
$$

and let A denote the 2×2 matrix

$$
A = \begin{bmatrix} 10 & 4 \\ 21 & 16 \end{bmatrix}.
\tag{5.13}
$$

As a matrix over $\mathbf{R}$, A is easily seen to be invertible with inverse

$$
A^{-1} = \begin{bmatrix} 10 & 13 \\ 14 & 14 \end{bmatrix}.
\tag{5.14}
$$

Now suppose we want to transmit the message "COME AT MID-NIGHT." Replacing the spacing between words with the symbol $\emptyset$ and grouping the resulting symbols as vectors of 2-components, we get

$$
\begin{bmatrix} C \\ O \end{bmatrix} \begin{bmatrix} M \\ E \end{bmatrix} \begin{bmatrix} \emptyset \\ A \end{bmatrix} \begin{bmatrix} T \\ \emptyset \end{bmatrix} \begin{bmatrix} M \\ I \end{bmatrix} \begin{bmatrix} D \\ N \end{bmatrix} \begin{bmatrix} I \\ G \end{bmatrix} \begin{bmatrix} H \\ T \end{bmatrix} \begin{bmatrix} . \\ \emptyset \end{bmatrix},
$$

where we have added a space (null) after the period to make the number of symbols in the message divisible by 2. Applying the mapping α of (5.12) to each of these vectors produces the following sequence of vectors over $\mathbf{R}$:

$$
\begin{bmatrix} 2 \\ 14 \end{bmatrix} \begin{bmatrix} 12 \\ 4 \end{bmatrix} \begin{bmatrix} 30 \\ 0 \end{bmatrix} \begin{bmatrix} 19 \\ 30 \end{bmatrix} \begin{bmatrix} 12 \\ 8 \end{bmatrix} \begin{bmatrix} 3 \\ 13 \end{bmatrix} \begin{bmatrix} 8 \\ 6 \end{bmatrix} \begin{bmatrix} 7 \\ 19 \end{bmatrix} \begin{bmatrix} 26 \\ 30 \end{bmatrix}.
$$

Multiplying each of these vectors by the matrix A gives the sequence

$$
\begin{bmatrix} 14 \\ 18 \end{bmatrix} \begin{bmatrix} 12 \\ 6 \end{bmatrix} \begin{bmatrix} 21 \\ 10 \end{bmatrix} \begin{bmatrix} 0 \\ 11 \end{bmatrix} \begin{bmatrix} 28 \\ 8 \end{bmatrix} \begin{bmatrix} 20 \\ 23 \end{bmatrix} \begin{bmatrix} 11 \\ 16 \end{bmatrix} \begin{bmatrix} 22 \\ 17 \end{bmatrix} \begin{bmatrix} 8 \\ 3 \end{bmatrix},
$$

which on applying the inverse α^{-1} of (5.12) becomes

$$
\begin{bmatrix} O \\ S \end{bmatrix} \begin{bmatrix} M \\ G \end{bmatrix} \begin{bmatrix} V \\ K \end{bmatrix} \begin{bmatrix} A \\ L \end{bmatrix} \begin{bmatrix} , \\ I \end{bmatrix} \begin{bmatrix} U \\ X \end{bmatrix} \begin{bmatrix} L \\ Q \end{bmatrix} \begin{bmatrix} W \\ R \end{bmatrix} \begin{bmatrix} I \\ D \end{bmatrix}.
$$

Thus, the transmitted ciphertext is "OSMGVKAL,IUXLQWRID". The decipherment is carried out in a similar fashion except that it uses (5.10).

Hill introduced his system well before the advent of high-speed digital computers and, at that time, its implementation for matrices of any size was quite awkward and tedious. Because of this, he went so far as to patent a mechanical device for doing the encipherment. Apparently, a prototype of Hill's device was never built, but a picture of it is shown in Kahn [4, p. 409].

It is interesting that the matrix A for Hill's mechanical encipherment machine was built into the machine (really hard-wired) and could not easily be changed. This meant that another machine based on the matrix A^{-1} would be required for decipherment. For this reason, Hill advocated using a so-called *involutory matrix* satisfying $A = A^{-1}$, in which case the same machine could be used for both enciphering and deciphering.

Example 5.4

Consider the matrix defined by

$$B = \begin{bmatrix} 18 & 13 \\ 30 & 13 \end{bmatrix}. \tag{5.15}$$

It is then easy to verify that as a matrix over the modulo 31 field, B satisfies the matrix equation $B = B^{-1}$ (or equivalently $B^2 = I$), so that B is involutory. Using the correspondence (5.12) and the matrix B of (5.15) in place of (5.13) in the enciphering equation (5.9), the encipherment of the message "COME AT MIDNIGHT." is found to be "BZUJNBTØKØGLFIBXVX" which is transmitted with spaces as "BZUJNBT K GLFIBXVX". The reader can verify that a second application of the transformation (5.9) recovers the plaintext.

The basic transformation used in the Hill system (5.9) can be depicted as

$$\mathcal{L}^n \xrightarrow{\alpha} \mathbf{R}^n \xrightarrow{A} \mathbf{R}^n \xrightarrow{\alpha^{-1}} \mathcal{L}^n,$$

where the mapping from $\mathbf{R}^n$ to $\mathbf{R}^n$ is just a linear transformation; i.e.,

$$y = Ax. \tag{5.16}$$

With this in mind, a number of variations to the basic Hill system are easy to describe. In the first place, one may use different correspondences for converting from letters to numbers. This idea is depicted by

$$\mathcal{L}^n \xrightarrow{\alpha} \mathbf{R}^n \xrightarrow{A} \mathbf{R}^n \xrightarrow{\beta^{-1}} \mathcal{L}^n.$$

Another idea is to use in place of (5.16) an affine transformation

$$y = Ax + b, \tag{5.17}$$

where b is a column vector.

To make the system even more secure, one could vary the linear transformation with the vector to be enciphered; that is, one might use a different linear transformation for enciphering each of the different plaintext vectors P_k so that encipherment of the kth vector would be carried out with a transformation of the form

$$y = A(k)x, \tag{5.18}$$

where the matrix $A(k)$ is a function of k. Other variations of the Hill system are described by Levine [5].

Of course the "sandwiched" transformation $\mathbf{R}^n \xrightarrow{A} \mathbf{R}^n$ need not be linear; i.e., it could be replaced by $\mathbf{R}^n \xrightarrow{f} \mathbf{R}^n$, where f is any one-to-one mapping from $\mathbf{R}^n$ to $\mathbf{R}^n$. These ideas and others have been considered in the literature and several are explored in the chapter exercises.

5.6 *Exercises and Projects*

1. Decipher the following newspaper cryptogram *(The Anderson (SC) Independent*, February 28, 1995):

AKZ WKYZL HVFZWAYHD AH XRGTYDC UYDN
YDEABSOZDAE YE AKGA YA XBHRHDCE AKZ
RYLZ HL AKZ XRGTZB. - CZHBCZ VZBDGBN EKGU

2. Decipher the following newspaper cryptogram *(The Greenville (SC) News*, April 4, 1995):

DIPBF SIP DICHT DIFM IJUF HPD DCVF QPW YPX-
CGM FLFWACBF SCGG BPPHFW PW GJDFW IJUF
DP QCHX DCVF QPW CGGHFBB. - FXSJWX BD-
JHGFM

3. How many different keys (permutations) are possible for a simple substitution cipher based on the standard 26-letter alphabet? How many of these could be used for a newspaper cipher which is required to have no fixed letters (other than spaces and punctuation marks)?

4. In standard English, the letters E and T occur about 12.7% and 9.1% of the time, respectively. Make your own estimate of these percentages using a short newspaper or magazine article. Also use the article to estimate the conditional probability that a given three-letter word is "THE".

5. Read Edgar A. Poe's "The Gold Bug" [8].

6. (Project for statisticians) Given a long portion of ciphertext, are there statistical tests that can be used to determine whether or not the ciphertext is the result of a simple substitution encipherment? (**Hint:** you might want to read about the *index of coincidence*. An elementary treatment can be found in [10].)

7. Decipher the remainder of cryptogram (5.6).

8. Let **R** denote the ring of integers modulo 26.

 a. How many different permutations of **R** are represented by functions of the form (5.4)?

 b. What is the subgroup of $\mathcal{P}(\mathbf{R})$ generated by the permutations of the form (5.4)?

 c. Among the functions of the form (5.4), find those which have no fixed points.

 d. Show that the function $f(x) = x^5$ defines a permutation of **R**.

 e. Find those exponents t for which the function $f(x) = x^t$ defines a permutation of **R**. Do these functions form a group?

9. For the case where α is the association (5.3) and $f(x)$ is the special instance of (5.4) with $a = 1$ and $b = 3$, show that the simple substitution defined by $\alpha^{-1} \circ f \circ \alpha$ is the Caesar cipher based on the cycle π of (5.2) with a key of $k = 3$.

10. Consider the modulo 26 function $y = f(x) = 9x + 14$ used in Example 5.2. What is the order of this mapping as a member of the group of permutations of $\mathbf{R} = \{0, 1, 2, \ldots, 25\}$?

11. Let **R** be a commutative ring (with identity) of order 26. Prove that **R** is isomorphic to the ring of integers modulo 26.

12. Let **F** denote the ring of integers modulo 31 (so that **F** is a field), let $\mathcal{L}$ denote the extended alphabet as given in (5.1), and let $\alpha : \mathcal{L} \to \mathbf{R}$ denote the correspondence (5.12). Encipher and then decipher the message "ATTACK AT DAWN."

 a. Use the function $f(x) = 8x + 24$.

 b. Use the function $f(x) = 2x^7$ (which is a permutation of **F**). Also show that the inverse of $f(x) = 2x^7$ is $f^{-1}(x) = 4x^{13}$.

13. Let $\mathbf{R} = \{0, 1, 2, 3\}$ denote the ring of integers modulo 4. Find a function from **R** to **R** which is not representable by a polynomial in $\mathbf{R}[x]$. Can two different polynomials over **R** represent the same function?

14. Suppose $\mathbf{R}$ is a finite commutative ring with an identity. Prove that every function from $\mathbf{R}$ to $\mathbf{R}$ is representable by a polynomial in $\mathbf{R}[x]$ if and only if $\mathbf{R}$ is a finite field.

15. Encipher the message WE WILL ATTACK AT DAWN using the standard Vigenére square.

   a. Use the key sequence ITS MY PARTY.

   b. Use a ciphertext autokey where the first letter is enciphered with the letter I and thereafter each ciphertext letter is used as the next key letter.

   c. Use a plaintext autokey where the first message letter W is enciphered using the key letter Z and thereafter each plaintext letter is enciphered using as key letter the previous plaintext letter (so that the second message letter E is enciphered using as key the first plaintext letter W, and so on).

   d. Use correspondence (5.3) and the numerical key stream $k_1, k_2, \ldots,$ where $k_1 = 2$ and $k_{n+1} = 5k_n + 17 \pmod{26}$ for $n \geq 1$. Also, find the period of this key stream: i.e., the smallest positive value of t such that $k_{n+t} = k_n$ for all n.

16. Decrypt the following message given that it has been enciphered using plaintext autokey as follows. The first letter was enciphered using I as the key letter, and each subsequent letter was enciphered using the previous plaintext letter as the key letter.

$$\text{XALIVGXBQTUNHYOCGAFBHOZVPQWGWLV}$$

17. Decrypt the following message given that it has been enciphered using ciphertext autokey as follows. The first letter was enciphered using Z as the key letter, and each subsequent letter was enciphered using the previous ciphertext letter as the key letter.

$$\text{BJYFJATXUNNHAOYCAIANBUPTKIAEGARV}$$

18. Consider the field $\mathbf{F} = \{0, 1\}$ modulo 2, and let the binary sequence $k_1, k_2, k_3, \ldots$ be defined by $k_1 = 1, k_2 = 1, k_3 = 0, k_4 = 0, k_5 = 0$, and $k_j = k_{j-3} + k_{j-5}$ for each $j \geq 6$. Find the period of the sequence $k_1, k_2, k_3, \ldots$ and describe how streams such as these might be used, say in conjunction with the ASCII code and the Vigenére system, to encipher messages.

19. Prove that an $n \times n$ matrix A over the ring of integers modulo 26 is invertible (respectively, involutory) if and only if it is invertible (respectively, involutory) mod 2 and mod 13.

20. Decipher the ciphertext OSMGVKAL,IUXLQWRID using the inverse matrix (5.14) and the correspondence (5.12).

21. A *transposition cipher* is one in which the letters of the plaintext are rearranged but not changed. For example, by interchanging every other letter in the word "TRANSPOSITIONS" one obtains the cipher-text "RTNAPSSOTIOISN". Show that interchanging every other letter is a special case of the basic 2×2 Hill encipherment, and describe the transposition ciphers that permute groups of n letters.

22. Show that a Vigenére cipher with key length n is a special case of the affine Hill cipher defined by (5.17).

23. As cryptanalyst, you have intercepted the following message from Alice to Bob:

> K , H Y H B G G H N . N K V ' H P N Q X . L G O . I
> B ' B O W J Ø E X G . N L Z H Y ? N

You know that in the recent past they have communicated using the basic 2×2 Hill system, the letter set (5.11), and the letter-ring correspondence (5.12). They have changed the enciphering matrix A of (5.9), from message to message, but it has been fixed for each message. Decipher the message above assuming that it is a "Dear John" letter; i.e., assume that "DEAR JOHN," appears in the plain text. (Bob's last name is John.)

24. In the simple substitution newspaper ciphers, no letter is enciphered into itself: i.e., $f(x) \neq x$ for all letters x. Assume that encipherment uses the basic Hill system and that the ring $\mathbf{R}$ is a finite field $\mathbf{F}$.

   a. Give necessary and sufficient conditions on the $n \times n$ matrix A in order that the transformation (5.9) have no fixed vectors. (**Hint:** look at eigenvalues.)

   b. Show if the matrix A of (5.9) is involutory, then there are always fixed vectors.

   c. Determine the number of fixed vectors of the matrix B of (5.15) and describe the cycle structure of the permutation it induces on the set S of all 2×1 column matrices (mod 31).

25. Prove that if A is an invertible matrix over a finite ring $\mathbf{R}$, then there is a positive integer m such that $A^{-1} = A^m$ and discuss any significance

this fact may have had to Hill when he patented his mechanical device for matrix encipherment. Find the m for the matrix (5.13).

26. (Project: *a two-message problem* [6, 7]) As cryptanalyst you have intercepted the following two ciphertext messages. Each of these have originated at enemy headquarters and have gone to two different outposts.

> Cipher 1: D J M E U H Z ? ' Y ? O L J C E M A Ø J V F
> Cipher 2: B F . A L U Ø G A , C . Ø Q C E , B C P Z N

From past history you know that the following assumptions are often valid for messages from the headquarters to the two outposts.

- The two ciphertext messages come from the same plaintext.

- Both messages are enciphered using the basic Hill system (5.9).

- The (unknown) enciphering matrices, call them A_1 and A_2, are each 2×2, the letter set $\mathcal{L}$ is given by (5.11), and the letter-ring correspondence is given by (5.12), where $\mathbf{R}$ is the field $\{0, 1, 2, \ldots, 30\}$ modulo 31.

- Both A_1 and A_2 are involutory $(A_i^{-1} = A_i)$.

Suppose the first three of these assumptions are valid.

a. Determine the product $M = A_1 A_2^{-1}$. (**Hint:** the equations $Y_1 = A_1 X$ and $Y_2 = A_2 X$ imply that $A_1^{-1} Y_1 = A_2^{-1} Y_2$.)

b. Devise a procedure to test the validity of the first three assumptions. (If the assumptions are true, then M can be determined in more than one way.)

c. If all four assumptions are valid, show that necessarily $\det(M) = \pm 1$ and that M is similar to its inverse. Are these conditions valid for the M you found in part (a)?

d. Prove that a square matrix M is the product of two involutory matrices (i.e., $M = XY$ where $X^2 = I$ and $Y^2 = I$) if and only if M is invertible and there exists an involutory solution to the equation $MX = XM^{-1}$.

e. Solve the equation $MX = XM^{-1}$ for the matrix M you found in part (a), expressing your answer in terms of the fewest number of arbitrary parameters.

f. Using a computer, determine which of the matrices you found in part (e) are involutory. (Note: there are a total of 994 involutory 2×2 matrices modulo 31, but only 62 of them satisfy the equation in part (e).)

h. Using a computer and the involutory solutions found in part (f), decipher Cipher 1 given earlier by trying all 62 putative matrices. (Only one should give good English.)

27. (Project)

a. Consider the modulo 31 sequence $t_0, t_1, t_2, t_3, \ldots$ which is defined recursively by $t_0 = 0$, $t_1 = 1$, and $t_{n+2} = t_{n+1} + 18t_n$ for $n \geq 0$. Write a computer program to find the period of this sequence.

b. Consider the modulo 31 matrix below which is defined in terms of a variable t. Show that this matrix is involutory for each t from the field $\mathbf{F} = \{0, 1, 2, \ldots, 30\}$ modulo 31. In view of (5.18) and part (a) above, describe how it might be used to generalize the stream cipher described in Exercise 5.15(d).

$$A(t) = \begin{bmatrix} a_{11}(t) & a_{12}(t) & a_{13}(t) \\ a_{21}(t) & a_{22}(t) & a_{23}(t) \\ a_{31}(t) & a_{32}(t) & a_{33}(t) \end{bmatrix},$$

where

$$a_{11}(t) = 2t^5 + 4t^4 + 27t^3 + 1$$
$$a_{12}(t) = 2t^6 + 4t^5 + 29t^4 + 4t^3 + 27t^2$$
$$a_{13}(t) = 6t^6 + 12t^5 + 19t^4$$
$$a_{21}(t) = 29t^4 + 27t^3 + 29t$$
$$a_{22}(t) = 29t^5 + 27t^4 + 29t^3 + 25t^2 + 30$$
$$a_{23}(t) = 25t^5 + 19t^4 + 25t^2$$
$$a_{31}(t) = 2t^2 + 2t$$
$$a_{32}(t) = 2t^3 + 2t^2 + 2t + 2$$
$$a_{33}(t) = 6t^3 + 6t^2 + 1$$

5.7 References

[1] W. Diffie and M. E. Hellman, "New directions in cryptography," *IEEE Transactions on Information Theory* **22**, 644–654 (1976).

[2] L. S. Hill, "Cryptography in an algebraic alphabet," *Amer. Math. Monthly* **36**, 306–312 (1929).

[3] L. S. Hill, "Concerning certain linear transformation apparatus of cryptography," *Amer. Math. Monthly* **38**, 135–154 (1931).

[4] D. Kahn, *The Code Breakers: The Story of Secret Writing*, Macmillan, New York, 1968.

[5] J. Levine, "Variable matrix substitution in algebraic cryptography," *Amer. Math. Monthly* **65**, 170–179 (1958).

[6] J. Levine and J. V. Brawley, "Involutory commutants with some applications to algebraic cryptography I," *J. Reine and Angewandte Math.* **224**, 20–43 (1966).

[7] J. Levine and J. V. Brawley, "Involutory commutants with some applications to algebraic cryptography II," *J. Reine and Angewandte Math.* **227**, 1–24 (1967).

[8] Edgar A. Poe, "The Gold Bug"; text can be found at the site <http://www.tgn.net/~pambytes/poe/goldbug.html>

[9] R. A. Rueppel, *Analysis and Design of Stream Ciphers*, Springer-Verlag, New York, 1986.

[10] A. Sinkov, *Elementary Cryptanalysis: A Mathematical Approach*, Random House & The L. W. Singer Company, New York, 1968.

[11] D. R. Stinson, *Cryptography: Theory and Practice*, CRC Press, Boca Raton, FL, 1995.

Chapter 6

Mathematical Models in Public-Key Cryptology

Joel Brawley
Shuhong Gao

Prerequisites: linear and modern algebra, elementary number theory

6.1 Introduction

Chapter 5 has described several of the classical models of cryptography in which the decryption key was the same as or easily derivable from the encryption key. This meant that the corresponding encryption and decryption algorithms were closely related in the sense that one could be easily deduced from the other. Such cryptographic systems are called *symmetric-key* or *conventional* systems, and their security relies exclusively on the secrecy of the keys. Other examples of private-key systems are the Data Encryption Standard (DES) [24] and IDEA [12], in which users of the system who share a secret key can communicate securely over an unsecure channel. In all of the private-key systems, two users who wish to correspond must have a common key *before* the communication starts. Also, in practice, establishing a common secret key can be expensive, difficult, and sometimes nearly impossible, especially in a large network where the users need not know each other.

In 1976, Diffie and Hellman [7] introduced a revolutionary new concept called *public-key cryptography* based on the simple observation that the encryption and decryption could be separated; i.e., they recognized that a knowledge of the encryption key (or equivalently, the encryption algorithm) need not imply a knowledge of the decryption key (or algorithm). In such a system, the encryption key can be made public, say in a public directory, while the decryption key can be kept secret. Anyone wishing to send a message to a person in the directory can simply look up the public encryption key for that person and use it to encrypt the message. Then, assuming the decryption key is known only to the intended receiver of the message, only that person can decrypt the message.

Of course in such a public-key system it must be computationally infeasible to deduce the decryption key (or the decryption algorithm) from the public key (or the public encryption algorithm), even when general information about the system and how it operates is known. This leads to the idea of one-way functions.

A function f is called a *one-way function* if for any x in the necessarily large domain of f, $f(x)$ can be efficiently computed but for virtually all y in the range of f, it is computationally infeasible to find any x such that $f(x) = y$.

Public-key cryptography requires a special set of one-way functions $\{E_k : k \in \mathcal{K}\}$ where $\mathcal{K}$, the so-called *key space*, is a large set of possible *keys*, and E_k is a map from a plaintext space $\mathcal{M}_k$ to a ciphertext space $\mathcal{C}_k$. The one-way nature of E_k implies that for virtually all ciphertexts $c = E_k(m)$ it is computationally infeasible to recover the plaintext m from a given k and c. However, since the legitimate recipient of the message must be able to recover m from c, more is required of these one-way functions. Specifically, each E_k must have an inverse D_k, and this inverse must be easily obtainable given some additional secret information d. The extra information d is called a *trap-door* of E_k and the functions E_k themselves are called *trap-door one-way functions*. It is also required that, with a knowledge of D_k, $m = D_k(c)$ be easy to compute for all c in the ciphertext space. Thus, a public-key cryptosystem consists of a family of trap-door one-way functions.

Before proceeding, we will make a few remarks on the length of messages. For a given key k, the function E_k usually acts only on plaintexts of fixed length whereas, in practice, a message can be of arbitrary length. However, the message can be cut into appropriate pieces, called blocks, so that E_k can act on each block. The whole message is then encrypted by encrypting each block individually. This operating mode is called the *Electronic Code Book* (ECB) mode. (Other operating modes include *Cipher Block Chaining* (CBC) mode, *Cipher Feedback* (CFB) mode, and *Output Feedback* (OFB) mode [24].) The point here is that the plaintext space (i.e., the domain of E_k) may be finite but a message of arbitrary length can be encrypted using E_k.

To summarize we give our first model of public-key cryptography.

DEFINITION 6.1 *A (deterministic) public-key cryptosystem consists of the following components:*

1. *A set $\mathcal{K}$ called the key space whose elements are called keys.*

2. *A rule by which each $k \in \mathcal{K}$ is associated with a trap-door one-way function E_k with domain $\mathcal{M}_k$ (the plaintext space) and range $\mathcal{C}_k$ (the ciphertext space).*

3. *A procedure for generating a random key $k \in \mathcal{K}$ together with a trap-door d for E_k and the inverse map $D_k : \mathcal{C}_k \longrightarrow \mathcal{M}_k$ such that*

$$D_k(E_k(m)) = m, \text{ for all } m \in \mathcal{M}_k.$$

The key space $\mathcal{K}$ is also called the *public-key space*, and the set of trap-doors d is called the *private-key space*. Relative to (3), it is also required that random keys $k \in \mathcal{K}$ and their corresponding trapdoors d be easy to generate.

In practice, the complete description of all the components (1)–(3) of a cryptosystem is public knowledge. A person (user) who wants to become a part of the communication network can proceed as follows:

- Use (3) to generate a random key $k \in \mathcal{K}$ and the corresponding trap-door d.

- Place the encryption function E_k (or equivalently the key k) in a public directory (say in the user's directory or home page), keeping d and the decryption function D_k secret.

Now suppose that Bob wants to send a message m to a user Alice. To do this, he simply looks up her public enciphering function E_{k_A} and computes $c = E_{k_A}(m)$ which he sends to Alice. On receiving c, Alice computes

$$D_{k_A}(c) = D_{k_A}(E_{k_A}(m)) = m,$$

thereby recovering the message. An eavesdropper might intercept c and can obtain E_{k_A} from public files, but cannot find m from c without knowledge of d_A (or equivalently D_{k_A}).

Actually, it is currently unknown as to whether one-way functions truly exist; indeed, a proof of the existence of such functions would settle the famous **P=NP** problem of computer science. However, there are a number of functions that are believed to be one-way. For example, it is assumed by experts that integer multiplication is a one-way function because it is very easy to multiply large integers, but it seems very hard to factor large numbers using the current knowledge and technology. This assumption is the basis for several public-key cryptosystems including the RSA and Rabin cryptosystems, discussed in Section 6.2.

In the above model of public-key cryptography, two identical plaintext messages m are always encrypted into the same ciphertext $c = E_k(m)$, and in the ECB mode, this feature can cause some leaking of information. For example, if the cryptosystem were used to encrypt personnel data and the salary fields were encrypted separately, then by simply looking at the ciphertexts one could identify people with the same salary. A natural question arises: Is it possible to design a public-key system in which such identical plaintexts are encrypted to different ciphertexts? Surprisingly, the answer is yes! The idea is to use random numbers (also referred to as redundant information or nonces).

To explain this idea more fully, suppose with each $k \in \mathcal{K}$ there is also associated a large set $\mathcal{R}_k$, called a *randomization set*, and a map

$$E_k : \mathcal{M}_k \times \mathcal{R}_k \longrightarrow \mathcal{C}_k.$$

In order to encrypt a plaintext $m \in \mathcal{M}_k$ with such a map, one picks a random number $r \in \mathcal{R}_k$ and computes the ciphertext

$$c = E_k(m, r).$$

Consequently, in the ECB mode of operation, a sequence of plaintext blocks $m_1, \ldots, m_t$ will be encrypted to a sequence $c_1, \ldots, c_t$, where $c_i = E_k(m_i, r_i)$ and each r_i is chosen independently and randomly from $\mathcal{R}_k$ for $i = 1, \ldots, t$. The set $\mathcal{R}_k$ is usually large, so the chance of picking the same r is small and the ciphertext values c_i will generally be different even when the corresponding plaintext blocks are identical.

Here again, just as in our first model of public-key cryptography, each E_k is required to be a one-way function with a trap door, but it need not be fully invertible since the recipient does not need to recover r. What is needed for decryption is that E_k be partially invertible; i.e., there needs to exist a function $D_k : \mathcal{C}_k \longrightarrow \mathcal{M}_k$ such that $D_k(E_k(m, r)) = m$, for all $m \in \mathcal{M}_k$ and all $r \in \mathcal{R}_k$. The function D_k is called a *partial inverse* of E_k since it recovers only part of the input to E_k. We call a one-way function E_k with this property a *partial-trap-door one-way function*, and we give our second model of public-key cryptography.

DEFINITION 6.2   *A **probabilistic public-key cryptosystem** consists of the following:*

  1. A set $\mathcal{K}$ called the key space whose elements are called keys.

2. *A rule by which each $k \in \mathcal{K}$ is associated with a partial-trap-door one-way function E_k with domain $\mathcal{M}_k \times \mathcal{R}_k$ and range $\mathcal{C}_k$. (Here, $\mathcal{M}_k$ is called the plaintext space, $\mathcal{R}_k$ the randomization set, and $\mathcal{C}_k$ the ciphertext space.)*

3. *A procedure for generating a random key $k \in \mathcal{K}$ together with a partial-trap-door d for E_k and the map $D_k : \mathcal{C}_k \longrightarrow \mathcal{M}_k$ such that*

$$D_k(E_k(m,r)) = m, \text{ for all } m \in \mathcal{M}_k, r \in \mathcal{R}_k.$$

Again, the elements $k \in \mathcal{K}$ are called the public keys and the partial-trap-doors d the private keys. Obviously, the deterministic model is a special case of the probabilistic model in which $\mathcal{R}_k$ has only one element. In order for the probabilistic model to be useful and secure, the following properties are needed.

P1. Given a public key k, it is easy to compute $E_k(m,r)$ for $m \in \mathcal{M}_k$ and $r \in \mathcal{R}_k$.

P2. Given a private key d, it is easy to compute $D_k(c)$ for $c \in \mathcal{C}_k$.

P3. Knowing k and $c \in \mathcal{C}_k$, it is infeasible to decide for any $m \in \mathcal{M}_k$ whether m can be encrypted to c under E_k. Thus, it is infeasible to determine D_k or d from the general information about the cryptosystem.

P4. It is easy to generate a random key $k \in \mathcal{K}$ and the corresponding private key d.

In the deterministic model it was required that it be infeasible to determine m from a knowledge of only E_k and c (as E_k is one-way). The corresponding requirement P3 for the probabilistic model is much stronger because if one cannot even decide whether a plaintext m can be encrypted to a given c, then certainly one cannot find a plaintext that can be encrypted to c. In this connection we note that in the deterministic model, it is trivial to decide if a plaintext $m \in \mathcal{M}_k$ can be encrypted to a given ciphertext $c \in \mathcal{C}_k$, as one can simply compute $E_k(m)$ and check whether or not it is c. Requirement P3 also implies that even when an adversary has a potentially matched pair (m, c) of plaintext and ciphertext, he or she cannot even verify that there exists $r \in \mathcal{R}_k$ such that $E_k(m,r) = c$. Therefore, a probabilistic cryptosystem can provide a higher level of security than a deterministic one.

The first probabilistic public-key cryptosystem was given by McEliece in 1978 (see [24]). In this system, the trap-door is based on the fact that the encoding process for error-correcting codes can be easy whereas decoding can be hard. In 1985 ElGamal [8] proposed a probabilistic system whose trap-door is based on the fact that exponentiation in a finite field (to be described later) is easy, but the inverse process, the so-called discrete logarithm problem, can be hard. The ElGamal system is a modification of the Diffie-Hellman key exchange scheme [7], whose security is also based on the discrete logarithm problem. Both the ElGamal and the Diffie-Hellman systems will be discussed in Section 6.3.

We close our general discussion of cryptographic models with a few remarks concerning a further generalization of property P3 which required that it be hard to decide whether a given plaintext m can be encrypted to a given ciphertext c. Even under this condition, there can still be leaking of information. More specifically, it can happen that the probability that m is encrypted to a given c is significantly different for different values of m; consequently, for a given c it is possible to infer some partial probabilistic information about the plaintext space. To avoid such leaking, one can replace P3 by the following stronger condition.

P3′. Given k and $c = E_k(m, r) \in \mathcal{C}_k$, it is infeasible to infer any partial information about m.

While we shall not describe any systems with property P3′, we do give a few references. The first probabilistic cryptosystem proven to satisfy P3′ was given by Goldwasser and Micali [10], and later Blum and Goldwasser [4] gave a more efficient system.

In the discussions above we have tacitly assumed that the adversary was passive in that he or she could only eavesdrop on a communication. However, if the adversary were active and could inject or alter messages, then some systems (e.g., the Rabin and ElGamal systems, discussed later) are vulnerable to an *adaptive chosen ciphertext attack*, in which the adversary is assumed so powerful that he or she can obtain the decryptions of many ciphertexts of his or her own making, though *not* the target ciphertext. Recently, Cramer and Shoup [6] proposed a cryptosystem that is secure against such an attack and is believed to be practical as well.

6.2   Cryptosystems Based on Integer Factorization

Given two primes, say $p = 863$ and $q = 877$, it is an easy process to multiply them by hand to get the product $n = 756851$. However, it is not nearly so easy to determine by hand the factors p and q from only a knowledge of the product 756851. In a similar fashion, if p and q are large, say 1,000 digits each, then a computer can readily find the 2,000 digit product (since multiplying two k-digit numbers requires at most $O(k^2)$ operations), but even the fastest of today's computers cannot generally determine the factors from only the product. This leads us to consider two central problems in the history of mathematics, namely the problems of (a) determining whether a given integer is a prime, and (b) determining the factorization into primes of a given integer. These two problems have been attacked by some of the best mathematicians of all time, including the great C. F. Gauss (1777–1855) who wrote [9]:

> The problem of distinguishing prime numbers from compos-
> ite numbers and of resolving the latter into their prime fac-
> tors is known to be one of the most important and useful in
> arithmetic. It has engaged the industry and wisdom of an-
> cient and modern geometers to such an extent that it would
> be superfluous to discuss the problem at length ... Further,
> the dignity of the science itself seems to require solution of
> a problem so elegant and so celebrated.

Gauss wrote these words some 175 years before primality testing and the integer factorization problem were applied to modern day cryptography, so as important as they were in Gauss' day, they are even more important today.

It is clear from his words that Gauss realized that primality testing was a different problem from that of integer factorization. Since his time, significant progress has been made on both of these problems. For example, there is an efficient probabilistic algorithm called the Rabin-Miller test (see [3, 5]) which can recognize a composite number of (say) 1,000 digits without ever factoring that number, and the primality of a number can also be determined efficiently [1, 2]. Over the years, much improved factoring algorithms [13] have also been developed, but despite this progress, factoring a general composite number with as few as (say) 200 digits is still out of reach of the fastest computers using the best algorithms known today.

It is not our purpose here to delve into the theory of primality testing and integer factorization (for which we refer the reader to [13, 20] for recent developments). Instead, we simply wish to emphasize that it is easy to generate and multiply large prime numbers but it is not generally possible to factor the resulting answer in reasonable time; that is, integer multiplication appears to be a one-way function. This belief forms the basis for several public-key cryptosystems. We will discuss two of these after reviewing several ideas from modular arithmetic.

Let n be a positive integer and let $\mathcal{Z}_n = \{0, 1, \ldots, n-1\}$ denote the ring of integers modulo n. It turns out that exponentiation in this ring is easy; i.e., for any $\alpha \in \mathcal{Z}_n$ and positive integer e, the computation of $\alpha^e \bmod n$ can be done efficiently. We demonstrate how this is done with an example.

Example 6.1

In order to compute α^{29}, first write 29 in binary form

$$29 = 1 \cdot 2^4 + 1 \cdot 2^3 + 1 \cdot 2^2 + 0 \cdot 2^1 + 1.$$

It then follows that

$$\alpha^{29} = \alpha^{1 \cdot 2^4} \cdot \alpha^{1 \cdot 2^3} \cdot \alpha^{1 \cdot 2^2} \cdot \alpha^{0 \cdot 2^1} \cdot \alpha^1 = \left(\left(\left((\alpha^2 \cdot \alpha)^2 \cdot \alpha \right)^2 \right) \alpha^0 \right)^2 \cdot \alpha.$$

Thus, only four squarings and three multiplications are needed to compute $\alpha^{29} \bmod n$. (Here, it is important that the reductions modulo n be done at each squaring or multiplication to avoid large intermediate integers.)

The above idea can be generalized to show that α^e can be computed with $\log_2(e) - 1$ squarings and at most $\log_2(e) - 1$ multiplications, where $\log_2(e)$ is the length of e in its binary representation. So, for any $\alpha \in \mathcal{Z}_n$ and $e > 0$, $\alpha^e \bmod n$ can be computed efficiently.

Now consider the reverse of the operation of exponentiation modulo n. Assuming n is specified, two different problems arise: (a) given α and $y \in \mathcal{Z}_n$, find an integer x (if one exists) such that $\alpha^x \equiv y \bmod n$; and (b) given e and $y \in \mathcal{Z}_n$, find x (if one exists) such that $x^e \equiv y \bmod n$. These two problems are intrinsically different and each of them leads to a public-key cryptosystem.

Problem (a) is called *the discrete logarithm problem* modulo n and is believed hard for almost all n. (For certain values of n, it is easy; e.g., when n has only small prime factors or when n is a prime but $n-1$ has only small prime factors.) We will discuss this problem and its relation to the ElGamal system in the next section.

Problem (b) asks for the computation of an eth root of an integer y modulo n. This is easy when the complete factorization of n is known but believed hard otherwise. For cryptographic purposes, the most important case is when n is the product of two large (distinct) primes and this is the case we shall develop here. (Our discussion can be readily generalized to the situation where n is square-free.) The RSA system arises when we examine this situation with $\gcd(e, \phi(n)) = 1$ and the Rabin system comes from considering the case in which e divides $\phi(n)$. Here $\phi(n)$ is the familiar Euler ϕ-function which equals the number of integers in $\mathcal{Z}_n$ that are relatively prime to n. When $n = pq$ with p and q distinct primes, it is given by $\phi(n) = (p-1)(q-1)$.

Assume then that $n = pq$ and let e be an integer with $\gcd(e, \phi(n)) = 1$. Then there exists an integer d such that $ed \equiv 1 \bmod \phi(n)$. Using Fermat's little theorem, it is straightforward to show that

$$x^{ed} \equiv x \bmod n, \text{ for all } x \in \mathcal{Z}_n. \tag{6.1}$$

This means that for any $x \in Z_n$, $x^d \bmod n$ is an eth root of x modulo n; hence an eth root of x can be computed efficiently provided d is known. Thus an important question is whether d can be computed efficiently.

The answer is yes if n can be factored, but no otherwise. To explain this statement, first assume that the complete factorization of n into primes is known. Then $\phi(n)$ can be computed quickly. By applying the extended Euclidean algorithm [3] to e and $\phi(n)$, it is easy to find d such that $ed \equiv 1 \bmod \phi(n)$. Conversely, assume that a number d is known with $ed \equiv 1 \bmod \phi(n)$. Then $\Phi = ed - 1$ is a multiple of $\phi(n)$. By Exercise 6.3, n can be easily factored using Φ. Hence computing d from e and n is equivalent to factoring n. Therefore, the factors of n provide a trap-door for inverting the function $P_{e,n} : \mathcal{Z}_n \longrightarrow \mathcal{Z}_n$ defined as

$$P_{e,n}(x) = x^e \bmod n,$$

where $x^e \bmod n$ denotes the smallest nonnegative integer congruent to x^e modulo n.

We can now describe the RSA cryptosystem, which bases its security on the belief that the class of functions $P_{e,n}$ are trap-door one-way functions.

DEFINITION 6.3   RSA cryptosystem

- *The public-key space $\mathcal{K}$ is the set of integer pairs (e, n) where n is a product of two large distinct primes, $1 < e < \phi(n)$ and $\gcd(e, \phi(n)) = 1$.*

- *For each $k = (e, n) \in \mathcal{K}$, the plaintext and ciphertext spaces are $\mathcal{M}_k = \mathcal{C}_k = \mathcal{Z}_n$.*

- *For each $k = (e, n) \in \mathcal{K}$, the encryption function is $E_k = P_{e,n}$.*

- *For each $k = (e, n) \in \mathcal{K}$, the corresponding private key is (d, n) where $ed \equiv 1 \bmod \phi(n)$, and the decryption function is $D_k = P_{d,n}$.*

We also need a rule for generating a random pair of public and private keys (e, n) and (d, n), but this is not difficult. We have mentioned that primality testing can be done efficiently and further there are many primes with a given number say of t digits (by the prime number theorem). This means that one can generate a t-digit prime as follows. Choose a random number of t digits and test it for primality. If it is not prime, repeat the procedure until a prime is obtained. In this way one can get a pair of primes p and q of any desired size. Then defining $n = pq$ and $N = (p-1)(q-1)$, one may choose a random integer $e \in \mathcal{Z}_N$ such that $\gcd(e, N) = 1$. For this e, it is a simple matter to compute d with $ed \equiv 1 \bmod N$. Then (e, n) is a public key and (d, n) is the corresponding private key.

We should point out that, even though computing d from e and n is equivalent to factoring n, it has not been proven that *inverting* $P_{e,n}$ is equivalent to computing d or factoring n, as there may exist some other method to compute eth roots modulo n without factoring n or computing d. This raises the following research question.

Open Problem. *Given a composite integer n and a positive integer e with $\gcd(e, \phi(n)) = 1$, prove or disprove that computing eth roots modulo n is equivalent to factoring n.*

We next consider the case in which e divides $\phi(n)$, used in the Rabin system. Here, the function $P_{e,n}$ is no longer a permutation on $\mathcal{Z}_n$ since for a given e and y, there can be several x that map to the same y under $P_{e,n}$. However, if e is small relative to n (say, $e \leq (\log n)^c$ for some constant c) then it can be proved that finding the inverse images of y under $P_{e,n}$ is equivalent to factoring n. Thus, if e is (say) one of $2, 3, 5, 6, 7$, then $P_{e,n}$ is a candidate trap-door one-way function where the factors of n again provide the trap-door for inverting $P_{e,n}$. Such trap-door one-way functions can be used to build up public-key cryptosystems; indeed, the Rabin system uses only the function $P_{2,n}$ and is described as follows.

DEFINITION 6.4   **Rabin cryptosystem**

- *The public-key space is $\mathcal{K} = \{n : n = pq$, where p and q are large distinct primes$\}$.*

- For each $n \in \mathcal{K}$, the plaintext space is a subset $\mathcal{M}_n \subset \mathcal{Z}_n$ such that $x_1^2 \not\equiv x_2^2 \bmod n$ for all different $x_1, x_2 \in \mathcal{M}_n$, and the ciphertext space $\mathcal{C}_n = \{x^2 \bmod n : x \in \mathcal{M}_n\}$.

- For each $n \in \mathcal{K}$, the encryption function is $E_n = P_{2,n}$, the private-key is the pair (p, q) such that $n = pq$, and the decryption function is described below.

To decrypt a ciphertext under the Rabin system, we need to describe how to compute square roots modulo n given the factors of n. The idea is to compute square roots modulo each prime factor of n separately and then use the Chinese remainder theorem to combine them to get a square root modulo n. Suppose that $c \in \mathcal{Z}_n$ has a square root modulo n. Then c also has square roots modulo p and q. Moreover, if r and s are some square roots of c modulo p and q, respectively, and if a and b are integers such that $ap + bq = 1$, then it is easy to check that $\pm aps \pm bqr$ is a square root of c modulo n for any choice of $+$ and $-$ signs. This gives four square roots of c. It can be shown that c has exactly four square roots if and only if $\gcd(c, n) = 1$ and c has at least one square root modulo n. To get the correct plaintext one just needs to check which of the roots is in $\mathcal{M}_k$.

It remains to show how to compute square roots modulo a prime p. But this is also easy, especially when $p \equiv 3 \bmod 4$, for then $r = c^{(p+1)/4} \bmod p$ is a square root of c whenever c is a *quadratic residue* in $\mathcal{Z}_p$: that is, $c \equiv x^2 \bmod p$ for some $x \in \mathcal{Z}_p$. If p is congruent to 1 modulo 4, square roots modulo p can also be computed efficiently but will not be described here (see [3, 5] for details). Hence computing square roots modulo $n = pq$ is easy when p and q are known.

We conclude this section by showing that if one can compute square roots modulo n then one can actually factor n; that is, we show that computing square roots modulo n is equivalent to factoring n. This also provides a nice example illustrating the power of randomness in computing. Suppose that there is an algorithm σ which, when presented a quadratic residue $x \in \mathcal{Z}_n$, outputs a square root of x in $\mathcal{Z}_n$, denoted by $\sigma(x)$. (One may think of σ as an algorithm, a black box, or an oracle.) Then n can be factored by the following simple algorithm. First, randomly and uniformly pick a nonzero element $a \in \mathcal{Z}_n$ and compute $x = a^2 \bmod n$. Next, input x to σ and get $b = \sigma(x)$. Then compute $h = \gcd(a - b, n)$. It can be shown that for each run of the algorithm, the computed number h is a proper factor of n with probability at least $1/2$. If the algorithm is run t times, then the probability of getting a factor of n is at least $1 - (1/2)^t$. Thus for $t = 10$, the chance of finding a factor of n is over 99.9%! Therefore n can be factored quickly.

As we have indicated, breaking the Rabin system is equivalent to factoring integers. This is the first example of a public-key cryptosystem with provable security against a passive adversary who can only eavesdrop. Compared to the RSA system, the encryption in the Rabin system is more efficient (only one square), and the decryption costs approximately the same as in RSA. However, the same proof above shows that the Rabin system is totally insecure against an active adversary who can mount a chosen ciphertext attack. Under this attack, an adversary chooses some (possibly many) ciphertexts and asks for the corresponding plaintexts, then deduces from them the secret key. The reader should be able to see why this attack works against the Rabin system. Because of this, the Rabin system is not used in practice.

6.3   Cryptosystems Based on Discrete Logarithms

Let $\mathbf{F}_q$ be a finite field of q elements so that $q = p^n$ for some prime p and integer n. It is well known that the multiplicative group of nonzero elements of $\mathbf{F}_q$, denoted by $\mathbf{F}_q^*$, is a cyclic group of order $q - 1$. Thus if α is a generator of this multiplicative group, then every nonzero element β in $\mathbf{F}_q$ is given by $\beta = \alpha^x$ for some integer x; in fact, for each β there is a unique integer in the range $0 \leq x < q - 1$ with this property. For a given x and α, the power α^x can be quickly computed by the square-and-multiply method as demonstrated in Example 6.1. The inverse problem, i.e., the problem of finding, for a given α and β, the x in the range $0 < x < q - 1$ satisfying $\beta = \alpha^x$, is the discrete logarithm problem; it is believed to be hard for many fields. Thus, exponentiation in finite fields is a candidate for a one-way function.

Example 6.2

For the prime $p = 1999$, the ring $\mathcal{Z}_p$ is a finite field and the nonzero elements $\mathcal{Z}_p^*$ of $\mathcal{Z}_p$ form a group G under multiplication modulo p:

$$G = \mathcal{Z}_p^* = \{1, 2, \ldots, p - 1\}.$$

Furthermore, the element $\alpha = 3$ is a generator of G, and is also known as a *primitive element* modulo p:

$$G = \{1, \alpha, \alpha^2, \ldots, \alpha^{p-2}\} \bmod p.$$

It is easy to compute that

$$3^{789} \equiv 1452 \bmod p.$$

However, it is not nearly so easy to determine that $x = 789$, given only that x is in the range from 0 to 1997 and satisfies the equation

$$3^x \equiv 1452 \bmod 1999.$$

A more realistic challenge is to find an integer x such that

$$3^x \equiv 2 \bmod p, \text{ where } p = 142 \cdot (10^{301} + 531) + 1.$$

We know a solution exists, but we don't know its value.

The above discussion can be generalized to any group G (whose operation is written multiplicatively). The *discrete logarithm problem* for G is to find, for given $\alpha, \beta \in G$, a nonnegative integer x (if it exists) such that $\beta = \alpha^x$. The smallest such integer x is called the discrete logarithm of β to the base α, and is written $x = \log_\alpha \beta$. In Example 6.2, $\log_3 1452 = 789$. Clearly, the discrete logarithm problem for a general group G is exactly the problem of inverting the exponentiation function $\exp : \mathcal{Z}_N \longrightarrow G$ defined by $\exp(x) = \alpha^x$ where N is the order of α.

The difficulty of this general discrete logarithm problem depends on the representation of the group. For example, consider G to be the cyclic group of order N. If G is represented as the additive group of $\mathcal{Z}_N$, then computing discrete logarithms in G is equivalent to solving the linear equation $ax \equiv b \bmod N$, where a, b are given integers; this can be easily done by using the extended Euclidean algorithm. If G is represented as a subgroup of the multiplicative group of a finite field as above or as a multiplicative group of elements from $\mathcal{Z}_m$ (where m may be composite or prime), then the problem can be "hard." For an elliptic curve group [14], the discrete logarithm problem seems to be harder. (As we have indicated, no one has been able to prove that these discrete logarithm problems are really hard, but they have been studied by number theorists for a long time with only limited success.) For recent surveys and a more detailed study of the discrete logarithm problem, the reader is referred to [15, 18, 19]. We now describe two cryptosystems whose security is based on the assumption that the discrete logarithm problem is hard.

The Diffie-Hellman key exchange scheme is a protocol for establishing a common key between two users of a classical cryptosystem. As we mentioned earlier, for a large network of users of a conventional cryptosystem, the secure distribution of keys can be complicated and logistic. In 1976, Diffie and Hellman [7] gave the following simple and elegant solution for this problem.

DEFINITION 6.5   Diffie-Hellman key exchange scheme. *Given the public group G and an element $\alpha \in G$ of order N, two parties, say Bob and Alice, establish a common key using the following steps:*

- *Alice picks a random integer $a \in \mathcal{Z}_N$, computes $A = \alpha^a$, and sends it to Bob.*

- *Bob picks a random integer $b \in \mathcal{Z}_N$, computes $B = \alpha^b$, and sends it to Alice.*

- *Alice computes $B^a = \alpha^{ba}$ and Bob computes $A^b = \alpha^{ab}$. Their common key is $k = \alpha^{ab} = \alpha^{ba}$.*

An eavesdropper, who knows G and α from the public directory, after intercepting A and B, is then faced with the following problem.

DEFINITION 6.6   **Diffie-Hellman problem.** *Let G be a group and let $\alpha \in G$. Given $A = \alpha^a$ and $B = \alpha^b$, compute $k = \alpha^{ab}$.*

If one can solve the discrete logarithm problem, then it is clear that one can solve the Diffie-Hellman problem; hence, the latter problem is no harder than the former. It is believed that the two problems are equivalent and, in fact, this equivalence has been established for some special cases. In any event, the Diffie-Hellman key exchange scheme is secure provided the Diffie-Hellman problem is hard.

The Diffie-Hellman key exchange scheme is widely used (with some variants) in practice to generate "session keys," for example, in secure Internet transactions. This scheme itself is not a public-key cryptosystem; however, ElGamal [8] showed that it could easily be converted into one. Note that if Alice publishes $k_A = \alpha^a$ but keeps a secret, then anyone, say Charlie, can share a common key with Alice in the same way that Bob did; i.e., Charlie can pick a random integer c, and compute $r = \alpha^c$ and $k_A^c = \alpha^{ac}$. Before sending r to Alice, Charlie can encrypt any message m he wishes by simply computing the product $r_1 = m \cdot k_A^c$. Then he sends the pair (r, r_1) to Alice. Alice can compute $k_A^c = r^a$ from r and her private key a, and so decrypt m.

DEFINITION 6.7   **ElGamal cryptosystem.** *Given the public group G (written multiplicatively) and an element $\alpha \in G$ of order N, let $G_1 = <\alpha>$, the subgroup of G generated by α.*

- *The key space is $\mathcal{K} = G_1$.*

- *For each $k \in \mathcal{K}$, the plaintext and ciphertext spaces are*

$$\mathcal{M}_k = G, \quad \mathcal{C}_k = G_1 \times G = \{(\beta_1, \beta_2) : \beta_1 \in G_1, \beta_2 \in G\}.$$

  The randomization set is $\mathcal{R}_k = \mathcal{Z}_N$.

- *For each $k \in \mathcal{K}$, the encryption function $E_k : \mathcal{M}_k \times \mathcal{R}_k \longrightarrow \mathcal{C}_k$ is given by $E_k(m, r) = (\alpha^r, k^r \cdot m)$.*

- *For each $k \in \mathcal{K}$, the corresponding private key is the integer $d \in Z_N$ such that $k = \alpha^d$, and the decryption function $D_k : \mathcal{C}_k \longrightarrow \mathcal{M}_k$ is given by $D_k(c_1, c_2) = c_2 \cdot (c_1^d)^{-1}$.*

It is easy to check that if $k = \alpha^d$, then $D_k(E_k(m,r)) = m$ for all $m \in M_k$ and $r \in \mathcal{R}_k$. To obtain a random key, one just chooses $d \in \mathcal{Z}_N$ at random and computes $k = \alpha^d$. In practice, one has to be extremely careful in choosing the group G so that the discrete logarithm problem is hard. Originally, Diffie and Hellman (1976) and ElGamal (1985) used the multiplicative group of $\mathcal{Z}_p$ for a large prime p. The above description of their systems is actually a natural generalization to an arbitrary group. The most studied groups for cryptographic purposes are multiplicative subgroups of $\mathcal{Z}_p$, $\mathcal{Z}_m$ (where m is a product of two large primes), $\mathbf{F}_{2^n}$, and elliptic curve groups over finite fields. While $\mathcal{Z}_p$ is currently the most popular choice, there is increasing interest in using elliptic curves over finite fields, particularly the fields $\mathbf{F}_{2^n}$ [14].

Note that breaking the ElGamal cryptosystem by a ciphertext-only attack is equivalent to solving the Diffie-Hellman problem. Thus the ElGamal cryptosystem is another example with provable security if the Diffie-Hellman problem is indeed hard. The major disadvantage of the system is the message expansion by a factor of two, but there are ways to improve it in both efficiency and security as we discuss next.

Observe that the multiplicative operation $k^r \cdot m$ in the encryption function E_k could be replaced by other operations. For example, if the elements in G are represented as binary strings of 0s and 1s, then we can let

$$E_k(m,r) = (\alpha^r, k^r \oplus m)$$

where m is any binary string and $\oplus$ is the bitwise "exclusive or" operation (XOR); e.g., $(1100) \oplus (0101) = 1001$. In this case, m does not have to be in G. If the elements in G are represented as binary strings of length ℓ, then the plaintext space $\mathcal{M}_k$ can be any subset of binary strings of length ℓ. Note that in the ElGamal cryptosystem, it is required that a plaintext be in the group. This is not trivial to achieve for some groups (e.g., elliptic curves), but the above approach solves this problem. Also, $\oplus$ is computationally cheaper than multiplication in a group.

However, the above alternative does not solve the problem of message expansion. One way around this is to use k^r as a key in a conventional cryptosystem, say, DES or IDEA [12]. That is, define

$$E_k(m,r) = (\alpha^r, \widetilde{E}_{k^r}(m)),$$

where $\widetilde{E}$ is any encryption function in a conventional cryptosystem. Here m can be a message of arbitrary length and $\widetilde{E}$ operates on m, say, with cipher block chaining (CBC) mode. With this modification, the cryptosystem is sometimes called a *hybrid cryptosystem*, which combines the advantages of both public-key and conventional cryptosystems. Such cryptosystems are practical but do not have provable security (a typical phenomenon for conventional cryptosystems). To improve security, one can use a cryptographically strong pseudo-random bit generator to expand k^r to a much longer string and then XOR it with m.

It should be noted that the ElGamal cryptosystem is completely insecure against an adaptive chosen ciphertext attack, mentioned earlier. Indeed, given an encryption (c_1, c_2) of a message m, one can ask for the decryption of $(c_1, c_2 \cdot \alpha)$, which is $\alpha \cdot m$, so m can be deduced immediately.

6.4   Digital Signatures

Suppose that you wish to transmit an electronic file. A natural question is how one can put a piece of information at the end of the file that serves the same role as a handwritten signature on a document. It turns out that the digital signature is one of the main applications of public-key cryptography.

Handwritten signatures have the following main features:

- The signature is unique and unforgeable. It is proof that the signer deliberately signed the document. It convinces the recipient of the document and any third party, say a judge, that it has been signed by the claimed signer.

- The signature is not reusable. It is part of the document and cannot be moved to another document. If the document is altered or the signature is moved to another document, then the signature is no longer valid.

How do we realize a signature digitally? Since it is easy to copy, alter, or move a file on computers without leaving any trail, one needs to be very careful in designing a signature scheme. In keeping with the above properties of a handwritten signature, a digital signature should be a number that depends on some secret known only to the signer and on the content of the message being signed. It must also be verifiable: i.e., the recipient of the message (or any unbiased third party) should be able to distinguish between a forgery and a valid signature without requiring the signer to reveal any secret information (private key). Thus in a signature scheme, we need two algorithms: one used by the person signing the message and the other used by the recipient verifying the signature. In the following, we describe two methods based on the RSA and the ElGamal cryptosystems [8, 21]. Incidentally, the recently adopted Digital Signature Algorithm (DSA) in the US Digital Signature Standard (DSS) [17] is a variation of ElGamal signature scheme, and we will describe it as well.

To describe the RSA digital signature scheme, note that the encryption function $E_k = P_{e,n}$ and the decryption function $D_k = P_{d,n}$ in the RSA system are commutative: that is,

$$D_k(E_k(x)) = E_k(D_k(x)) \equiv x^{ed} \equiv x \bmod n, \text{ for all } x \in \mathcal{Z}_n.$$

Suppose that a user Alice has public key $k = (e, n)$ and private key (d, n) for the RSA cryptosystem. Then Alice can use her private key to encrypt a message (or a file) $m \in \mathcal{Z}_n$ and use the ciphertext $s = D_k(m) = m^d \bmod n$ as her signature for the message m. Anyone, seeing the message m and the signature s, can compute $m_1 = E_k(s)$ and accept the signature if and only if $m_1 = m$. This proves that Alice indeed signed the message m, since an adversary trying to forge a signature for Alice on a message m would have to solve the equation $s^e \equiv m \bmod n$ for $s \in \mathcal{Z}_n$ (which is presumably hard). So if Bob shows m and s to a judge and if $E_k(s) = m$, the judge should be convinced that no one but Alice could have signed the statement.

There is one catch though — and this occurs when all or a significant fraction of the elements in Z_n represent valid messages. In this case, one could easily forge a signature as follows. Pick $s \in \mathcal{Z}_n$ at random and compute $m = E_k(s)$ where k is Alice's public key. Then with high probability, m is a valid message and in this case, since $E_k(s) = m$ holds, s is a valid signature of Alice for m. To avoid this possibility in practice, one adds some redundant information to the message. Namely, we require the message to have some additional structure (e.g., it should be in some standard format). Thus, a random element in $\mathcal{Z}_n$ will be a valid message with only vanishing probability. This comment also applies to the DSA and the ElGamal signature schemes discussed below.

The ElGamal cryptosystem cannot, as it stands, be used to generate signatures, but it can be modified to suit signature purposes. In this case, the signature scheme is probabilistic in that there are many possible valid signatures for every message and the verification algorithm accepts any of the valid signatures as authentic. Suppose that p is a large prime for which computing discrete logarithms in $\mathcal{Z}_p$ is infeasible, and $\alpha \in \mathcal{Z}_p$ is a primitive element. Also suppose that Alice chooses a random integer $a \in \mathcal{Z}_{p-1}$ as her private key and $\beta = \alpha^a \bmod p$ as her public key. To sign a message $m \in \mathcal{Z}_{p-1}$, Alice can

- Pick a random $k \in \mathcal{Z}_{p-1}$, with $\gcd(k, p - 1) = 1$.

- Compute γ and δ where

$$\gamma = \alpha^k \bmod p, \quad \delta = (m - a\gamma)k^{-1} \bmod (p - 1).$$

- Sign the message m with (γ, δ).

Since $\beta^\gamma \gamma^\delta \equiv \alpha^{a\gamma + k\delta} \equiv \alpha^m \bmod p$, anyone can verify Alice's signature:

- Get Alice's public key β (α and p are public).

- Compute $e_1 = \beta^\gamma \gamma^\delta \bmod p$ and $e_2 = \alpha^m \bmod p$.

- Accept the signature as valid only if $e_1 = e_2$.

The US Digital Signature Standard (DSS) was adopted on December 1, 1994. In DSS, a digital signature algorithm (DSA) is proposed and it is a variation of the ElGamal signature scheme. We describe DSA briefly as follows. Choose primes p and q with $q|(p-1)$ and

$$2^{159} < q < 2^{160}, \quad 2^{L-1} < p < 2^L.$$

That is, q has 160 bits and p has L bits where $512 \le L \le 1024$ and L is a multiple of 64. Suppose $\alpha \in \mathcal{Z}_p$ has order q: i.e., $\alpha \not\equiv 1 \bmod p$ but $\alpha^q \equiv 1 \bmod p$. A user, say Alice, has a random nonzero integer $a \in \mathcal{Z}_q$ as her private key and $\beta = \alpha^a \bmod p$ as her public key. To sign a message $m \in \mathcal{Z}_q$, Alice can

- Pick a random nonzero $k \in \mathcal{Z}_q$.

- Compute $\gamma = (\alpha^k \bmod p) \bmod q$, $\delta = (m + a\gamma)k^{-1} \bmod q$.

- Sign the message m with (γ, δ).

To verify Alice's signature (γ, δ) for the message m, the receiver can

- Get Alice's public key β.

- Compute

$$e_1 = m\delta^{-1} \bmod q, \; e_2 = \gamma\delta^{-1} \bmod q, \; \gamma_1 = (\alpha^{e_1}\beta^{e_2} \bmod p) \bmod q.$$

- Accept the signature as valid only if $\gamma_1 = \gamma$.

To see why this works, note that

$$k\delta \equiv m + a\gamma \bmod q.$$

Thus

$$\alpha^{k\delta} \equiv \alpha^{m+a\gamma} \bmod p \equiv \alpha^m \beta^\gamma \bmod p.$$

Since q is prime and $\delta \not\equiv 0 \bmod q$, we see that the map $x \mapsto x^\delta$ in the multiplicative group generated by α is a permutation. Thus

$$\alpha^k \equiv \alpha^{m\delta^{-1}} \beta^{\gamma\delta^{-1}} \bmod p.$$

Reducing both sides modulo q (after they are reduced modulo p), we have $\gamma = \gamma_1$.

Note that in the above signature scheme, the size of a signature equals (in RSA) or doubles (in both ElGamal and DSA) the size of the message being signed. This can be awkward in practice, especially if the message being signed is long. One way to overcome this problem is to first hash the message to a string of fixed size and then sign the hashed value of the message. Together with DSS, the National Institute of Standards and Technology (NIST) has also published a Secure Hash Standard (SHS) [17]. In SHS, a Secure Hash Algorithm (SHA) is proposed, which maps a message of arbitrary length to a binary string of length 160. We will not describe the details here, and the interested reader is referred to the references.

Public-key cryptography has many other applications, including identification, authentication, authorization, data integrity, and smart cards. In fact, NIST proposed in February 1997 a standard for entity authentication using public-key cryptography; the reader can consult the website <http://csrc.ncsl.nist.gov/fips/>. For further study we recommend the books [16, 22, 23, 24], and for the early history of cryptography, we suggest [11]. Computational number theory is nicely covered in [3, 5, 20]. Additional information on cryptographic methods, algorithms, and protocols can be found at <http://www.ssh.fi/tech/crypto/>.

6.5 *Exercises and Projects*

1. Develop an algorithm for computing $\alpha^e \bmod n$ using the square-and-multiply method indicated in Example 6.1. Implement your algorithm and test it for $n = 12345$; $\alpha = 123$; and $e = 0, 111, 12344, 54321$.

2. Let $n = 863 \cdot 877 = 756851$ and let $e = 5$. Given that 863 and 877 are primes, find $\phi(n)$ and compute d such that $ed \equiv 1 \bmod \phi(n)$.

3. The following problems relate to the discussion of the RSA system.

   a. Given $n = pq = 591037$ and $\phi(n) = 589500$, determine p and q by first determining a quadratic equation that p satisfies and then solving it using the quadratic formula.

b. Let M be a given multiple of $\phi(n)$. Write $M = 2^e m$ where m is odd. Prove that for random $a, b \in \mathcal{Z}_n$, the probability that $\gcd(n, a^{m2^i} - b^{m2^i})$ is a proper factor of n for some $0 \le i \le e$ is at least $1/2$. As a consequence, show that n can be factored efficiently when a multiple of $\phi(n)$ is given.

4. You are an RSA cryptosystem user with public key $n = 756851$ and $e = 5$. Suppose that a number in $\mathcal{Z}_n$ is always written as a 6 digit number (padding zeros in front if necessary). Then the numbers in $\mathcal{Z}_n$ represent a triple of letters under the correspondence $00 \leftrightarrow A$, $01 \leftrightarrow B$, ..., $25 \leftrightarrow Z$; e.g., $1719 = 001719 \leftrightarrow ART$. Your private key is the number d computed in Exercise 6.2. Decrypt the following ciphertext:

375365	752560	389138	193982	283519	350016	92892	86995
604644	125895	706746	323635	574615	226430	533566	419464

5. Suppose that p is a prime congruent to 3 modulo 4 and c is a quadratic residue modulo p. Prove that $x = c^{(p+1)/4} \bmod p$ is a square root of c modulo p.

6. Suppose Bob is a Rabin cryptosystem user with public key $n = 5609$ and private key $(p, q) = (71, 79)$. With each number in $\mathcal{Z}_n$ representing two letters as in Exercise 6.4, decrypt the following ciphertext:

924	642	2299	223	5374	121	2217	4474	719	839	5060
1474	3607	3763	2015	3586	3204	5060	10	2017	169	5101
446	4837	288	2217	4474	719	839	5060	1474	3988	

7. Investigate how to choose a convenient plaintext space for Rabin's cryptosystem. That is, for $n = pq$, find a subset M_n of $\mathcal{Z}_n$ such that (a) $x_1^2 \not\equiv x_2^2 \bmod n$ for all different $x_1, x_2 \in M_n$; (b) for any $x \in \mathcal{Z}_n$, it is easy to decide whether $x \in M_n$; and (c) M_n should be large, say of size $O(n)$.

8. Decrypt the following ElGamal ciphertexts. The parameters are $p = 3119$, $\alpha = 7$, $\beta = 1492$, and $d = 799$. Each number in $\mathcal{Z}_p$ represents two letters as in Exercise 6.4.

(1139, 1035)	(79, 1438)	(1489, 2725)	(2928, 87)	(691, 888)
(3010, 1012)	(1316, 1734)	(1790, 1385)	(2775, 1267)	(1807, 2319)
(2910, 2668)	(142, 238)	(123, 1994)	(916, 2055)	(3053, 2491)
(810, 247)	(1674, 2521)	(617, 1798)	(2705, 144)	(776, 650)
(1440, 311)	(1620, 713)	(938, 572)	(2209, 968)	(1037, 45)

9. Let $p = 877$ and $\alpha = 2$. Alice uses the ElGamal signature scheme, and her public key is $\beta = 253$.

   a. Verify that $(137, 217)$ is a valid signature of Alice for the message $m = 710$.

   b. Suppose your secret key is $a = 133$. Sign the message $m = 606$.

10. Suppose that Bob uses the DSA with $q = 103$, $p = 10 \cdot q + 1 = 1031$, $\alpha = 14$, $a = 75$, and $\beta = 742$. Determine Bob's signature on the message $x = 1001$ using the random value $k = 49$, and verify the resulting signature.

6.6 References

[1] L. M. Adleman and M.-D. A. Huang, *Primality Testing and Abelian Varieties over Finite Fields*, Lecture Notes in Mathematics, Volume 1512, Springer-Verlag, Berlin, 1992.

[2] A. O. L. Atkin and F. Morain, "Elliptic curves and primality proving," *Math. Comp.* **61**, 29–68 (1993).

[3] E. Bach and J. Shallit, *Algorithmic Number Theory, Volume I: Efficient Algorithms*, MIT Press, Cambridge, MA, 1996.

[4] M. Blum and S. Goldwasser, "An efficient probabilistic public-key cryptosystem which hides all partial information," *Advances in Cryptology — CRYPTO 84*, Lecture Notes in Computer Science, Volume 196, Springer-Verlag, Berlin, 1985, 289–299.

[5] H. Cohen, *A Course in Computational Algebraic Number Theory*, Graduate Texts in Mathematics 138, Springer-Verlag, Berlin, 1993.

[6] R. Cramer and V. Shoup, "A practical public key cryptosystem provably secure against adaptive chosen ciphertext attack," *Advances in Cryptology — CRYPTO 98*, Lecture Notes in Computer Science, Volume 1462, Springer-Verlag, Berlin, 1998, 13–25.

[7] W. Diffie and M. E. Hellman, "New directions in cryptography," *IEEE Trans. Info. Theory* **22**, 644–654 (1976).

[8] T. ElGamal, "A public key cryptosystem and a signature scheme based on discrete logarithms," *IEEE Trans. Info. Theory* **31**, 469–472 (1985).

[9] C. F. Gauss, *Disquisitiones Arithmeticae*, Braunschweig, 1801. English Edition, Springer-Verlag, New York, 1986.

[10] S. Goldwasser and S. Micali, "Probabilistic encryption," *Journal of Computer and System Science* **28**, 270–299 (1984).

[11] D. Kahn, *The Codebreakers: The Story of Secret Writing*, Macmillan, New York, 1968.

[12] X. Lai, "On the design and security of block ciphers," *ETH Series in Information Processing*, Volume 1, Hartung-Gorre Verlag, Konstanz, Switzerland, 1992.

[13] A. K. Lenstra and H. W. Lenstra, Jr., *The Development of the Number Field Sieve*, Lecture Notes in Mathematics, Volume 1554, Springer-Verlag, Berlin, 1993.

[14] A. J. Menezes, *Elliptic Curve Public Key Cryptosystems*, Kluwer, Boston, 1993.

[15] A. J. Menezes, I. F. Blake, X. Gao, R. C. Mullin, S. A. Vanstone, and T. Yaghoobian, *Applications of Finite Fields*, Kluwer, Boston, 1993.

[16] A. J. Menezes, P. C. van Oorschot, and S. A. Vanstone, *Handbook of Applied Cryptography*, CRC Press, Boca Raton, FL, 1996.

[17] National Institute for Standards and Technology, *Secure Hash Standard* (SHS), FIPS 180-1 (1995); *Digital Signature Standard* (DSS), FIPS 186-1 (1998).

[18] A. M. Odlyzko, "Discrete logarithms in finite fields and their cryptographic significance," *Advances in Cryptology — EUROCRYPT 84*, Lecture Notes in Computer Science, Volume 209, Springer-Verlag, Berlin, 1985, 224–314.

[19] A. M. Odlyzko, "Discrete logarithms and smooth polynomials," in *Finite Fields: Theory, Applications, and Algorithms*, G. L. Mullen and P. J.-S. Shiue (Eds.), Contemporary Mathematics, Volume 168, American Mathematical Society, Providence, RI, 1994, 269–278.

[20] C. Pomerance (Ed.), *Cryptography and Computational Number Theory*, Proc. Symp. Appl. Math., Volume 42, American Mathematical Society, Providence, RI, 1990.

[21] R. L. Rivest, A. Shamir, and L. Adleman, "A method for obtaining digital signatures and public-key cryptosystems," *Communications of the ACM* **21**, 120–126 (1978).

[22] B. Schneier, *Applied Cryptography*, 2nd ed., Wiley, New York, 1995.

[23] G. J. Simmons (Ed.), *Contemporary Cryptography: The Science of Information Integrity*, IEEE Press, New York, 1992.

[24] D. R. Stinson, *Cryptography: Theory and Practice*, CRC Press, Boca Raton, FL, 1995.

Chapter 7

Nonlinear Transverse Vibrations in an Elastic Medium

Philip B. Burt

Prerequisites: differential equations

7.1 Introduction

Physical systems have been modeled by ordinary and partial differential equations since calculus was developed. The construction of an appropriate model involves the elimination or minimization of effects deemed to be negligible, and results in one or more dependent functions expressed in terms of the independent and controllable variables. This is a mathematical model for the physical system. The solution of the differential equations leads to expressions for the dependent functions that require initial or boundary values to complete the model. These solutions can then be compared with empirical information to test the model.

The development of solutions for differential equations is also a subject with an extensive history. The problem here is to devise techniques which, in themselves, do not contribute to the inaccuracy inherent in comparing the model with experimental data. For example, the solution of a complex system of differential equations using a computer is subject to computational fictions that can obscure the relevance of a model. Nevertheless, some type of approximation is almost always necessary. This is a result of the well-known theorem of introductory differential equations that "exact equations are uninteresting and interesting solutions are inexact." (It will be left to the reader to track down the source and proof of this theorem.)

Linear second-order differential equations have the most frequent application to physical problems. The body of literature on such equations is enormous and many techniques have been devised for approximating solutions. These techniques comprise a tribe dubbed *perturbation theory* since, for the most frequently encountered examples, there is within the problem a set of parameters which can be varied continuously and which, when zero, reduce the problem to one with an exact solution. In this context, exact solutions are those in which all boundary values can be inserted, the resulting functions evaluated over a range of interesting values of the independent variables, and comparison with data made. This, in turn, implies that the physical system is controllable over the range of the parameters.

A vast universe of physical problems is excluded from the class of perturbative problems. In these systems, parameters characterizing the input enter the analysis intrinsically. That is, any description that omits these parameters is unphysical. For such systems, described by nonlinear differential equations, the techniques applicable to linear systems including perturbation theory may be wholly inapplicable. New approaches must be devised. In this chapter one such technique will be presented.

The physical problem to be described consists of a string embedded in an elastic medium. Transverse waves can be generated on the string. These waves will be affected by the elastic properties of the medium. For large amplitude waves, nonlinear effects are also qualitatively significant. A novel analysis of this wave motion along with a systematic approximation using the Ricatti equation will constitute the solution method. As an interesting aside, application of these ideas to problems in nuclear physics will also be mentioned.

7.2 A String Embedded in an Elastic Medium

The modeling of a string with ends fixed between two supports is a classic described in many introductory physics and calculus textbooks [6]. A small amplitude transverse displacement of the string results in waves described by the *wave equation*

$$\frac{\partial^2 Y}{\partial x^2} - \frac{1}{v^2}\frac{\partial^2 Y}{\partial t^2} = 0, \tag{7.1}$$

where Y is the amplitude of the displacement, x is the position along the string, v is the wave speed (a characteristic of the string determined by its mass density and elastic properties), and t is the time. Boundary conditions for fixed ends are typically that Y vanish at both ends of

the string. Problems with other boundary conditions are also of great
interest and are considered below.

For example, Figure 7.1 shows solutions to the wave equation (7.1)
with fixed endpoints. The initial solution displayed at the top evolves to
the middle function (shown as dashed) and later to the lower function.
The solution periodically returns to the initial function.

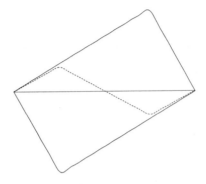

FIGURE 7.1   **Solutions of the wave equation (7.1)**

Figure 7.2 shows solutions of the vibrating string problem with infinite length. The initial function displayed at the center splits into
left-traveling and right-traveling waves (shown as dashed curves).

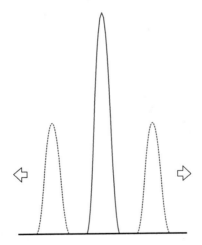

FIGURE 7.2   **Solutions for a vibrating string of infinite length**

The wave equation describes the string in a "vacuum," a medium whose properties do not enter the problem. However, an elastic medium will offer resistance to the string motion. As a first generalization of (7.1), we add a term which is linear in Y to obtain the new equation

$$-m^2 Y + \frac{\partial^2 Y}{\partial x^2} - \frac{1}{v^2}\frac{\partial^2 Y}{\partial t^2} = 0. \tag{7.2}$$

The coefficient $-m^2$ is characteristic of the medium and describes its elastic properties. As the displacement of the string increases, the elasticity of the medium restores it to the equilibrium condition.

Interesting solutions of (7.1) or (7.2) can be found in which the time and space behavior separate. That is, the function Y can be written in the form

$$Y(x,t) = X(x)T(t).$$

For such solutions, the function $T(t)$ satisfies an ordinary differential equation of the form

$$\frac{d^2 T}{dt^2} + (m^2 + k^2)v^2 T = 0, \tag{7.3}$$

where k is a separation constant arising from the boundary conditions satisfied by $X(x)$.

Separable solutions are appropriate for *standing waves* such as may be found when the ends of the string are fixed. However, *traveling waves* are also interesting. These do not separate, but may be found by superposition of separable solutions $Y = A_k e^{i(kx-\omega t)}$ for different k. There is an interesting difference between superpositions of separable solutions of (7.1) and (7.2). A wave packet consisting of superpositions of solutions for different k will have a form as a function of position that distorts as time progresses. The reason for this distortion is that the phase speed of the wave solutions of (7.1) is v, while that of the solutions of (7.2) is ω/k, given by

$$\left(\frac{\omega}{k}\right)^2 = v^2\left(1 + \frac{m^2}{k^2}\right),$$

which differs for the various terms in the superposition. Clearly the case $m = 0$ corresponds to the constant wave speed in (7.1). The source of the distortion of the wave packet is clear. Contributions for different k travel with different phase speeds so that as time progresses, the phase front for a wave component will travel further than the front for a larger value of k. The phenomenon described here is known as *dispersion*. It is ubiquitous in physical systems and, in fact, is the rule rather than the exception. The solutions that interest us most are traveling waves as would be seen on a string of infinite length.

The description of the response of the elastic medium in terms of the string displacement Y is an idealization that is itself an exception. In fact, a nonlinear response is most likely. This may be viewed as an amplitude dependent m^2 or more generally as a function whose lowest order term is m^2Y. If this function is taken to be a series in positive powers of Y, it can be easily seen that only odd powers will occur. The reason is that the force must restore the string to its original undisturbed position, whether the displacement is positive or negative. An even power of Y with a negative coefficient will always lead to a negative acceleration of the string irrespective of the sign of Y. Thus we are led to a generalization of the wave equation having the form

$$\frac{1}{v^2}\frac{\partial^2 Y}{\partial t^2} - \frac{\partial^2 Y}{\partial x^2} + m^2 Y + bY^3 = 0. \tag{7.4}$$

Solutions appropriate for the infinite string can be found in Burt [2]. Equation (7.4) is a particular nonlinear *Klein-Gordon equation* [2]. It has been relatively well studied and has special solutions of known form. For example, when $b < 0$ there is a traveling wave solution of the form $Y = \alpha \tanh \beta(x - ct)$; see Exercise 7.6.

An important class of solutions is again given by traveling wave solutions, i.e., solutions in a variable $z = kx \pm k_0 vt$. These solutions can be constructed in various ways. The method we will use in this chapter is called the *method of base equations*. This method relates solutions of (7.4) to solutions of an associated linear differential equation called the *base equation*. For this particular case, the base equation is (7.2), the linear Klein-Gordon equation. Base equation techniques have been related to Backlund transformations and are applicable to a wide variety of linear and nonlinear equations. The most extensive discussion of the technique is found in the Clemson University dissertation of J. L. Reid [10]. Other reviews may be found in Burt [2] and Ranganathan [9].

7.3 An Approximation Technique for Nonlinear Differential Equations

We want to analyze nonlinear differential equations using a method that does not rely on expansions in powers of the coefficient of the nonlinearity (also known as a coupling constant). A systematic method has been developed which uses the fact that linear second-order differential equations can be converted to the first-order Ricatti equation [11]. For example, (7.3) can be rewritten as a first-order equation through the introduction of a new dependent variable

$$p = \frac{1}{T} \frac{dT}{dt}$$

to obtain

$$\frac{dp}{dt} + p^2 + V^2 = 0, \tag{7.5}$$

where $V^2 = (m^2 + k^2)v^2$. This is a special example of the Ricatti equation. Solutions of this equation can be converted to solutions of (7.3) and vice versa. The reader should note that the solutions of (7.3) are sine waves and thus vanish every half period. Thus the transformation equation is valid only for t in intervals of half-period length, and indeed the solution of (7.5) has the form

$$p(t) = V \tan (C - Vt),$$

with $C = \tan^{-1} \left(\frac{T_0'}{VT_0} \right)$. Notice that solutions here depend only on the ratio of T_0' to T_0.

In the same way, the differential equation[1]

$$\frac{d^2 R}{dz^2} + f(R) = 0$$

can be transformed by the change of variable

$$p = \frac{1}{R} \frac{dR}{dz}$$

to the form

$$\frac{dp}{dz} + p^2 + \frac{1}{R} f(R) = 0.$$

The change of variable can also be put into the form $\frac{1}{R} dR = p\,dz$, which leads to the expression $R = K e^{\int^z p(s)\,ds}$, where K is determined by the value of the solution at an initial value of z. Substitution gives

$$\frac{dp}{dz} + p^2 + \frac{1}{K} e^{-\int^z p(s)\,ds} f\left(K e^{\int^z p(s)\,ds} \right) = 0.$$

The specific example with

$$f(R) = \frac{m^2 R}{k_0^2 - k^2} + \frac{bR^3}{k_0^2 - k^2}$$

---

[1]A solution of the model $\frac{1}{v^2} \frac{\partial^2 Y}{\partial t^2} - \frac{\partial^2 Y}{\partial x^2} + af(Y) = 0$ of the form $Y = R(kx - \omega t)$ gives this equation. Here $z = kx - \omega t$ and $a = \frac{\omega^2}{v^2} - k^2$. The form of the argument $kx - \omega t$ is inherited from the Fourier series/transform method of solving the classical wave equation with initial condition $Y(x, 0)$ and $Y_t(x, 0)$.

is interesting since we may find exact solutions and this will guide us in describing a method for other functions $f(R)$. (For $b/[k_0^2 - k^2] > 0$, the solution of an initial value problem containing this differential equation is periodic and the period is dependent on the initial values.) The equation for p becomes

$$\frac{dp}{dz} + p^2 + \frac{m^2}{k_0^2 - k^2} + \frac{b}{k_0^2 - k^2} K^2 \, e^{2 \int^z p(s)\, ds} = 0. \qquad (7.6)$$

For convenience we take

$$M^2 = \frac{m^2}{k_0^2 - k^2},$$

$$B = \frac{b}{k_0^2 - k^2}.$$

Approximate solutions of the integro-differential equation (7.6), or those for other functions $f(R)$, are derived by choosing an approximate function for $p(s)$ in the integrand. The integral is evaluated and the resulting equation solved for the next approximation. This is reinserted in the integral, a new equation derived and solved, and so on. At no point is an expansion in the parameter describing the strength of the nonlinearity made. This method has been used to generate exact solutions, and second iterations on the Lane-Emden equation have also been calculated; see Burt and Pickett [5] and Burt [3]. In the next section a specific assumption will be made which brings (7.6) back to a Ricatti equation.

7.4 Base Equation Solution of Ricatti Equation

The integral equation (7.6) can be approximated in a way that leads back to a Ricatti equation. In the integrand, assume

$$p(s) = V \tan (C - V s), \qquad (7.7)$$

with $C = \tan^{-1} \left(\frac{R_0'}{V R_0} \right)$, so that the exponential term becomes

$$e^{2 \int^z p(s)\, ds} = e^{2 \int^z \frac{d}{ds} \ln[\cos (C - V s)]\, ds}$$

and the approximate p satisfies the Ricatti differential equation

$$\frac{dp}{dz} + p^2 + M^2 + BK^2 \left| \cos (C - V z) \right|^2 = 0,$$

where $K = R_0 \Big/ \sqrt{R_0^2 + \left(\frac{R_0'}{V} \right)^2}$. If we change variables using

$$p(z) = \frac{1}{U} \frac{dU}{dz}, \qquad (7.8)$$

then $U(z)$ satisfies

$$\frac{d^2U}{dz^2} + (M^2 + BK^2 \left| \cos(C - Vz) \right|^2) U = 0.   \qquad (7.9)$$

We don't know how to solve this equation so we replace it with an averaged equation. The average value of $\left| \cos(C - Vz) \right|^2$ over a half period π/V is $\pi/(2V)$. The differential equation (7.9) is replaced with

$$\frac{d^2U}{dz^2} + \left(M^2 + BK^2 \frac{\pi}{2V}\right) U = 0.   \qquad (7.10)$$

The solution of (7.10) is converted to $p = \frac{U'}{U}$ and used in the integrand of (7.6) to find another Ricatti differential equation for p, etc.

The choice of the trial function $p(s)$ to place in the integral term of (7.6) is the key to this method. A good choice will lead back (approximately) to the same function for specified choices of R_0 and R_0'. An alternate choice for p to that of (7.7) is

$$e^{2 \int^z p(s)ds} = p^2 \cos^2(C - Vz),$$

which gives the Ricatti equation

$$\frac{dp}{dz} + \sigma p^2 + M^2 = 0,   \qquad (7.11)$$

with $\sigma = 1 + BK^2 \cos^2(C - Vz)$. Again using an averaged value for the cosine term we get (7.11), this time with $\sigma = 1 + BK^2 \frac{\pi}{2V}$. If we change again to new variables by (7.8) we obtain

$$\frac{d^2U}{dz^2} + (\sigma - 1)\frac{1}{U}\left(\frac{dU}{dz}\right)^2 + M^2U = 0.   \qquad (7.12)$$

Equation (7.12) can be solved in terms of solutions of the base equation

$$\frac{d^2u}{dz^2} + \sigma M^2 u = 0.   \qquad (7.13)$$

These solutions have the form

$$U = (Au_1 + u_2)^{1/\sigma},   \qquad (7.14)$$

where u_1 and u_2 are independent solutions of (7.13) and A is an arbitrary constant. If for convenience u_2 is chosen so $u_2(0) = 0$, then A is determined by the value of R_0. Such solutions have been referred to as a *weak Backlund transformation* of (7.12); for example, see [11].

The reason for making this circuitous choice for p is to use the solution in (7.14) as a starting point for a more general discussion where the string is driven by a space- and time-dependent nonlinear force. This will require B to be space and time dependent. Other choices of p in the integrand of a generalized (7.6) will lead to new approximate differential equations and subsequent iterations. Other examples are given in [4].

An interesting and important application of the nonlinear Klein Gordon equation and its generalizations is to the interaction of certain subnuclear particles (pi mesons) with themselves and with constituents of the nucleus. The fact that solutions of the nonlinear equations can be obtained from solutions of related linear equations (in this case, the Klein-Gordon equation) with the same coefficients in the linear term is then of utmost importance since the nonlinear theories are non-perturbative, that is intrinsically nonlinear. Extensive discussion of this application can be found in reference [2].

To repeat, the traveling wave solutions of (7.4) are known. Methods of obtaining these solutions (such as using the Ricatti base equations) may be studied to learn ways of finding traveling wave solutions of more general equations, such as

$$\frac{1}{v^2}\frac{\partial^2 Y}{\partial t^2} - \frac{\partial^2 Y}{\partial x^2} + f(Y) = 0.$$

Finding ways to choose trial functions $p(s)$ in the integral term of the Ricatti type differential-integral equation such as (7.6) so that resulting solutions of this equation are better trial functions is a worthy problem.

7.5   Exercises and Projects

1. Construct solutions to the linear model for a vibrating string. That is, using superposition of separable solutions of the wave equation, construct the function that satisfies

$$\frac{\partial^2 Y}{\partial x^2} - \frac{1}{v^2}\frac{\partial^2 Y}{\partial t^2} = 0$$

$$Y(0,t) = Y(a,t) = 0, \ t \geq 0$$

$$Y(x,0) = f(x), \ Y_t(x,0) = g(x) \ \text{for} \ 0 \leq x \leq a.$$

Hint: If $Y(x,t) = X(x)T(t)$ satisfies the partial differential equation and the boundary conditions, then $X(x)$ is some multiple of $\sin(\lambda_n x)$ for $\lambda_n = (n\pi/a)^2$ and this gives $T'' + \lambda_n v^2 T = 0$. Suppose we define $F(x) = f(x)$ for $0 \leq x \leq a$ and require F to be odd and periodic with period $2a$; further define $G(s) = \int_0^s g(x)\,dx$ for $0 \le s \le a$ and require G to be even and periodic with period $2a$. Then

$$Y = \frac{1}{2}[F(x - vt) + F(x + vt)] + \frac{1}{2v}[G(x - vt) + G(x + vt)];$$

see [8, pp. 396–406]. As a special case suppose $f(x)$ is a multiple of $\sin\left(\frac{n\pi x}{a}\right)$ for some positive integer n and $g(x)$ is another multiple. Then $Y(x,t)$, given as a product of $\sin\left(\frac{n\pi x}{a}\right)$ and a function of t, is the solution of the problem. Such a solution would correspond to a simple vibration in place. This is called a *standing wave*.

2. (Project) Consider a perfectly flexible elastic string that vibrates in a plane, but do not assume that each particle of the string moves along a line perpendicular to a coordinate axis. Construct a mathematical model for the motion of the string. See [8, p. 400].

3. A graphics package, such as MATLAB [7], Mathematica [13], or Maple [12], can be used in the following problems.

a. Plot $Y(x,0), Y(x,1), Y(x,2)$ for x between 0 and 2 for

$$Y(x,t) = \sum_{n=1}^{25} B(n) \sin\left(\frac{n\pi x}{2}\right) \cos\left(\frac{n\pi t}{2}\right),$$

where

$$B(n) = \frac{2}{3} \int_0^{1.5} x \sin\left(\frac{n\pi x}{2}\right) dx + 2 \int_{1.5}^2 (2 - x) \sin\left(\frac{n\pi x}{2}\right) dx.$$

These graphs show the solution of (7.1) for $v = 1$ and particular initial conditions. What other values for t will give the same graph?

b. Plot $Y(x,0), Y(x,0.5), Y(x,1), Y(x,1.5), Y(x,2)$ for x between 0 and 2 for

$$Y(x,t) = \sum_{n=1}^{25} B(n) \sin\left(\frac{n\pi x}{2}\right) \cos\left(\frac{\sqrt{1 + n^2}\,\pi t}{2}\right),$$

using the $B(n)$ given in part (a). These graphs show the solution of (7.2) for $m = v = 1$ and the same initial conditions.

4. Suppose u_1 and u_2 are independent solutions of $\frac{d^2u}{dz^2} + g(z)u = 0$ and $U = (Au_1 + u_2)^{1/\sigma}$. Show $U(z)$ satisfies

$$\frac{d^2U}{dz^2} + (\sigma - 1)\frac{1}{U}\left(\frac{dU}{dz}\right)^2 + \frac{g(z)}{\sigma}U = 0.$$

For the case $g(z) = \sigma M^2$ and $\sigma = 3$, plot a family of U solutions. That is, plot solutions for various values of A.

5. Consider the initial value problem

$$\frac{d^2Y}{dt^2} + M^2Y + 2B^2Y^3 = 0$$

$$Y(0) = \sigma, \quad \frac{dY}{dt}(0) = 0.$$

Multiply the differential equation by $\frac{dY}{dt}$ and integrate, then separate the variables and integrate again to obtain

$$\int_\sigma^Y \frac{dy}{\sqrt{C - (M^2y^2 + B^2y^4)}} = \pm t,$$

where $C = M^2\sigma^2 + B^2\sigma^4$. To integrate the term on the left, observe that the expression within the radical can be expressed as

$$X = C - M^2y^2 - B^2y^4 = -B^2\left(y^4 + \frac{M^2}{B^2}y^2\right) + C$$

$$= -B^2\left(y^4 + \frac{M^2}{B^2}y^2 + \frac{M^4}{4B^4}\right) + C + \frac{M^4}{4B^2}$$

$$= D^2 - B^2\left(y^2 + \frac{M^2}{2B^2}\right)^2,$$

where $D^2 = C + \frac{M^4}{4B^2}$. Factorization gives

$$X = \left[D + B\left(y^2 + \frac{M^2}{2B^2}\right)\right]\left[D - B\left(y^2 + \frac{M^2}{2B^2}\right)\right].$$

Our integrand now has the form

$$\frac{1}{\sqrt{B}}\int_\sigma^Y \frac{dy}{\sqrt{\gamma^2 + By^2}\sqrt{\sigma^2 - y^2}} = \pm t,$$

where $\gamma^2 = D + \frac{M^2}{2B}$ (note $B\sigma^2 = D - \frac{M^2}{2B}$). Standard tables [1, p. 596] give

$$Y = \sigma\, \text{cn}\left(Bt\sqrt{\sigma^2 + \frac{\gamma^2}{B}}, \mu\right),$$

with $\mu^2 = \frac{\sigma^2}{\sigma^2 + (\gamma^2/B)}$ and where cn is an elliptic function. Recall that

$$\gamma^2 = M^2\sigma^2 + B^2\sigma^4 + \frac{M^4}{4B^2} + \frac{M^2}{2B}.$$

This gives a family of periodic solutions, as σ varies.

6. A computer algebra package, such as Maple [12] or Mathematica [13], can be used in the following problems.

   a. If $B < 0$ show that $y = \dfrac{M}{\sqrt{-B}} \tanh \left(\dfrac{M}{\sqrt{2}} z \right)$ is a solution of

   $$\frac{d^2 y}{dz^2} + M^2 y + B y^3 = 0.$$

   [This solution may be derived by the same method used in Exercise 7.5 for the case $C = -\frac{M^4}{2B}$ in the expression for $(\frac{dy}{dz})^2$.]

   b. For the solution given in part (a), calculate

   $$p = \frac{1}{y} \frac{dy}{dz} \quad \text{and} \quad \frac{dp}{dz} + p^2 + M^2.$$

   The result of the latter expression is what should result when the correct choice of p is made in the integral term in (7.6) corresponding to this function $p(z)$.

   c. For the solution given in Exercise 7.5, calculate

   $$p = \frac{1}{y} \frac{dy}{dz} \quad \text{and} \quad \frac{dp}{dz} + p^2 + M^2.$$

   The result of the latter expression is what should result when the correct choice of p is made in the integral term in (7.6) corresponding to this function $p(z)$.

7. Equation (7.2) can be generalized to two space dimensions to describe the propagation of waves on an infinite membrane. Particular solutions to this equation are

$$Y = \sqrt{-\frac{M^2}{B} \frac{U^2 - 1}{U^2 + 1}},$$

where $U = \sum_{i=1}^{N} e^{\alpha_i (K_i x + L_i Y - \omega_i t)}$. Find all conditions that α_i, K_i, L_i, and ω_i must satisfy. (See reference [2].)

7.6 References

[1] M. Abramowitz and I. A. Stegun, *Handbook of Mathematical Functions*, 10th printing, National Bureau of Standards, United States Department of Commerce, Washington, D.C., 1972.

[2] P. B. Burt, *Quantum Mechanics and Nonlinear Waves*, Harwood Academic Press, Chur, 1981.

[3] P. B. Burt, "Nonperturbative solutions of the Lane-Emden equation," *Physics Letters* **A109**, 133–135 (1985).

[4] P. B. Burt, "Non-perturbative solution of nonlinear field equations," *II Nuova Cimento* **B100**, 43–52 (1987).

[5] P. B. Burt and T. J. Pickett, "Nonlinear solution construction of some nonlinear field equations," *Letters to Nuova Cimento* **44**, 473–75 (1985).

[6] W. E. Gettys, F. J. Keller, and M. J. Skove, *Physics, Classical & Modern*, McGraw-Hill, New York, 1992.

[7] The MathWorks, Inc., *Using MATLAB Version 5*, The MathWorks, Inc., Natick, MA, 1997. <http://www.mathworks.com/>

[8] A. L. Rabenstein, *Introduction to Ordinary Differential Equations*, 2nd ed., Academic Press, New York, 1972.

[9] P. V. Ranganathan, "Homogeneous solution of some nonlinear partial differential equations," *Journal of Mathematical Analysis and Applications* **136**, 357–367 (1988).

[10] J. L. Reid, *Exact Solution of Some Nonlinear Problems of Mathematical Physics and to Some Nonlinear Differential Equations*, Ph.D. dissertation, Clemson University, Clemson, SC (1974).

[11] C. Rogers and W. F. Ames, *Nonlinear Boundary Value Problems in Science and Engineering*, Academic Press, Boston, 1989.

[12] Waterloo Maple, Inc., *Maple V Learning Guide*, Springer-Verlag, New York, 1997. <http://www.maplesoft.com/>

[13] S. Wolfram, *The Mathematica Book*, 3rd ed., Cambridge University Press, Cambridge, 1996. <http://www.wri.com/>

Chapter 8

Simulating Networks with Time-Varying Arrivals

Marie Coffin
Peter C. Kiessler

Prerequisites: probability, statistics

8.1 Introduction

Simulation is a popular modeling tool for studying systems that exhibit random behavior. The arrival times, service times, and routing schemes in queueing networks are often best modeled as random phenomena. Several texts [17, 19] deal with simulation of queueing networks, and much computer software, such as SLAM [13] and SIMAN [12], is designed primarily for simulating these systems. This chapter introduces simulation by considering a particular queueing network (a university registration system) and introducing the tools necessary for modeling this system. These tools come from many different areas of statistics, probability, and operations research.

The registration system is introduced in Section 8.2. There we discuss how to model this system as a queueing network. In the process of building the network, we describe the random behavior that the arrival process, service times, and routing scheme exhibit. The later sections are devoted to examining methods for simulating these entities.

Section 8.3 serves as brief introduction to random number generation. Section 8.4 discusses some of the statistical issues that arise in the modeling of random systems. It is important to note that for the registration model, a fair amount of data are available to help us build the network. In Section 8.5 we discuss the arrival processes considered for the network. Since the registration process itself lasts only a single day, a finite point process is an appropriate model. A fairly complete discussion of such processes is found in Daley and Vere-Jones [2]. As will be seen, finite point processes are extensions of Poisson processes that exhibit nonstationary behavior. In Section 8.6 we discuss modeling queueing systems as generalized semi-Markov schemes. These structures present a queueing network as a flow chart for building the simulation.

8.2   The Registration Problem

Many concepts of queueing networks will be illustrated by considering various aspects of a real-life queueing problem, namely, the Clemson University registration process. The registration task is an important one, because registration for all students must be completed in a short period of time.

The registration process consists of six procedures: registration form distribution, Guaranteed Student Loan (GSL) signing, dropping and adding of classes, fee assessment, fee payment, and turning in registration forms. This can be accomplished in several ways. In one scheme, the different tasks may be located on different parts of campus. Alternatively, each college could organize registration in its own way. A third alternative would impose a centralized registration for all students; this last alternative has been modeled by Ashby [1] and Kirschman [8].

Before building a queueing network, the elements of the network must be identified. *Customers* are entities that obtain service in the network. *Service stations* are locations where work is completed for (or on) the customers. *Service discipline* is the order in which customers are served. Each service station has *servers* to perform a task, and a *waiting room* where customers wait to be processed. The paths followed by the customers among the service stations are called *routes*.

In the Clemson Registration Network example, the customers are the students who need to register for classes. Although there are more than 15,000 students enrolled at Clemson University, not all the students register on the same day. Some students participate in late enrollment after classes start. We assume that 15,000 students must register in a 7.5-hour day. Customers can enter the network during the first 5.5 hours (330 minutes). A nonstationary arrival process (Section 8.5.2) is probably an appropriate arrival model, because arrival rates tend to be smaller during the early morning and late afternoon hours, and larger in the middle of the day.

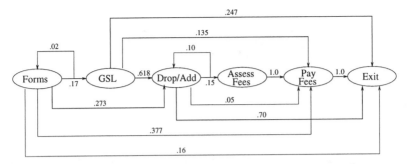

FIGURE 8.1   **Diagram of the registration network**

Figure 8.1 is a schematic diagram of the queueing network, showing the possible routes a student could follow during registration [1]. The network consists of six service stations or queues.

- *Queue 1*: registration form distribution;

- *Queue 2*: GSL signing;

- *Queue 3*: dropping and adding classes;

- *Queue 4*: fee assessment;

- *Queue 5*: fee payment;

- *Queue 6*: turning in completed registration forms.

All customers enter the network at Queue 1 and depart from Queue 6, although the routes through the network may vary. A total of 77 servers are allocated to the network, as shown in Table 8.1. The waiting room for each queue is a line — students wait in line until it is their turn for service. The service discipline in each queue is first in, first out. The mean and variance for Queues 1–5 in Table 8.2 were obtained from observational data [1]. The mean and variance for Queue 6 were simply assumed to be 0.25 and 0.0, respectively. We assume the workers are capable of working at any queue, and thus may be re-allocated for optimal performance.

8.3   *Generating Random Numbers*

Computer simulation of stochastic events depends on the ability to generate random events with properties that match the properties of the mathematical model. These events are usually expressed in the form of random numbers. Often when people use the phrase "random numbers"

Table 8.1   Allocation of (77) servers

Queue	Queue Description	Number of Servers
1	Form Distribution	22
2	GSL Signing	4
3	Drop/Add	15
4	Fee Assessment	14
5	Fee Payment	14
6	Form Turn-in	8

Table 8.2   Mean, variance of service times for queues

Queue	Mean Service Time per Server (min)	Variance per Server (min^2)
1	0.46590	0.1315
2	0.85165	0.1892
3	1.10900	0.4892
4	0.75370	0.4160
5	0.72482	0.2146
6	0.25000	0.0000

they mean uniformly distributed random numbers, but random numbers can follow any distribution. We assume the reader is familiar with the usual density functions [16].

Most computer languages, and many calculators, have the capacity to generate random numbers (more properly called "pseudo-random numbers") from a uniform distribution. Some computer languages also have the ability to generate pseudo-random numbers from other distributions, but this capability varies from language to language. In this section, methods will be discussed for generating pseudo-random numbers from a desired distribution, assuming a set of uniformly distributed pseudo-random numbers is available. A more thorough treatment of methods for generating pseudo-random numbers is given in [17].

Formally, the random variable U has the *uniform* (0,1) distribution, written $U \sim \mathrm{UNIF}(0,1)$, if $P(U \leq u) = u$ for $u \in [0,1]$. (Note: by convention, we use U to denote a random variable, whereas the notation $u \sim \mathrm{UNIF}(0,1)$ indicates that u is a realization of the random variable U.) Uniformly distributed random numbers are the basis for random numbers generated from any other distribution.

8.3.1   Random Numbers from Discrete Distributions

The random variable Y has a *discrete distribution* if, for some countable set $\{y_1, y_2, \ldots\}$, there exists a set of nonnegative probabilities $\{p_1, p_2, \ldots\}$

where $p_i = P(Y = y_i)$, such that $\sum_{i=1}^{\infty} p_i = 1$. Such a distribution may be finite (simply define $p_i \equiv 0$ for all $i > n$), or countably infinite. Random numbers that obey such a distribution can easily be generated from uniform random numbers by the following algorithm.

ALGORITHM 8.1: Generating random numbers, discrete case

1. *Generate $u \sim$ UNIF$(0, 1)$.*
2. *If $u < p_1$ then $Y = y_1$*
   else $Y = y_k$ where $\sum_{i=1}^{k-1} p_i \leq u < \sum_{i=1}^{k} p_i$.

The validity of Algorithm 8.1 follows directly since

$$P(Y = y_i) = P\left(\sum_{j=1}^{i-1} p_j \leq u < \sum_{j=1}^{i} p_j\right)$$

$$= \sum_{j=1}^{i} p_j - \sum_{j=1}^{i-1} p_j = p_i,$$

where the second equality holds because $u \sim$ UNIF$(0, 1)$.

8.3.2   Random Numbers from Continuous Distributions

A random variable Y is *continuous* if its distribution function F defined by $F(y) = P(Y \leq y)$ is a continuous function. The following theorem and its corollary [3], which apply to any distribution function, provide an approach for generating random numbers from continuous distributions.

THEOREM 8.1

If Y has distribution function F, then the random variable $U = F(Y) \sim$ UNIF$(0, 1)$.

COROLLARY 8.1

If $U \sim$ UNIF$(0, 1)$ and $G(u) = \min\{x : u \leq F(x)\}$, then $Y = G(U)$ has distribution function F.

As a result, when F can be written in closed form, one can generate random numbers from this distribution by first generating uniform random numbers U, and then letting $Y = \min\{y : F(y) \geq U\}$. The complicated form of this expression is necessary because in general F is not a one-to-one function. If F is one-to-one, this expression simplifies to $Y = F^{-1}(U)$. (Note that Step 2 of Algorithm 8.1 is a simple application of this, since for a discrete distribution, F is a step function.)

Some continuous distributions, such as the gamma distribution [16], do not have a distribution function that can be written in closed form. For these distributions, the following algorithm can generate pseudo-random numbers by using uniform random numbers and numerical integration.

ALGORITHM 8.2: Generating random numbers, continuous case

1. Let f be the density function of Y: i.e., $P(Y \leq y) = \int_{-\infty}^{y} f(t)\,dt$.
2. Choose an error bound η.
3. Generate $u \sim \text{UNIF}(0, 1)$.
4. Search for y such that $\left| \int_{-\infty}^{y} f(t)\,dt - u \right| \leq \eta$.
5. Then, except for the small error η, y has density f.

8.3.3 Shortcuts

Sometimes one can generate pseudo-random numbers more easily by applying one's knowledge of probability. For example, if $Y_1, Y_2, \ldots, Y_k$ are independent and identically distributed (iid) exponential random variables with mean 5, then $Z = \sum_{i=1}^{k} Y_i$ is a gamma random variable with scale parameter 5 and shape parameter k.

Similarly, if $Y_1, Y_2, \ldots, Y_n$ are iid according to any distribution with mean μ and standard deviation σ, and n is large enough, then $\overline{Y}$ is approximately normally distributed, with $\text{E}(\overline{Y}) = \mu$ and $\text{Var}(\overline{Y}) = \sigma^2/n$. The size of n needed for $\overline{Y}$ to be approximately normal depends on the distribution of the Y_i. For $Y_i \sim \text{UNIF}(0, 1)$, $n = 3$ is sufficient.

8.3.4 Generating Dependent Random Numbers

The above strategies are all valid for generating independent random numbers, appropriate for simulating random sampling from a population. However, in some situations, it is desirable to generate random numbers that are dependent. Some methods will be described here for generating random numbers with a specified dependence structure.

8.3.4.1 *Dependent Normal Random Numbers*

It is particularly easy to generate normal random numbers with a specific positive correlation, since any linear combination of independent normal random variables is also a normal random variable. Algorithm 8.3 produces dependent normal random variables $Y_1, Y_2, \ldots, Y_n$ having a specified correlation $\mathrm{Corr}(Y_i, Y_j) = \rho$, where $0 < \rho < 1$. Unfortunately, in many applications, normally distributed random numbers are not what is needed. There are very few other distributions for which one can generate dependent sets of numbers in this fashion.

ALGORITHM 8.3: Generating dependent normal variables

1. Let Z and $X_1, X_2, \ldots, X_n$ be independent standard normal random variables.
2. Define $c = \sqrt{\rho/(1-\rho)}$.
3. For $i = 1, 2, \ldots, n$ let $Y_i = \frac{c}{\sqrt{c^2+1}} Z + \frac{1}{\sqrt{c^2+1}} X_i$.

8.3.4.2 *Markov Chains*

A general approach to generating dependent random variables uses the joint distribution function of the random variables. One method of generating the joint distribution function is to generate the random vector sequentially using conditional distributions. This method is particularly appealing when the conditional distributions are easy to manipulate, as is the case for Markov chains.

Let $\{X_n : n = 0, 1, \ldots\}$ be a Markov chain having a finite state space given by $\{1, 2, \ldots, k\}$, one-step transition matrix P, and initial distribution π^o; see Chapter 13. Recall that each row of P sums to 1, so each row of this matrix is a probability distribution on $\{1, 2, \ldots, k\}$.

A realization of the Markov chain may be simulated as follows. First initialize the Markov chain; that is, generate the random variable X_0 using the distribution π^o as in Section 8.3.1. Now proceed inductively. Suppose the random variables $X_0, X_1, \ldots, X_n$ have been generated, realizing the values $i_0, i_1, \ldots, i_n$, respectively. The Markov property states that X_{n+1} is conditionally independent of $(X_0, X_1, \ldots, X_{n-1})$ given X_n:

$$P(X_{n+1} = i_{n+1} \mid X_0 = i_0, X_1 = i_1, \ldots, X_n = i_n) = p_{i_n, i_{n+1}}.$$

Thus, the random variable X_{n+1} is generated as in Section 8.3.1 using row i_n of P.

Example 8.1

Consider the Markov chain $\{X_n : n = 0, 1, \ldots\}$ having state space $\{1, 2, 3\}$ and one-step transition matrix

$$P = \begin{bmatrix} 1/3 & 1/6 & 1/2 \\ 1/6 & 1/2 & 1/3 \\ 1/2 & 1/3 & 1/6 \end{bmatrix}.$$

Assume that the Markov chain starts in state 1. In a simulation of this Markov chain, the next state can be found by generating a uniform random number u and then making the transition to the next state based on the distribution appearing in the first row of P. For example, if $u = 0.635$ is the random number generated, then state 3 is next visited since $\frac{1}{3} + \frac{1}{6} \le u < 1$. The process then repeats, starting from state 3 and using row 3 of P.

In Example 8.1 the statistic

$$Y_N = \frac{1}{N} \cdot |\{n \in \{1, \ldots, N\} : X_n = 1\}|$$

calculates the proportion of time that the Markov chain visits state 1 during the first N transitions. It follows that since $\{X_n : n = 0, 1, \ldots\}$ is *ergodic* [14], Y_N converges with probability one to the limiting probability of the Markov chain being in state 1. Since the one-step transition matrix in this example is *doubly stochastic* (each row and each column sums to 1), this probability equals $\frac{1}{3}$. Thus, the statistic Y_N is a consistent estimator of the limiting probability. Moreover, using ideas from regenerative processes, Iglehart and Shedler [6] calculate a central limit theorem for Y_N. In this way one may calculate confidence intervals for the limiting probabilities.

The method described above may be modified to generate sample paths of regular continuous-time Markov chains. Let $\{X(t) : t \ge 0\}$ be a continuous-time Markov chain having the state space $\{1, 2, \ldots, k\}$. Let Y_n be the nth state visited by the process and let T_n be the time of the nth jump of the process. Then

$$P(Y_{n+1} = i_{n+1}, T_{n+1} - T_n > t \mid Y_0 = i_0, T_0 = 0, \ldots, Y_n = i_n, T_n = t_n)$$
$$= p_{i_n, i_{n+1}} e^{-\nu_{i_n} t},$$

where ν_{i_n} is defined below.

From this equation we deduce that the process moves from state to state according to a Markov chain and spends an exponential amount of time in a state having a parameter (ν_{i_n}) that depends only upon the current state of the process.

A realization from $\{X(t) : t \geq 0\}$ can be generated as follows. First generate the initial state according to the initial distribution. Then generate the time the process spends in this state using the appropriate exponential distribution. Given that $Y_0, T_1, Y_1, T_2 - T_1, \ldots, Y_n$ have been generated, generate $T_{n+1} - T_n$ according to an exponential distribution having parameter ν_{Y_n} and generate Y_{n+1} using the method for discrete-time Markov chains.

Example 8.2

Consider a continuous-time Markov chain $\{X(t) : t \geq 0\}$ having state space $\{1, 2, 3, 4\}$, generator

$$Q = \begin{bmatrix} -1/2 & 1/4 & 1/4 & 0 \\ 1/8 & -3/8 & 1/8 & 1/8 \\ 0 & 1/4 & -1/2 & 1/4 \\ * & * & * & * \end{bmatrix},$$

and initial distribution $\pi^o = [\frac{1}{2}, \frac{1}{4}, \frac{1}{4}, 0]$. We would like to simulate the operation of this process in order to calculate the amount of time it takes the Markov chain to reach state 4. Therefore, we do not need to specify the fourth row of the generator Q, which gives the instantaneous rates of transition out of each state. By first sampling from the initial distribution π^o, an initial state is selected (say state 1). The amount of time spent in this state is exponentially distributed with parameter $\frac{1}{2}$. The next state visited by the process is then obtained by randomly selecting from states $2, 3, 4$ proportionally to $\frac{1}{4}, \frac{1}{4}, 0$, respectively. Running this simulation for a large number of iterations allows one to build a histogram of the required distribution function.

The distribution function given in Exercise 8.2 is an example of a *phase-type distribution*, which is a generalization of a gamma distribution with an integer shape parameter. Suppose that a continuous-time Markov chain has generator

$$Q = \begin{bmatrix} -1/5 & 1/5 & 0 & 0 \\ 0 & -1/5 & 1/5 & 0 \\ 0 & 0 & -1/5 & 1/5 \\ * & * & * & * \end{bmatrix}$$

and $\pi^o = [1, 0, 0, 0]$. Then starting from state 1 it is not difficult to show that the amount of time it takes the Markov chain to reach state 4 is the sum of three independent exponential random variables having mean 5; that is, the transition time from state 1 to state 4 has a gamma distribution with shape parameter 3 and scale parameter 5. Phase-type distributions have proven to be very useful in stochastic modeling.

Other aspects of random number generation will be considered in Sections 8.4 and 8.5.

8.4   Statistical Tools

Suppose a set of data has been collected representing the output of a process. The next step would be to build a mathematical model of the underlying process. Ideally, the model should incorporate all of the available kinds of knowledge: information obtained from the data, other information about how the process behaves, and assumptions one is willing to make about the process. Generally, if less of one kind of information is available, more of another kind will be needed. If we don't have very much data, we will need to know a lot about the underlying process, and/or be willing to make strong assumptions about it. Conversely, if we don't know much about the process, we will need to collect a lot of data. How do we incorporate the data into the model?

Suppose that instead of collecting actual data, one wishes to simulate the behavior of a particular model. This could be done by generating random numbers, probably via computer, as discussed in the previous section. How can one tell whether the generated numbers actually look like the product of the model in question? What does it mean for random numbers to "look like" the product of a particular model? These are all statistical questions, which we will attempt to answer in this section.

Given a sample $y_1, y_2, \ldots, y_n$, one can model the density function that underlies the sample using a three-step process.

- Using a graph of the data (and perhaps outside information), choose a parametric family of density functions.

- Using the data, estimate the parameter(s) of the density function.

- By comparing the data and the estimated density function, assess whether or not the chosen density function is suitable.

8.4.1   Choosing a Parametric Model

The most useful graphs for this situation are *histograms* and *probability plots*. A histogram is one form of the familiar "bar chart." If the sample size is not too small, the histogram can give a fair idea of the underlying distribution. For example, the family of exponential density functions is defined by

$$f(y) = \lambda e^{-\lambda y}, \ y \geq 0$$

and has the form illustrated in Figure 8.2. Given a histogram like the one in Figure 8.3, one might suppose that for some choice of λ, the exponential density would be a good model.

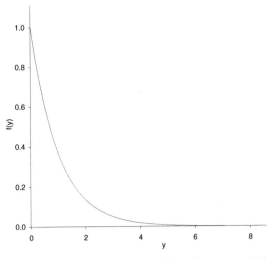

FIGURE 8.2 **Exponential distribution with $\lambda = 1$**

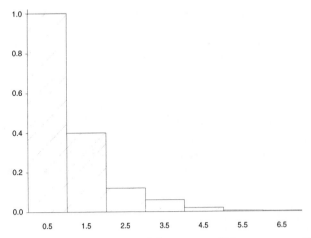

FIGURE 8.3 **Histogram of data, possibly exponential**

To get a more exact idea of whether or not the chosen density fits the data, one could make a probability plot. If $y_{(1)}, y_{(2)}, \ldots, y_{(n)}$ are the *ordered* data points, then each $y_{(i)}$ is an estimate of $Q(\frac{i-0.5}{n})$, the

$\frac{i-0.5}{n}$ quantile of the underlying true density. Now, one can also calculate $Q(\frac{i-0.5}{n})$, $i = 1, \ldots, n$ from the proposed distribution. A plot of the sample quantiles against the quantiles of the proposed model gives a probability plot. If the data actually came from the proposed distribution, the probability plot should look like a straight line. Small random deviations from a straight line are not a concern (this would be expected from a finite sample), but curvature in the plot would indicate that the distribution was not correctly chosen. For plotting purposes, it will not matter what value(s) of the parameter(s) are used to calculate the population quantiles, so some value is chosen for convenience. For the exponential density given earlier, choosing $\lambda = 1$ simplifies the calculations.

In statistics courses, the normal probability plot is commonly discussed. However, a probability plot can be constructed for any proposed density function. Deviations from a straight line should be carefully considered; frequently the nature of the deviations will point the way to a more appropriate model.

A parametric family is chosen not only to resemble the data; each parametric model has certain implications that may or may not correspond to the real situation. For example, the exponential distribution has the *memoryless property*: $P\left(Y \in (t, t+s) \,|\, Y \geq t\right) = P\left(Y \in (0, s)\right)$. In practical terms, this means that knowing $Y \geq t$ provides no additional information about where Y is likely to fall in (t, ∞). In some situations the memoryless property fits our intuition, while in others it seems contrary to common sense. This and other implications should be taken seriously when a model is being chosen.

Example 8.3

The length of time that a light bulb lasts is often modeled with an exponential density. If you have two working light bulbs, one new and one a year old, the memoryless property implies that neither one of them is "better than" the other. Each has a 50/50 chance of being the first to fail.

8.4.2   Parameter Estimation

Once a parametric family has been chosen, data will be used to estimate the parameter(s) of the model. Perhaps the most common method of estimation for such a situation is *maximum likelihood estimation*: this yields the parameter estimate that is "most likely," given the data. More formally, for an *iid* sample $y_1, y_2, \ldots, y_n$, the likelihood is defined by

$$L(\lambda \mid y_1, y_2, \ldots, y_n) = \prod_{i=1}^{n} f(y_i \mid \lambda)$$

and the maximum likelihood estimate (MLE) is the value of λ that maximizes L.

In practice, we may not have an *iid* sample, and n may not be fixed (as was assumed here). When observations occur over time (and the time of the next occurrence is unknown), having a fixed sample size means that you don't know when the experiment will be finished, and you have no guarantee that it will be finished in a timely fashion. Instead, one may choose to run the experiment for a fixed length of time T, and take as many observations as occur in that time. In statistical terms, the data are *censored*, and this particular type of censoring is called *Type II right censoring*. As a general rule, data of this type are not *iid*, and the sample size N is a random variable. In such a case, the correct likelihood function is

$$L(\lambda \mid y_1, \ldots, y_n) = \sum_{k=0}^{\infty} \prod_{i=1}^{n} f(y_i \mid \lambda) P(N = k).$$

Maximum likelihood estimates have an invariance property. Namely, if $g(\lambda)$ is any function of the parameter λ and if $\hat{\lambda}$ is the MLE of λ, then the MLE of g is simply $\hat{g} = g(\hat{\lambda})$. Note, however, that the MLE may have other, less desirable properties. For example, there is no guarantee that $\hat{\lambda}$ will be an unbiased estimate of λ, and even if it is, it does not follow that $\hat{g}$ is an unbiased estimate of g.

8.4.3 Goodness of Fit

Once the parameters of the model have been estimated, one can make a formal test of whether or not the parametric family was a good choice. Analytically, one can formally test in several ways the null hypothesis

H_0: the data represent a random sample from the distribution F.

The χ^2 *goodness of fit test* is one simple and widely-used procedure.

This test involves partitioning the domain of the distribution into intervals $I_1, \ldots, I_k$, chosen by the practitioner, where I_i has endpoints a_i and b_i. Then compare the number of observations d_i in each interval to the number e_i predicted by the proposed distribution, where e_i is computed as $e_i = F(b_i) - F(a_i)$. The test statistic

$$X^2 = \sum_{i=1}^{k} (d_i - e_i)^2 / e_i$$

has approximately a $\chi^2(k-2)$ distribution. If the data actually come from the distribution F, X^2 should be small, and if the data come from some other distribution, X^2 should be large. Thus, one rejects H_0 if X^2 exceeds a suitable quantile (such as the .90, .95, or .99 quantile) of the $\chi^2(k-2)$ distribution.

Some care must be taken in choosing the intervals $I_1, \ldots, I_k$. The test statistic X^2 will have an approximately $\chi^2(k-2)$ distribution, provided $e_i \geq 5$ for all i. If this condition does not hold, the practitioner should choose different endpoints for the intervals, or simply combine several intervals into one. The easiest way to ensure appropriate intervals is to choose $n/5$ intervals so that $P(I_i) = 5/n$. That is, $a_1 = -\infty$, $b_1 = Q(5/n) = a_2$, and, in general, $a_i = b_{i-1}$, $b_i = Q(5i/n)$. Then each $e_i = nP(I_i) = n \times 5/n = 5$.

The χ^2 goodness of fit test was originally designed for observations taken in groups. This is why, if exact values are observed, one must artificially divide them into intervals before applying the test. There are many other tests that do not require this. The intuitively appealing *Cramer-von Mises test* is related to the probability plot discussed in Section 8.4.1. The Cramer-von Mises test statistic is

$$W^2 = \sum_{i=1}^{n} \left(F(y_{(i)}) - \frac{i - 0.5}{n} \right)^2 + \frac{1}{12n}.$$

Essentially, the test statistic measures the distance between the sample and population quantiles. Once again, if the data come from the proposed distribution, we would expect W^2 to be small. Formally, one rejects H_0 if W^2 exceeds some suitable quantile of the distribution. In practice, it is more convenient to scale W^2 by the sample size, so that one set of quantiles works for any n. The scaled quantiles are:

Function	Quantile				
	0.85	0.90	0.95	0.975	0.99
$(W^2 - 0.4/n + 0.6/n^2)(1 + 1/n)$	0.284	0.347	0.461	0.581	0.743

An introduction to goodness of fit testing is found in [9, Chapter 9] and [3, Section 9.12].

8.4.4 Density Estimation

At times it is desired to generate random numbers from (approximately) the same distribution as a given set of data. As mentioned above, this can be done by choosing a parametric model for the data, estimating the parameters of the model, perhaps performing goodness of fit tests to make sure the parametric model is plausible, and then generating data from the model in the manner described in Section 8.3. A simpler approach is possible, which is also useful for situations in which none of the usual parametric families seem suitable for the data.

Given data $y_1, y_2, \ldots, y_n$ from some unknown distribution F, assumed to have density f, how can we estimate f directly from the data? Actually, many such estimates exist. One obvious estimate is

$$\hat{f}(x) = \begin{cases} \frac{1}{n} & \text{if } x \in \{y_1, y_2, \ldots, y_n\} \\ 0 & \text{otherwise.} \end{cases} \tag{8.1}$$

Generating m random numbers from (8.1) is equivalent to sampling m times with replacement from $\{y_1, y_2, \ldots, y_n\}$. Because one is sampling with replacement, m may be larger than n. (This *resampling* scheme is the basis of the statistical techniques of *bootstrapping* and *jackknifing*.)

One disadvantage of generating random numbers via resampling is that only those values observed in the original data set will appear in the sample, and repeated values will necessarily be more common in the generated sample than in the population. Because of this, it may be desirable to generate random samples from a continuous (rather than discrete) $\hat{f}$. A relative frequency histogram is one such continuous density estimate. Because of the simple nature of $\hat{f}$, it is easy to find $\widehat{F}$ and generate random numbers according to Corollary 8.1. If boundary values are chosen so there are no empty intervals, $\widehat{F}$ will be invertible.

A main disadvantage of the histogram estimate is that the "shape" of the histogram depends in part on the interval widths and the choice of boundaries for intervals. In practice, different choices of interval boundaries may lead to substantially different histograms. An estimate that does not have this problem is the *moving histogram*

$$\hat{f}(x) = \frac{1}{2hn} \cdot |\{i \in \{1, \ldots, n\} : y_i \in (x - h, x + h)\}|,$$

where h is some small number and $2h$ is the interval width. The form of $\hat{f}$ is somewhat more complicated than before. As a result, it is difficult to write $\widehat{F}$ and its inverse in closed form, although these quantities are easy to find on a computer.

The *kernel estimator* is more sophisticated than the moving histogram estimator — kernel estimators with arbitrary degrees of smoothness may be constructed. The general form of such an estimate is

$$\hat{f}(x) = \frac{1}{nh} \sum_{i=1}^{n} K\left(\frac{x - y_i}{h}\right)$$

where K, the *kernel function*, may be any density function; for example, the normal density is often a popular choice. For a given set of data, random samples may be generated according to the kernel estimate via the methods in the previous section.

The details of density estimation, in particular the choice of h, the necessary assumptions, and the asymptotic performance, are omitted here. The interested reader should consult [18, 20] for more information.

8.5   Arrival Processes

8.5.1   The Poisson Process

The most common assumption about arrival streams in queueing models is that the arrival process forms a stationary *Poisson process.*

DEFINITION 8.1   A **Poisson process** with rate λ is a stochastic process $\{N_t : t \geq 0\}$ whose sample paths are right continuous and nondecreasing, and whose values are nonnegative integers satisfying

- $N_0 = 0;$

- For each $t > 0$, N_t has a Poisson distribution having mean $\lambda t;$

- $\{N_t : t \geq 0\}$ has stationary and independent increments.

Stationarity implies that the distribution of the number of arrivals in an interval depends on the length of the interval, but not on its location. The following alternative description of a Poisson process gives an idea of when it is an appropriate model for an arrival process. Let N be a random variable having a Poisson distribution with parameter λT and let $Y_1, Y_2, \ldots$ be a sequence of independent, uniform $[0, T]$ random variables that are independent of N. For $0 \leq t \leq T$, let N_t be the number of Y_n, $n \leq N$, that are less than or equal to t. Then the process $\{N_t : 0 \leq t \leq T\}$ is a Poisson process having rate λ.

DEFINITION 8.2   The (N, Y) **characterization** of a *Poisson process over a finite horizon* $[0, T]$ *with rate* λ *is a Poisson random variable* N *with mean* λT *and a sequence* $Y_1, Y_2, \dots$ *of independent uniform* $[0, T]$ *random variables which are also independent of* N.

To demonstrate how this characterization of a Poisson process can be used to determine when the Poisson model is appropriate, consider the following scenario. There is a large population of potential arrivals to a store. On any given day, only a small portion of the population enters the store. Given that a person decides to enter the store, the time at which the person arrives at the store is uniformly distributed over the day. (Here a day can represent the hours the store is open.) Thus, the number of arrivals in a day has a binomial distribution with a small probability of success and a large number of trials; hence, its distribution is approximately Poisson. This, combined with the uniform distribution assumption of the unordered arrival times, shows the Poisson model is appropriate in this setting.

The Poisson process is an example of a renewal process since the times between arrivals form a sequence of independent and exponentially distributed random variables having mean $1/\lambda$; see [15, Section 5.3.3]. This fact is useful in several ways.

- One can simulate a sample path from a Poisson process with a given λ by drawing a pseudo-random sample from an exponential distribution and using the sample points as interarrival times of the process.

- Given a set of arrival times, one can assess whether or not they were generated by a Poisson process by assessing whether or not they look like *iid* exponential random variables.

- Given a set of arrival times that are known (or assumed) to be generated by a Poisson process, the arrival times can be used to estimate the rate λ of the process.

One application of the Poisson process is to generate random numbers from an arbitrary continuous distribution of a nonnegative random variable. Let F be such a distribution function having density function f. The *hazard function* h is

$$h(x) = \frac{f(x)}{1 - F(x)}.$$

The distribution function can then be written as

$$F(t) = 1 - \exp\left(\int_0^t h(s)\, ds\right).$$

Assume that the hazard function is bounded by a constant $\lambda < \infty$. Let $T_1, T_2, \ldots$ be the successive arrival times in a Poisson process having arrival rate λ and let $Y_1, Y_2, \ldots$ be a sequence of $[0, 1]$ uniform random variables. Let

$$\kappa = \inf\left\{n > 0 : \frac{h(T_n)}{\lambda} \leq Y_n\right\}.$$

Then T_κ has distribution function F; see [7, pp. 66–68].

8.5.2   The Nonstationary Poisson Process

The (N, Y) characterization of a Poisson process states that given N arrivals over a time interval $[0, T]$, the unordered arrival times are independent and uniformly distributed. In the example of arrivals to a store, this implies that once a customer enters the store, he or she is equally likely to enter during any time period. This assumption is perhaps too strong in many applications. The *nonstationary Poisson process* is a model that relaxes the uniform arrival time assumption.

DEFINITION 8.3   *A nonstationary Poisson process with mean function $m(t)$ is a stochastic process $\{N_t : t \geq 0\}$ whose sample paths are right continuous and nondecreasing, with nonnegative integer values satisfying*

- $N_0 = 0$;

- *For $t \geq 0$, $N(t)$ has a Poisson distribution with mean $m(t)$;*

- $\{N_t : t \geq 0\}$ *has independent increments.*

Let N be a random variable having a Poisson distribution with parameter λT and let $Y_1, Y_2, \ldots$ be a sequence of independent random variables having distribution F that is concentrated on the interval $[0, T]$. Suppose that N and $Y_1, Y_2, \ldots$ are independent. For $0 \leq t \leq T$, let N_t be the number of $n \leq N$ for which $Y_n \leq t$. Then $\{N_t : t \geq 0\}$ is a nonstationary Poisson process with mean function $\lambda F(t)\, T$. This statement requires proof and is left to the reader as an exercise.

8.5.3   The General Model

The nonstationary Poisson process is constructed from the Poisson process by relaxing the assumption that $Y_1, Y_2, \ldots$ are uniformly distributed over the interval $[0, T]$. The general model is now constructed by relaxing the assumption that N, the number of arrivals in $[0, T]$, has a Poisson distribution. Thus N is a nonnegative integer-valued random variable whose distribution is given by the probabilities $p(n)$, $n = 0, 1, 2, \ldots$. The estimation problem is identical to that encountered in survival analysis. Furthermore, generating sample paths for this process is accomplished by adapting the approaches used in simulating Poisson and nonstationary Poisson processes.

There are some disadvantages to this approach. First, many arrival processes typically used in queueing models (e.g., renewal and more generally Markov renewal processes) do not fit directly into this structure. A second disadvantage is that our time horizon is finite. This inhibits our ability to develop asymptotic results. We return to these issues later.

Let $\{N_t : 0 \leq t \leq T\}$ be the counting process; that is,

$$N_t = \sum_{n=1}^{N} 1_{[0,t]}(Y_n),$$

where for a set A

$$1_A(t) = \begin{cases} 1 & \text{if } t \in A \\ 0 & \text{otherwise.} \end{cases}$$

The process $\{N_t : 0 \leq t \leq T\}$ is an example of a *time inhomogeneous Markov chain*. One nice property of this process is that its transition function can be expressed explicitly.

THEOREM 8.2

The process $\{N_t : 0 \leq t \leq T\}$ *is a continuous-time Markov chain having transition function*

$$p_{j,j+k}(t, t+s) =$$
$$\frac{\sum_{n=j+k}^{\infty} \frac{n!}{j!k!(n-j-k)!} F(t)^j (F(t+s) - F(t))^k (1 - F(t+s))^{n-j-k} p(n)}{\sum_{n=j}^{\infty} \frac{n!}{j!(n-j)!} F(t)^j (1 - F(t))^{n-j} p(n)}$$

PROOF   For $0 \leq t_1 < t_2 < \cdots < t_n = t$ and $0 \leq j_1 \leq j_2 \leq \cdots \leq$

$j_n = j$ it follows from the definition of conditional expectation that

$$P(N_{t+s} - N_t = k \mid N_t = j, N_{t_{n-1}} = j_{n-1}, \ldots, N_{t_1} = j_1) =$$
$$\frac{P(N_{t_1} - N_0 = i_1, \ldots, N_{t_n} - N_{t_{n-1}} = i_n, N_{t+s} - N_t = k)}{P(N_{t_1} - N_0 = i_1, \ldots, N_{t_n} - N_{t_{n-1}} = i_n)}$$

where $i_0 = j_0 = 0$ and $i_1 = j_1 - j_0, i_2 = j_2 - j_1, \ldots, i_n = j_n - j_{n-1}$. Since the arrival times $Y_1, Y_2, \ldots$ are independent of N, the numerator on the right-hand side of the above equation equals

$$\sum_{m=j+k+1}^{\infty} \frac{m!}{i_1! i_2! \cdots i_n! k! (m-j-k)!} \prod_{l=1}^{n} (F(t_l) - F(t_{l-1}))^{i_l} \cdot$$
$$(F(t+s) - F(t))^k (1 - F(t+s))^{m-j-k} p(m)$$

and the denominator equals

$$\sum_{m=j+1}^{\infty} \frac{m!}{i_1! i_2! \cdots i_n! (m-j)!} \prod_{l=1}^{n} (F(t_l) - F(t_{l-1}))^{i_l} \cdot$$
$$(F(t+s) - F(t))^k (1 - F(t+s))^{m-j} p(m).$$

Cancelling common terms in the numerator and the denominator gives

$$P(N_{t+s} - N_t = k \mid N_t = j, N_{t_{n-1}} = j_{n-1}, \ldots, N_{t_1} = j_1) =$$

$$\frac{\displaystyle\sum_{m=j+k+1}^{\infty} \frac{m!}{k!(m-j-k)!} (F(t+s) - F(t))^k (1 - F(t+s))^{m-j-k} p(m)}{\displaystyle\sum_{m=j+1}^{\infty} \frac{m!}{(m-j)!} (1 - F(t+s))^{m-j} p(m)}$$

$$= \frac{\displaystyle\sum_{m=j+k+1}^{\infty} \frac{m!}{j!k!(m-j-k)!} F(t)^j (F(t+s) - F(t))^k (1 - F(t+s))^{m-j-k} p(m)}{\displaystyle\sum_{m=j+1}^{\infty} \frac{m!}{j!(m-j)!} F(t)^j (1 - F(t+s))^{m-j} p(m)}$$

$$= P(N_{t+s} - N_t = k \mid N_t = j).$$

It follows that $\{N_t : t \geq 0\}$ is a Markov process and has the proposed transition function, completing the proof. ∎

Using Theorem 8.2 it is now possible to generate the interarrival times in the process $\{N_t : t \geq 0\}$ as they are needed. Let $T_0 = 0$ and for $n = 1, \ldots, N$ let T_n be the time of the nth arrival. By Theorem 8.2

$$P(T_{N_t+1} > t + s \mid N_t = j) = p_{j,j}(t, t+s).$$

The above equation follows from the observation that $T_{N_t+1} > t + s$ if and only if $N_{t+s} - N_t = 0$. A somewhat more sophisticated argument using the strong Markov property shows that

$$P(T_{j+1} - T_j > s \mid T_j = t) = p_{j,j}(t, t + s).$$

Notice that, in general, the distribution of time between the jth and $(j + 1)$st arrival depends on both j and the time t at which the jth arrival occurs. Generating interarrival times using the Poisson process method requires knowing the hazard function for the interarrival time distribution. Let $h_{j,t}$ denote the hazard function for the time between arrivals j and $j + 1$, given that arrival j occurs at time t. Then

$$h_{j,t}(s) = - \frac{1}{p_{j,j}(t, t + s)} \frac{d}{ds} p_{j,j}(t, t + s).$$

The registration problem at Clemson is one where the model for the arrival process given above seems appropriate. There are roughly 15,000 students that need to register over a finite time period. To build the model, we must determine the length of the time horizon, the number of students that need to register, and their individual arrival times for registration.

Suppose the registration process actually occurs over a single day, starting at 8 a.m. and closing at 5 p.m., so the time horizon is 9 hours. Suppose, however, that by 5 p.m. all the students should have completed registration. In this case, there is a time prior to 5 p.m. at which arrivals are no longer accepted into the system. For example, this time may be fixed at 4:30 p.m. or may be a variable that is determined by the analysis.

The number of arrivals that occur over the period is a random variable N. It may be the case that (say) 15,580 students have been accepted and only those students can participate in the registration process. Then at most 15,580 students will register. It also likely that there are students who will not attend registration; either they will register late or they will not attend classes.

Suppose that in any year, S students are accepted, where S is the realization of a random variable, and $N \le S$. By collecting observations on N and S from n previous years, it might be possible to find a model that would allow us to predict N from S. We now study alternative models relating N to S.

Perhaps the simplest model is that N is approximately a constant proportion of S. That is, N/S has some distribution with mean μ and nonnegative variance σ^2. Then the proportion μ can be estimated via $\hat{\mu} = \frac{1}{n} \sum_{i=1}^{n} N_i/S_i$. (Note that we have not specified a distribution for

N, S, or N/S. The sample mean given here is the uniform minimum variance unbiased estimator of μ provided $\mu < \infty$ and N/S has a density; see [10, Section 2.4]).

Another simple model would be that N is approximately a linear function of S: namely, $N = \beta_0 + \beta_1 S + \epsilon$, where $E(\epsilon) = 0$. Then simple linear regression would allow us to estimate β_0 and β_1 (call these estimates $\hat{\beta}_0$ and $\hat{\beta}_1$), and the predicted value of N would be given by $\widehat{N} = \hat{\beta}_0 + \hat{\beta}_1 S = \hat{\beta}_0 + \hat{\beta}_1$ (15,580). One disadvantage of this model is that it is counterintuitive to have $\beta_0 \neq 0$. However, since the observed values of S and N will be large, this disadvantage has little or no practical significance. Another disadvantage is that there is no guarantee that the resulting $\widehat{N} \leq S$. It is left to the reader as an exercise to determine conditions on $\hat{\beta}_0$ and $\hat{\beta}_1$ that ensure $\widehat{N} \leq S$.

As an alternative to least squares estimation, one could use *restricted maximum likelihood*; that is, find the values of $\hat{\beta}_0$ and $\hat{\beta}_1$ that maximize the likelihood function, subject to the constraint that $\widehat{N} \leq S$. This is simplest to do if one assumes a parametric form for ϵ, and the usual assumption is $\epsilon \sim N(0, \sigma^2)$.

Yet a third model is specified by

$$N = \ln(S^\beta) \tag{8.2}$$

which (for a suitable choice of β) "grows" more slowly than the linear model discussed above. This model might be suitable if it appears that, when S increases, the *proportion* of students who do not register also increases.

Once we have predicted N, the next step is to determine when the arrivals occur. One assumption, perhaps a little unrealistic, is that all students choose their arrival times independently of each other. Then it remains to determine the distribution of the arrival time. History of the registration process may suggest a model for the distribution.

8.6　Queueing Models

8.6.1　Model Building

A convenient structure for simulating queueing systems is a *generalized semi-Markov scheme* (*GSMS*). References [4] and [19] discuss such schemes in detail. We show, via several examples, how to model queueing systems using a GSMS. The first example is that of a single server queue with balking. The second example is that of a single server queue having more than one class of arrivals. The third example is the registration model discussed earlier.

8.6.1.1 A Single Server Queue with Balking

Consider a queueing system in which customers arrive one at a time for service. A single server processes customers in order of their arrival. Customers arriving for service choose either to enter the queue or not depending upon the queue's size. Customers who do not enter the queue are said to *balk*.

Let us now construct a machine, a GSMS, that will run the dynamics of a single server queue with balking. First, the machine needs to know how many customers are in the system (the system state); here the state space S is the set of nonnegative integers. The evolution of this system is governed by two activities: arrivals and service completions. Thus the *activity set* is $A = \{a, s\}$, where a denotes arrivals and s denotes service completions. In some states of the system, only certain activities are possible; these activities are "active," whereas activities that cannot take place are "passive." For example, when the queueing system is empty, the server is idle; in this case the activity s is passive. Let $E(k)$ denote the active activities when the system is in state k. Thus $E(0) = \{a\}$ and $E(k) = \{a, s\}$ for $k > 0$. Events occur when activities are completed. When an event corresponds to the completion of activity a, that event will be an arrival. Likewise, when an event corresponds to the completion of activity s, that event will be a service completion.

Suppose that the current state is k and the next event is an arrival. In this case activity a generates the next event. Let $p(k + 1; a, k)$ denote the probability that the customer chooses to enter the system and let $p(k; a, k)$ denote the probability that the customer balks. Set $p(j; a, k) = 0$ for all other j. Now suppose that the system is in state $k > 0$ and the next event is a service completion. The next state is $k - 1$, so set $p(k - 1; s, k) = 1$ and $p(j; s, k) = 0$ for all $j \neq k$. Notice that for every pair (k, α) where $k \in S$ and $\alpha \in E(k)$, $\sum_{j \in S} p(j; \alpha, k) = 1$.

We have established mechanisms to update the system state and activities when an event (an arrival or a service completion) occurs. The next step is to determine the times and order in which the events occur. Let $\{w_a(n)\}$ denote the sequence of interarrival times and $\{w_s(n)\}$ denote the sequence of service times. Given these two sequences and $(S, A, \{E(k) : k \in S\}, \{p(\cdot; \alpha, k) : k \in S, \alpha \in E(k)\})$, we can describe the evolution of the system. Perhaps the most difficult aspect of queueing theory is the amount of notation that is needed to keep track of everything that is happening. We will need to keep track of three entities: the state Y_n of the system at the time of the nth activity, the vector

$[C_n(a), C_n(s)]$ of current clock readings, and the vector $[k_n(a), k_n(s)]$ counting how many arrivals and service completions have occurred.

In addition to the above we must keep track of activities that are new and activities that are old. At the nth event an activity is *old* if it is active and it is not the activity that triggers the next event. The activity is *new* if it becomes active at the nth event. This can occur in two ways: either it is inactive and becomes active, or it is active and is the activity that generates the next event (remaining active in the new state). Thus the new activities at the nth event are precisely those activities α for which $k_n(\alpha) = k_{n-1}(\alpha) + 1$.

Let us consider the evolution of such a system with the initial state $Y_0 = 0$. Then $E(0) = \{a\}$. Since only a is active, we must choose the interarrival time and thus set $k_0(a) = 1$ and $C_0(a) = w_a(1)$. Since s is passive $k_0(s) = 0 = C_0(s)$. The first event will then be an arrival that occurs at time $w_a(1)$. This arrival will join the system with probability equal to $p(1; a, 0)$ and balk with probability $p(0; a, 0)$. Suppose that the arrival joins the system. Then $Y_1 = 1$. Since $E(1) = \{a, s\}$, both activities are active. Note that at the first event there are no old activities and that both activities are new. Thus, $k_1(a) = 2$, $k_1(s) = 1$, and so the new clock readings are $C_1(a) = w_a(2)$ and $C_1(s) = w_s(1)$. The second event occurs at the minimum of $C_1(a)$ and $C_1(s)$. Let C_1^* denote this minimum — suppose this is $C_1(s)$. Then the next event is a service completion and the new system state is $Y_2 = 0$. In this case, a is an old activity and there are no new activities. Thus, $k_2(a) = k_1(a) = 2$ and $k_2(s) = k_1(a) = 1$. The clock times are updated as follows. Since activity a is old $C_2(a) = C_1(a) - C_1^*$ and $C_2(s) = 0$. The system continues to evolve in this fashion for an arbitrary length of time.

Notice that we have not yet placed any probability structure on the the interarrival and service time sequences. Placing a probability law on these sequences is the final step. While very general laws are possible, we will restrict ourselves to types similar to the following. The sequence of interarrival times derive from a finite process considered in the previous section. The service times form a sequence of *iid* random variables independent of the arrival process. Once a probability law is placed upon the sequences $\{w_a(n)\}$ and $\{w_s(n)\}$, the sequence $\{(Y_n, C_n, k_n)\}$ is a random process called a *generalized semi-Markov process (GSMP)*.

8.6.1.2   A Single Server Queue with Different Arrival Types

In the previous queueing example (with balking), the state space was quite simple; S was the set of nonnegative numbers. In many queueing scenarios the state space will be much more complex. The example presented here is one in which the elements of S are finite sequences.

Consider a queueing system in which customers arrive one at a time for service. An arriving customer is either a Type One customer or a Type Two customer. There is a single server who processes customers in the order of their arrival. The service requirement of a customer may depend upon the customer's type.

There are many methods of building a GSMS for this example. We choose the following. The set of activities is $A = \{a_1, a_2, s_1, s_2\}$. Thus, activities correspond to the two different types of arrivals and service requirements. In order to properly update the system we must keep track not only of the number of each type of customer but also their position in the queue. Thus an element of S is of the form $(j_1, j_2, \ldots, j_l)$, where l is the number of customers in the system and j_i is the type of customer in position i. Here we assume the customer in position 1 is the one currently receiving service, the customer in position 2 the one next in line, and so on. There is also a state 0 corresponding to the system being empty. The activities a_1 and a_2 are always active. Given a state $j = (j_1, j_2, \ldots, j_l)$, we have that $E(j) = \{a_1, a_2, s_{j_1}\}$. We leave it as an exercise for the reader to complete the description of this system.

8.6.1.3   The Registration Model

For this system there will be 78 activities, one for arrivals and one for each of the 77 servers. Let a denote the activity corresponding to arrivals. For $j = 1, \ldots, 6$ let l_j denote the number of servers allocated to node j. Let s_{ji}, for $j = 1, \ldots, 6$ and $i = 1, \ldots, l_j$, be the activity corresponding to server completions at node j by server i. The state space for this system is $S = \prod_{j=1}^{6}(\{0, 1, \ldots\} \times \{0, 1\}^{l_j})$. A state describes not only the number of customers at each node in the network but also which servers are busy. For example, if $l_1 = \cdots = l_6 = 2$ then the state

$$((5, 1, 1), (1, 0, 1), (0, 0, 0), (2, 1, 1), (1, 1, 0), (3, 1, 1))$$

indicates that 5 customers are in queue 1 with both servers busy, 1 customer is in queue 2 being served by the second server, the third queue is empty, the fourth queue has 2 customers, the fifth has 1 customer being served by the first server, and the sixth has 3 customers. (In this case, when there are at least 2 customers at a queue, both the servers are busy.) For a given state k the passive activities correspond to those servers who are idle when the system is in that state. For example, in the state given above, the passive activities are $s_{21}, s_{31}, s_{32}, s_{52}$ and hence the active activities are $a, s_{11}, s_{12}, s_{22}, s_{41}, s_{42}, s_{51}, s_{61}, s_{62}$.

Let P be the 7×7 stochastic matrix given by

$$P = \begin{bmatrix} .02 & .17 & .273 & 0 & .377 & .16 & 0 \\ 0 & 0 & .618 & 0 & .135 & .247 & 0 \\ 0 & 0 & .1 & .15 & .05 & .7 & 0 \\ 0 & 0 & 0 & 0 & 1 & 0 & 0 \\ 0 & 0 & 0 & 0 & 0 & 1 & 0 \\ 0 & 0 & 0 & 0 & 0 & 0 & 1 \\ 1 & 0 & 0 & 0 & 0 & 0 & 0 \end{bmatrix}.$$

For $i, j = 1, \ldots, 6$ let p_{ij} be the probability that a customer leaving node i goes next to node j, p_{i7} the probability that a customer leaves the network from node i, and p_{7j} the probability that a customer enters the network at node j. Thus, for this network all customers enter the network at node 1 and leave the network from node 6.

To construct the transition probabilities $p(\tilde{k}; \alpha, k)$ we assume that a customer entering a queue in which there are free servers chooses from among them with equal probability. Again let us consider the example above where $l_1 = \cdots = l_6 = 2$ and the state is

$$k = ((5, 1, 1), (1, 0, 1), (0, 0, 0), (2, 1, 1), (1, 1, 0), (3, 1, 1)).$$

Suppose that the next event is triggered by a completion of activity s_{42}. Since $p_{45} = 1$ we then have $p(\tilde{k}; s_{42}, k) = 1$, where

$$\tilde{k} = ((5, 1, 1), (1, 0, 1), (0, 0, 0), (1, 1, 0), (2, 1, 1), (3, 1, 1)).$$

Suppose now that activity s_{11} triggers the next event. Since $p_{13} = 0.273$, then $p(\tilde{k}; s_{11}, k) = 0.1365$ if

$$\tilde{k} = ((4, 1, 1), (1, 0, 1), (1, 1, 0), (2, 1, 1), (1, 1, 0), (3, 1, 1))$$

or

$$\tilde{k} = ((4, 1, 1), (1, 0, 1), (1, 0, 1), (2, 1, 1), (1, 1, 0), (3, 1, 1)).$$

The remaining transition probabilities are generated in a similar fashion.

To put a probability law on $\{w_\alpha(n) : n \geq 1, \alpha \in A\}$, where A is the set of activities, we assume that the arrival process and all of the service time sequences are mutually independent. The arrival process will be a finite point process and the service times of server i at node j form a sequence of *iid* random variables. The generation of the arrival times via simulation has been discussed earlier. Service times can be generated using one of the techniques in Section 8.3 and observational data.

8.6.2   Output Analysis

So far the discussion has focused on building a simulation model and not the analysis of its output. One must be careful in choosing the performance measures, performing the statistical analysis of the output, and making conclusions based upon this output. The registration model is an example of a finite horizon model because registration lasts for a fixed amount of time. In other applications the time horizon may be infinite. Finite horizon models suggest different performance measures and lead to different conclusions than do infinite horizon models.

For example, suppose that we wish to analyze the average number of customers in the system in the registration example. There are several possible candidates for the random variable that represents this quantity. For one choice, recall that $Y_n = (Y_n(1), \ldots, Y_n(6))$ is the vector whose jth element is the number of customers at node j at the time of the nth event. Set $\widetilde{Y}_n = \sum_{j=1}^{6} Y_n(j)$ and let

$$\widetilde{Y} = \frac{1}{N} \sum_{n=1}^{N} \widetilde{Y}_n,$$

where N is the total number of events in a run of the simulation. The random variable $\widetilde{Y}$ is the average number of customers in the network during the time horizon and is one possible choice for the average number of customers in the system. Another possible candidate is the random variable

$$\widetilde{Q} = \frac{1}{T} \int_0^T \widetilde{Q}(s) \, ds,$$

where $\widetilde{Q}(s)$ is the total number of customers in the system at time s and T is the length of time that the simulation runs. The random variables $\widetilde{Y}$ and $\widetilde{Q}$ are different random variables and as such may have different means. The random variable $\widetilde{Y}$ is a customer average while the random variable $\widetilde{Q}$ is a time average.

The next step is to examine statistical properties of the random variable $\widetilde{Y}$ or $\widetilde{Q}$. For example, the mean or variance of $\widetilde{Y}$ or $\widetilde{Q}$ might be of interest. In the finite horizon case, it is fairly simple to obtain an estimate for the mean of $\widetilde{Y}$: simply run the simulation M times, each time getting a new value for $\widetilde{Y}_m$. Let $\overline{Y} = \frac{1}{M} \sum_{m=1}^{M} \widetilde{Y}_m$. Now $\widetilde{Y}_1, \ldots, \widetilde{Y}_M$ is a sequence of *iid* observations, so the statistical analysis of $\overline{Y}$ is fairly routine.

In an infinite horizon model, only one iteration of the simulation may be required. Let us consider a scenario where this is the case. Suppose that the arrivals form a stationary Poisson process with parameter λ and that service times at node j are exponentially distributed having mean μ_j^{-1}. Suppose in addition that the input rate is less than the service rate at each of the nodes. Then the network under consideration is a *Jackson network* [15]. In this case the process $\{Y_n\}$ is a discrete-time ergodic Markov chain. Hence the process has a limiting distribution ν that is also a stationary distribution. Then as $N \to \infty$ (one needs to carefully define what it means for the random variable N to approach infinity), it follows that

$$\widetilde{Y} \to \sum (j_1 + \cdots + j_6)\nu(j_1, \ldots, j_6).$$

In Jackson networks, rather than the process $\{Y_n\}$, one studies the continuous-time process $\{Q(t) : t \geq 0\}$ where $Q(t) = (Q_1(t), \ldots, Q_6(t))$ and $Q_j(t)$ is the number of customers at node j at time t. The process $\{Q(t) : t \geq 0\}$ is a continuous-time Markov chain which has a limiting distribution π that is also a stationary distribution. Moreover as $T \to \infty$ one has

$$\widetilde{Q} \to \sum (j_1 + \cdots + j_6)\pi(j_1, \ldots, j_6).$$

Even in this case it is not necessarily true that $\widetilde{Y}$ and $\widetilde{Q}$ are equal.

The average queue length is just one of many performance measures that one may choose to analyze. If one wishes to analyze worst-case scenarios, it might be more appropriate to consider the maximum queue length. Another quantity often considered is the total amount of time a customer spends in the system. One might also be interested in the relationships between the different queues in the network. In this case it may be appropriate to estimate the correlation between the lengths of two queues.

8.7   Exercises and Projects

1. Write a computer program to generate 1000 random numbers from the following distribution.

y	$P(Y = y)$
1	0.40
2	0.21
5	0.37
12	0.01
100	0.01

2. Write a computer program to generate 1000 random numbers from a Poisson distribution.

3. Write a computer program to generate 1000 random numbers from an exponential distribution with mean 5.

4. Write a computer program to generate 1000 random numbers from a gamma distribution with shape parameter 3 and scale parameter 2.

5. Write a computer program to generate 1000 random numbers from a normal distribution with mean 100 and standard deviation 10.

6. Write a computer program to generate 1000 random numbers from a gamma distribution with shape parameter 3 and scale parameter 2 from exponential random variables. Compare the execution time to the execution time for Exercise 8.4.

7. Write a computer program to generate 1000 random numbers from a normal distribution with mean 100 and standard deviation 10 from uniform random numbers, as described in Section 8.3. Compare the execution time to the execution time for Exercise 8.5.

8. Simulate the Markov chain $\{X_n : n = 0, 1, \ldots\}$ having state space $\{1, 2, 3\}$, one-step transition matrix

$$
P = \begin{bmatrix} 1/3 & 1/6 & 1/2 \\ 1/6 & 1/2 & 1/3 \\ 1/2 & 1/3 & 1/6 \end{bmatrix},
$$

and assume the Markov chain starts in state 1. Calculate the statistic

$$
Y_N = \frac{1}{N} \cdot |\{n \in \{1, \ldots, N\} : X_n = 1\}|.
$$

9. Consider a continuous-time Markov chain $\{X(t) : t \geq 0\}$ having state space $\{1, 2, 3, 4\}$, generator

$$
Q = \begin{bmatrix} -1/2 & 1/4 & 1/4 & 0 \\ 1/8 & -3/8 & 1/8 & 1/8 \\ 0 & 1/4 & -1/2 & 1/4 \\ * & * & * & * \end{bmatrix},
$$

and initial distribution $\pi^o = [\frac{1}{2}, \frac{1}{4}, \frac{1}{4}, 0]$. Write a simulation that calculates the amount of time until the Markov chain reaches state 4. (To do this you do not need the fourth row of the generator.) Run the simulation for 1000 iterations and build a histogram of the resulting distribution function.

10. Make a normal probability plot of the data from Exercise 8.7. Does it appear that $n = 3$ was large enough for $\overline{Y}$ to be approximately normal?

11. Repeat Exercise 8.7, but using $n = 2$. Then make a normal probability plot for this set of data. In what way does the plot show that the data are not normally distributed?

12. Using the data from Exercise 8.3, make normal and exponential probability plots. Comment on them.

13. Using the data from Exercise 8.7, calculate both X^2 and W^2, and use them to test whether or not your pseudo-random data actually appear to come from a normal distribution.

14. Repeat Exercise 8.13, using the data from Exercise 8.11.

15. The data given below [1] are 100 actual service times (in seconds) for students obtaining registration forms. In order to simulate the registration queueing network, many more than 100 such service times would need to be generated. Using the data and both the histogram method and (8.1), generate pseudo-random samples of 10,000 service times. Compare the two methods in terms of speed and goodness of the results (i.e., do the generated samples, in fact, resemble the original sample?).

101.5	27.0	11.9	22.0	12.7	48.8	20.5	7.5	48.6	8.7
21.9	16.7	15.2	17.6	14.1	20.4	14.0	38.3	5.0	15.6
18.9	24.5	17.4	15.1	54.2	22.2	21.7	29.0	36.2	22.8
78.3	21.3	23.2	18.4	12.9	13.2	69.0	20.8	17.0	47.2
60.5	12.3	140.6	14.9	12.0	34.0	14.4	28.1	12.9	25.4
23.0	24.1	18.6	22.7	13.6	26.6	114.7	19.3	38.5	34.0
29.0	31.4	16.6	25.4	16.5	73.2	27.0	51.2	28.5	4.8
50.7	59.0	23.3	48.0	18.1	9.8	36.3	16.5	17.0	14.7
14.2	20.5	18.5	19.5	14.7	23.4	20.2	18.8	15.0	17.6
10.2	20.2	31.4	11.1	23.7	13.1	29.1	31.8	17.2	28.5

16. Using the data from Exercise 8.15, calculate and plot moving histograms for $h = 0.1, 0.5, 1.0$. What happens as the value of h increases?

17. Using the data from Exercise 8.15 and a moving histogram with $h = 0.5$, generate a random sample of 100 service times. Compare it to the pseudo-random samples generated in Exercise 8.15, in terms of speed and smoothness.

18. Use the (N, Y) characterization to simulate a sample path of a Poisson process having rate $\lambda = 5$ per hour over an 8-hour period.

19. Use the output from Exercise 8.18 to make a probability plot and perform a goodness of fit test to see if the interarrival times from that example are exponentially distributed.

20. Simulate the sample path of a Poisson process having rate $\lambda = 5$ per hour over an 8-hour period by generating random numbers from an exponential distribution having mean $1/\lambda$.

21. Use the random numbers generated in Exercise 8.20 to find the MLE of the mean of the exponential distribution.

22. Show that the MLE found in Exercise 8.21 is an unbiased estimate of the mean of the exponential distribution.

23. Find the MLE of λ, in both the exponential and Poisson cases.

24. The data [5] in Table 8.3 represent times that 41 successive vehicles on a major highway (travelling north) passed a fixed point. Assess the reasonableness of an underlying Poisson process for these data and estimate λ.

Table 8.3 Vehicle passing times on a highway (at H hours, M minutes, S seconds)

Vehicle	Time (H=22) M	Time (H=22) S	Vehicle	Time (H=22) M	Time (H=22) S
1	34	38	22		52
2		50	23		53
3		52	24	38	11
4		58	25		20
5	35	00	26		25
6		19	27		26
7		24	28		47
8		58	29		48
9	36	02	30		49
10		03	31		54
11		07	32		57
12		15	33	39	11
13		22	34		16
14		23	35		19
15		44	36		23
16		50	37		28
17	37	01	38		29
18		09	39		32
19		37	40		48
20		43	41		50
21		47			

(From D. J. Hand, F. Daly, A. D. Lunn, K. J. McConway, and E. Ostrowski, *A Handbook of Small Data Sets*, Chapman & Hall, London, 1994. Reprinted with permission of CRC Press LLC.)

25. Use the Poisson process approach to generate random numbers from a phase-type distribution. A phase-type distribution is a generalization of the mixture of exponentials. Essentially, such a distribution is the first passage time in a continuous-time Markov chain.

26. Let N be a random variable having a Poisson distribution with parameter λT and let $Y_1, Y_2, \ldots$ be a sequence of independent random variables having distribution F that is concentrated on the interval $[0, T]$. Suppose that N and $Y_1, Y_2, \ldots$ are independent. For $0 \le t \le T$, let N_t be the number of $n \le N$ for which $Y_n \le t$. Prove that $\{N_t : t \ge 0\}$ is a nonstationary Poisson process with mean function $\lambda F(t) T$. As part of the proof show that for each $t \in [0, T]$, N_t as constructed above has a Poisson distribution with mean $\lambda F(t) T$.

27. Use the transition function given in Theorem 8.2 to calculate $P(N_{t+s} - N_t = k \mid N_t = j)$ for both the Poisson process and the nonstationary Poisson process.

28. Find the asymptotic distribution of the $\hat{\mu}$ given at the end of Section 8.5.3, and use this to form a $(1 - \alpha)100\%$ confidence interval for μ.

29. Assume the linear model given in Section 8.5.3.

   a. What conditions on $\hat{\beta}_0$ and $\hat{\beta}_1$ would ensure $\widehat{N} \le S$?

   b. Consider the constrained model $\beta_0 \equiv 0$. What is the least squares estimate of β_1 in this case? Is it possible to have $\widehat{N} > S$?

   c. Find expressions for the restricted maximum likelihood estimates of β_0 and β_1, assuming $\epsilon \sim N(0, \sigma^2)$.

30. What conditions on β in (8.2) would ensure $N \le S$?

31. Find an expression for the restricted MLE of β in (8.2).

32. (Project) Using the information in Section 8.2 and the methodology of this chapter, build a simulation of the registration model. Choose several performance measures to analyze. Once the simulation is running, perform some sensitivity analysis. For example, compare your simulation to one where the arrival process forms a Poisson process. Or, attempt to improve system performance by rearranging the servers.

8.8 References

[1] K. L. Ashby, "A Model of Registration at Clemson University," Technical Report, Department of Mathematical Sciences, Clemson University, 1990.

[2] D. J. Daley and D. Vere-Jones, *An Introduction to the Theory of Point Processes*, Springer-Verlag, New York, 1988.

[3] E. J. Dudewicz and S. N. Mishra, *Modern Mathematical Statistics*, Wiley, New York, 1982.

[4] P. Glasserman and D. D. Yao, *Monotone Structure in Discrete Event Systems*, Wiley, New York, 1994.

[5] D. J. Hand, F. Daly, A. D. Lunn, K. J. McConway, and E. Ostrowski, *A Handbook of Small Data Sets*, Chapman & Hall, London, 1994.

[6] D. L. Iglehart and G. S. Shedler, *Regenerative Simulation of Response Times in Networks of Queues*, Lecture Notes in Control and Information Sciences, Springer-Verlag, New York, 1979.

[7] V. V. Kalashnikov, *Mathematical Methods in Queuing Theory*, Kluwer, Boston, 1994.

[8] J. S. Kirschman, "An Analysis of Registration at Clemson University," Technical Report, Department of Mathematical Sciences, Clemson University, 1991.

[9] J. F. Lawless, *Statistical Models and Methods for Lifetime Data*, Wiley, New York, 1982.

[10] E. L. Lehmann, *Theory of Point Estimation*, Wiley, New York, 1983.

[11] M. F. Neuts, *Matrix Geometric Solutions in Stochastic Models: An Algorithmic Approach*, Johns Hopkins University Press, Baltimore, 1981.

[12] C. D. Pegden, R. E. Shannon, and R. P. Sadowski, *Introduction to Simulation using SIMAN*, McGraw-Hill, New York, 1995.

[13] A. A. Pritsker, *Introduction to Simulation and SLAM II*, Wiley, New York, 1986.

[14] S. I. Resnick, *Adventures in Stochastic Processes*, Birkhäuser, Boston, 1992.

[15] S. M. Ross, *Introduction to Probability Models*, 5th ed., Academic Press, New York, 1993.

[16] S. M. Ross, *A First Course in Probability*, 3rd ed., Macmillan, New York, 1988.

[17] R. Y. Rubinstein, *Simulation and the Monte Carlo Method*, Wiley, New York, 1981.

[18] D. W. Scott, *Multivariate Density Estimation: Theory, Practice, and Visualization*, Wiley, New York, 1992.

[19] G. S. Shedler, *Regeneration and Networks of Queues*, Springer-Verlag, New York, 1987.

[20] B. W. Silverman, *Density Estimation for Statistics and Data Analysis*, Chapman and Hall, London, 1986.

Chapter 9

Mathematical Modeling of Unsaturated Porous Media Flow and Transport

Christopher L. Cox
Tamra H. Payne

Prerequisites: differential equations, numerical analysis, scientific computing

9.1 Introduction

Knowledge of the movement of chemicals and water in the soil has taken on ever-increasing importance since the dawn of the agricultural revolution. As contaminants from industrial spills or dumping and leaks in underground pipelines and storage tanks find their way into the water supply, and as agricultural scientists look for ways to improve crop production while minimizing the deleterious effects of herbicides, pesticides, and fertilizers, important questions involving the fate of these chemicals need to be addressed. In recent years, these concerns over groundwater pollution and resource management have resulted in numerous modeling efforts. Although extensive modeling has been done to access the fate of chemicals that reach the groundwater, comparatively little is known about how these chemicals move in the drier soil above the water table, the *vadose zone.*

This chapter focuses on modeling the movement of chemicals in the vadose zone. A fundamental modeling problem is to predict the movement of chloride through a uniform soil. In particular, we may want to predict the movement of chlorinated water as it moves from the soil surface to the water table. Thus, we are interested in predicting the concentration of chloride at various depths in the soil over time. The domain for such a problem is shown in Figure 9.1.

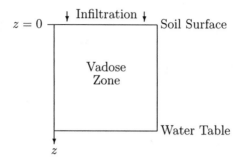

FIGURE 9.1   **Domain for unsaturated flow**

Our goal is to present a series of steps by which a numerical solution to the equations governing chemical movement in the vadose zone can be developed. The equations governing unsaturated porous media flow and transport are presented in Section 9.2, together with relevant definitions from soil physics. In Section 9.3 the constant-coefficient convection-dispersion equation in one dimension is considered. Both an analytical solution and a finite element solution are discussed. Section 9.4 shows how the traveling wave solution to Richards' equation can be coupled with the convection-dispersion equation (in one space dimension) with constant coefficients and/or closed-form functions for the coefficients. Finally, more complicated models and important modeling considerations are mentioned in Section 9.5.

9.2   Governing Equations

Partial differential equations (PDEs) are often the most effective way to mathematically describe a natural process, such as the movement of water and chemicals in the soil. Simple models for the diffusion of a quantity such as a chemical or heat in a continuous medium are developed in elementary courses (from conservation laws) and may result in a single partial differential equation of the form

$$\frac{\partial u}{\partial t} = D\frac{\partial^2 u}{\partial z^2} - V\frac{\partial u}{\partial z}$$

with conditions specified on u at two values of the spatial variable z (called *boundary conditions*), and a condition specifying u at some time t_0 (called an *initial condition*). The quantity D is a *diffusion coefficient*, whereas the quantity V measures the velocity of the flow in which the diffusion is taking place. Large values of D cause a rapid diffusion of u into the surrounding medium. We study a more involved model in which the diffusion coefficient depends on z and involves the diffusive flow of two quantities: a chemical contaminant carried by water and the water itself. Such a model will result in two partial differential equations.

9.2.1 The Convection-Dispersion Equation

The PDE that governs the movement of solutes in unsaturated porous media (i.e., the movement of chemicals in the soil) is known as the *convection-dispersion equation* and has the following form in one dimension [9]

$$\frac{\partial(\theta C)}{\partial t} = \frac{\partial}{\partial z}\left[\theta D_c(\theta)\frac{\partial C}{\partial z} - qC\right], \tag{9.1}$$

where

C = concentration of chemical (the dependent variable),

θ = water content,

t = time,

z = depth (positive downward),

q = flux,

$D_c(\theta)$ = apparent chemical diffusion coefficient.

Water content is a unitless quantity defined as the mass of water per unit mass of soil or the volume of water per unit volume of soil. In general the *flux* is the time rate of transport of a quantity across a given area. In our case, we will be interested in the time rate of transport of water across a given area of soil. By Darcy's law [5], the flux q satisfies

$$q = -K\nabla H = -K\left[1 + \frac{d\psi}{dz}\right], \tag{9.2}$$

where

$K(\theta)$ = hydraulic conductivity,

H = total hydraulic head,

ψ = tension head.

Hydraulic conductivity is the flux of water per unit gradient of hydraulic potential. The hydraulic conductivity quantifies the ease with which water flows through the soil; the higher the value of $K(\theta)$, the more easily water flows. For a given soil, $K(\theta)$ increases rapidly with increasing θ.

Since the soil is porous it draws in water and holds it, just like a sponge. The *tension head* ψ quantifies how strongly a soil retains water. A very moist soil will lose water fairly easily. However, more effort is required to remove water from a drier soil. The relationship between θ and ψ is known as a *soil characteristic curve*. Characteristic curves for soils are usually determined experimentally using a pressure plate apparatus or similar measuring device [8]. Here a saturated soil is subjected to increasing tension (suction). For each tension, the water content of the soil is measured. From these data, a plot of θ versus ψ (a soil characteristic curve) is produced. Figure 9.2 shows a soil characteristic curve for a Greary silt loam. The characteristic curve was produced using field data from [8] and MATLAB's `spline` routine [13].

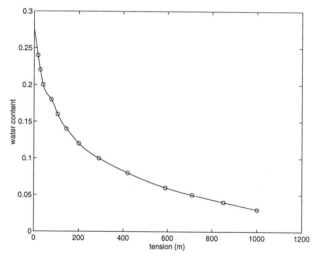

FIGURE 9.2   **Characteristic curve for a Greary silt loam**

Movement of chemicals through the soil can be accomplished by four processes [9]:

- chemical diffusion in the liquid phase — i.e., movement of chemicals in response to an aqueous concentration gradient [15];

- diffusion in the gas phase;

- convection (mass flow) of the chemical as the result of movement of water in which the chemical is dissolved;

- convection of the chemical in the vapor phase.

For unreactive chemicals such as chloride, diffusion in the gas phase and convection in the vapor phase are unimportant. Thus, we will consider only processes 1 and 3 in the subsequent analysis.

Field and laboratory studies have shown that the chemical diffusion term can take the form [9]

$$D_p(\theta) = D_{ol} a e^{b\theta},$$

where D_{ol} is the diffusion coefficient for the chemical in a pure liquid phase, $b = 10$, and $0.005 \leq a \leq 0.01$. Meanwhile, the convection term can take the form

$$D_m(q) = \lambda |v|,$$

where λ is a positive constant (which depends on the soil system) called the *dispersivity* and the soil water velocity is given by

$$v = \frac{q}{\theta}.$$

Thus we have

$$\begin{aligned} D_c(\theta) &= D_p(\theta) + D_m(q) \\ &= D_{ol} a e^{b\theta} + \lambda |v|. \end{aligned} \tag{9.3}$$

9.2.2   Richards' Equation

To model the movement of dissolved chemicals through unsaturated soil, (9.1) can be coupled with Richards' equation. Richards' equation governs the movement of water through unsaturated soil and has the following form in one dimension [11]

$$\frac{\partial \theta}{\partial t} = \frac{\partial}{\partial z}\left[D(\theta)\frac{\partial \theta}{\partial z}\right] - \frac{\partial K(\theta)}{\partial z}, \tag{9.4}$$

where $D(\theta)$ is the *soil water diffusivity*. Soil water diffusivity is defined as the hydraulic conductivity divided by the differential water capacity or the flux of water per unit gradient of moisture content in the absence of other force fields.

Several closed-form functions can be found that closely model the relationship between θ and K or D, as observed in field and laboratory experiments. In the manner of [1], we use either

$$K(\theta) = \theta^3 \quad \text{and} \quad D(\theta) = 2\theta, \tag{9.5}$$

or

$$K(\theta) = 100\theta^8 \quad \text{and} \quad D(\theta) = 100\theta^4. \tag{9.6}$$

Notice that as water content increases, D and K increase since water flows more easily through a moist soil.

Equations (9.1) and (9.4) can be coupled so that, for each time step, (9.4) is used to predict the water movement in the soil and the resulting water content θ is used in the solution of (9.1). An equation relating $d\theta/d\psi$, D, and K is available [4]:

$$\frac{d\theta}{d\psi} = D(\theta)/K(\theta). \tag{9.7}$$

Thus, once θ has been found using (9.4), we can rewrite (9.7) as

$$\frac{d\psi}{dz} = [K(\theta)/D(\theta)]\frac{d\theta}{dz}$$

for use in (9.2).

Only under extremely simplifying assumptions can the PDEs that govern the movement of water and chemicals in the soil be solved analytically. Otherwise, a numerical method is needed. Numerical approximation methods include the *finite difference method* and the *finite element method*. The finite element method will be the solution method used in this chapter.

9.3   Constant-Coefficient Convection-Dispersion

Consider a constant-coefficient version of (9.1):

$$\frac{\partial C}{\partial t} = D\frac{\partial^2 C}{\partial z^2} - V\frac{\partial C}{\partial z}. \tag{9.8}$$

One analytical solution for this equation is given in [16]:

$$C_A(z,t) = 0.5\,\text{erfc}\left[\frac{z - Vt}{2\sqrt{Dt}}\right] + 0.5e^{Vz/D}\,\text{erfc}\left[\frac{z + Vt}{2\sqrt{Dt}}\right], \tag{9.9}$$

where

$$\text{erfc}(x) = 1 - \frac{2}{\sqrt{\pi}}\int_0^x e^{-u^2}\,du.$$

Figure 9.3 shows $C_A(z,t)$ for $V = 877.9$, $D = 1$, and $t = 0.0001$.

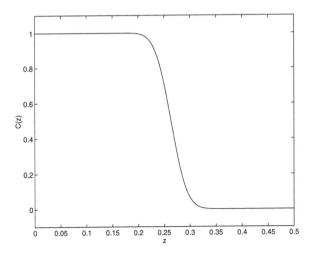

FIGURE 9.3   **Plot of $C_A(z,t)$ for $t = 0.0001$**

To formulate our initial/boundary value problem, we need to set the domain for z, the time interval, and initial and boundary conditions. A convenient choice uses $z \in [0,1]$ and $t \in [0.0001, 0.001]$, with

$$C(z, 0.0001) = C_A(z, 0.0001) \quad \forall z \in [0,1]$$
$$C(0,t) = C_A(0,t) \quad \forall t \in (0.0001, 0.001] \qquad (9.10)$$
$$C(1,t) = C_A(1,t) \quad \forall t \in (0.0001, 0.001]$$

since the solution of (9.8) subject to (9.10) is simply $C_A(z,t)$. Then we can use $C_A(z,t)$ as a benchmark for our numerical solution method. In particular, we can use the finite element method [7] to solve (9.8) subject to (9.10) and compare the finite element solution to $C_A(z,t)$. This assures the correctness of the finite element code before applying it to problems of a more complex nature.

We now give a brief description of the finite element method for a problem governed by a partial differential equation of the form

$$\frac{\partial u}{\partial t} = L(u(z,t)), \quad z_a \le z \le z_b, \quad t > 0, \qquad (9.11)$$

plus appropriate boundary and initial conditions, where $L(u)$ is a linear or nonlinear second-order partial differential operator with respect to z.

A standard finite element solution of (9.11) is a continuous piecewise-linear function

$$u(z,t) = \sum_{j=1}^{n} a_j(t)\phi_j(z). \tag{9.12}$$

Here $a_j(t)$ is the magnitude of the approximation at the nodal point z_j, $1 \le j \le n$, and $\phi_j(z)$ is the corresponding continuous piecewise-linear basis function satisfying

$$\phi_j(z_i) = \delta_{ij}, \quad 1 \le i, j \le n.$$

A typical basis function is shown in Figure 9.4.

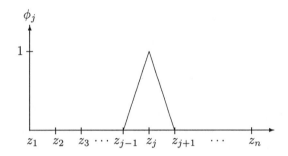

FIGURE 9.4   **Finite element basis function**

The weak form of (9.11) is found by multiplying both sides of (9.11) by a test function $v(z)$, and integrating both sides of the resulting equation over the spatial domain. Integration by parts is used to produce an equation with derivatives of no higher than first order. Into this equation we substitute the form (9.12) for u and require that the equation be satisfied for $v = \phi_i$, $i = 1, \ldots, n$. The result is a system of ordinary differential equations of the form

$$A(y)\dot{y} = g(y), \tag{9.13}$$

where A is a block tridiagonal matrix and $y = [a_1, \ldots, a_n]^T$. This system can be solved using a Gear-type ordinary differential equation solver such as DASSL [14].

The finite element solution of (9.8) and (9.10) with $V = 877.9$ and $D = 1$ is shown in Figure 9.5. The initial curve is plotted along with solutions at the times indicated on the graph.

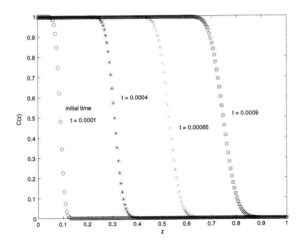

FIGURE 9.5   **Finite element solution: constant coefficients**

9.4   *Coupling the Equations*

9.4.1   *Traveling Wave Solution*

Recall that (9.4) governs the movement of water in unsaturated soil. It can be shown that as $t \to \infty$, the solution to (9.4) becomes a *traveling wave*, that is, a fixed shape moving at a constant speed [10]. The governing equation for the traveling wave solution is derived by substituting $U(z - ct)$ into (9.4) for θ. This produces

$$-c\frac{dU}{d\xi} = \frac{d}{d\xi}\left[D(U)\frac{dU}{d\xi}\right] - \frac{d}{d\xi}K(U), \qquad (9.14)$$

where $\xi = z - ct$ is the *traveling wave coordinate* and U satisfies the asymptotic conditions

$$\lim_{\xi \to -\infty} U(\xi) = \overline{u}, \quad \lim_{\xi \to \infty} U(\xi) = \underline{u}, \quad \lim_{|\xi| \to \infty} U'(\xi) = 0.$$

The constants $\overline{u}$ and $\underline{u}$ are upper and lower bounds on U, respectively. In (9.14), c is the constant velocity of the traveling wave, given by

$$c = \frac{K(\overline{u}) - K(\underline{u})}{\overline{u} - \underline{u}}. \qquad (9.15)$$

Integrating both sides of (9.14) over the interval $[\xi, \infty)$ results in the first-order equation

$$\frac{dU}{d\xi} = \frac{1}{D(U)}\left[K(U) - K(\underline{u}) + c(\underline{u} - U)\right]. \qquad (9.16)$$

For $\underline{u}$ and $\overline{u}$, we choose values that correspond to θ_{res}, the residual water content in air-dried soil, and θ_{sat}, the water content of the soil at saturation, respectively. Figure 9.6 shows the solution to (9.16) using (9.5) with boundary and initial conditions

$$\overline{u} = 0.3805, \quad \underline{u} = 0.15, \quad U_0 = 0.38. \tag{9.17}$$

The solution was obtained using the MATLAB adaptive Runga-Kutta ODE solver ode45 [13].

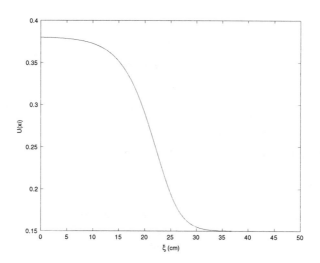

FIGURE 9.6   **Traveling wave**

9.4.2   Solving the Convection-Dispersion Equation

9.4.2.1   Constant Flux and Linear Diffusion

Now we substitute the traveling wave solution for θ in (9.1). Before we can solve for C, we need to specify q, $D_c(\theta)$, and determine our boundary and initial conditions. Following the example in [2], consider the problem where

$$t \in [4, 154], \quad z \in [0, 70], \quad D_c = 0.07\,\theta\,\frac{\text{cm}^2}{\text{min}}, \quad q = -0.026\,\frac{\text{cm}}{\text{min}}$$

and the initial and boundary conditions are given by

$$\begin{aligned}
C(z, 4) &= 1, & 0 \le z \le 10 \\
C(z, 4) &= 0, & 10 \le z \le 70 \\
C(0, t) &= 1, & C(70, t) = 0, \quad t > 4.
\end{aligned} \tag{9.18}$$

We can obtain values for θ from the traveling wave solution after we
have specified $\bar{u}$, $\underline{u}$, K, and D.

Using these conditions, we can solve (9.16) and associate this curve
with time $t_0 = 4$, denoting the solution as U_{t_0}. Then, if we translate
U_{t_0} at a constant speed given by (9.15), we will be able to determine
the water content for each z and t value in our domain. Figure 9.7
shows the solution to (9.1), (9.16), and (9.18), using $K(\theta)$ and $D(\theta)$
given by (9.6). The top graph shows the traveling wave solution, and
the bottom graph shows the associated convection-dispersion equation
solution using constant flux and linear diffusion terms. For both figures,
the curves (from left to right) indicate solutions at increasing times; that
is, the curve to the far left is the solution at the initial time, and curves
to the right indicate later times.

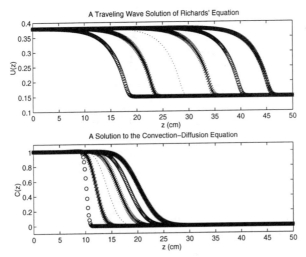

FIGURE 9.7   Coupled solutions: linear coefficients

Now we change the boundary conditions on C to

$$
\begin{aligned}
C(z,4) &= 0, \quad 0 \leq z \leq 70 \\
C(0,t) &= 4.35\,\theta(0,t) - 0.65 \\
C(70,t) &= 0, \quad t > 4
\end{aligned}
\tag{9.19}
$$

and allow the traveling wave to slowly emerge into the soil profile. Physically, this could correspond to a situation in which water contaminated
with an unreactive chemical is being applied to an initially dry, homogeneous soil. The emergence of the traveling wave solution through the
soil surface might correspond to a steadily increasing flow of chemical

solution. The movement of the traveling wave solution into the soil profile ($z = 0$) is shown in Figure 9.8. The vertical line at $z = 0$ represents
the soil surface. Thus, for each curve, we use the portion of the curve
to the right of $z = 0$ (the thick portion).

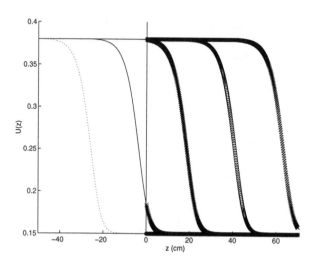

FIGURE 9.8   **Emergence of the traveling wave**

Figure 9.9 and Figure 9.10 show the finite element solution to this
problem at 30-minute time intervals using the values for D and K given
by (9.5) and (9.6), respectively. Again, in each figure, the top graph
shows the traveling wave solution, and the bottom graph shows the
associated solution of the convection-dispersion equation.

9.4.2.2   Nonlinear Coefficients

Now we use the quantities specified in (9.3) and (9.2) for D_c and q,
respectively. The constants are set as follows:

$$\lambda = 2, \quad D_{ol} = 10^{-5}, \quad a = 10, \quad b = 7.5 \times 10^{-3}.$$

In addition, the equations in (9.5) are used for D and K. Again we
use the boundary conditions given by (9.19). Figure 9.11 displays the
results.

A comparison of Figure 9.9 with Figure 9.11 is warranted at this point.
Notice that we see more convection (mass flow of the chemical due to
movement of the water) in Figure 9.11.

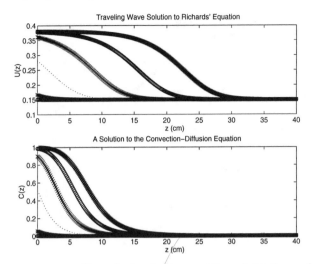

FIGURE 9.9   **Coupled solutions: D and K from (9.5)**

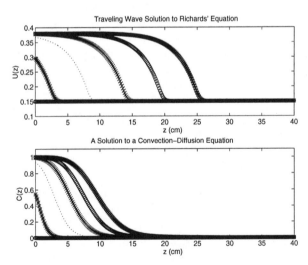

FIGURE 9.10   **Coupled solutions: D and K from (9.6)**

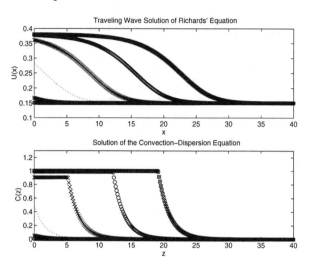

FIGURE 9.11   **Coupled solutions: nonlinear forms for q and D_c**

9.5   Summary and Suggestions for Further Study

This chapter began with a description of the equations governing flow
and transport through unsaturated porous media, in one spatial dimension. The goal was to demonstrate how these equations may be solved
numerically. A sequence of solutions for progressively more difficult
problems was presented, beginning with the constant coefficient transport problem for which an analytical solution exists, and culminating
with a traveling wave solution of the flow equation incorporated with
the transport equation with nonlinear coefficients. The examples in this
chapter illustrate that modeling unsaturated porous media flow is an interdisciplinary challenge, requiring knowledge of science, mathematical
analysis, and computing.

Several other aspects of porous media flow and transport modeling
merit attention.

1. **More Efficient Solution.** For the preliminary work described in
this chapter, the main goal was to obtain solutions with the desired
characteristic behavior. Thus, a high amount of resolution was used in
the spatial discretization. Numerical experiments can be used to provide
insight into how much resolution is necessary. Adaptive mesh procedures
could also be considered.

2. **Coupling Richards' Equation and the Convection-Dispersion
Equation.** A next step, after working with the traveling wave solution
to Richards' equation, could be to solve (9.4) to obtain values of θ for

(9.1). This problem will likely require the use of a stiff differential equation solver for the system of ODEs (9.13) resulting from applying the finite element method to (9.4).

3. **Using Real Field Data.** Instead of using closed-form equations for K and D, such as (9.5) or (9.6), experimental data can be used. For example, if field data are available for K and D over a reasonable range of water contents and the water content curve for a soil type has been determined, these data can be fit to obtain all of the information required to solve (9.4) and consequentially (9.1). Field data are not generally smooth, and thus some smoothing of the data may be required.

4. **Modeling Layered Systems.** All of the problems discussed so far have assumed a homogeneous soil: i.e., a soil in which the hydraulic conductivity is not a function of location. Although some soils can be approximated as homogeneous, all soils have some degree of heterogeneity. A more accurate approximation can sometimes be obtained by assuming the soil is layered: i.e., the soil has layers in which the hydraulic conductivity is not a function of location, but the hydraulic conductivity may change from layer to layer [12].

5. **Modeling Heterogeneous Systems.** For some problems, the assumption of a homogeneous or layered system is not a good approximation. In this situation, a different $K(\theta)$ relationship may hold for each node. Insufficient experimental data are often the limiting factor for analyzing these problems.

6. **2-D and 3-D Problems.** Each of the problems discussed above can be extended to two and three dimensions. When higher-dimensional problems are considered, the computational costs of solving these problems increase dramatically. As a result, the role of computational efficiency becomes paramount.

7. **Reactive Chemicals.** Many of the contaminants that find their way into water supplies are reactive and present in more than one phase. For these chemicals, modeling must also take into account phase equilibria as well as reactions between the contaminants and the soil system.

8. **Numerical Oscillations and Dispersion.** The solutions of (9.1) and (9.4) involve steep fronts (particularly for advection-dominated flows). These steep fronts are difficult to handle using numerical solution techniques. In particular, finite element solutions for problems involving steep fronts can exhibit unwarranted oscillatory behavior. Upwinding [3] can reduce these oscillations, but at the risk of smeared fronts (i.e., dispersion).

As the above list indicates, the examples presented in this chapter only touch on a few aspects of modeling unsaturated porous media flow and transport. This field of study offers the modeler a wide variety of interesting problems. Further information can be found by consulting the interdisciplinary Water Resources Research journal at the website <http://www.agu.org/pubs/agu_jourwrr.html>, as well as the modeling papers at <http://www.hwr.arizona.edu/avail_pubs.html> and <http://www.ticam.utexas.edu/Groups/SubSurfMod/home.html>.

9.6   Exercises and Projects

1. Graph $C(z,t) = C_A(z,t)$ for $V = 877.9$, $D = 1$, $z \in [0,1]$, and $t = 2, 4, 6$. What happens when you change V? Change D?

2. Use a symbolic algebraic package, such as Maple [6], to verify that $C_A(z,t)$ solves (9.9).

3. Verify that (9.14) and (9.16) both follow from (9.4).

4. Compare Figures 9.9 and 9.10. How did changing the functions used for K and D affect the solutions shown in these figures? Why?

5. (Project) Here we consider the traveling wave solution.

   a. Use an ODE solver to the find the solution of (9.16), (9.17), and (9.5) on the interval [0, 70]. Graph the resulting solution.

   b. If the wave found in part (a) corresponds to $t = 0$, show (on the same graph) the location of the wave at $t = 90$.

   c. Graph dU/dt for the traveling wave. Why would this function be important in the solution of (9.1) coupled with a traveling wave solution to (9.4)?

   d. Write a computer program that causes the traveling wave to emerge into the soil profile as shown in Figure 9.8. Plot the emergence of the curve at several times and indicate these times on your plot (**Hint:** the wave should move at a constant velocity given by (9.15).)

   e. Redo part (a) where K and D are now given by (9.6). How does your solution change? Why?

6. (Project) Write a computer program to obtain a finite element solution to (9.8) and (9.10). Compare your approximate solution to the analytical solution $C_A(z,t)$. How do these answers compare? If they differ, explain why.

9.7 References

[1] J. Bear, *Dynamics of Fluids in Porous Media*, Wiley, London, 1972.

[2] E. Bresler, "Simultaneous transport of solutes and water under transient unsaturated flow conditions," *American Geophysical Union* **9**, 975–986 (1973).

[3] G. F. Carey and J. T. Oden, *Finite Elements: Fluid Mechanics, Vol. 6* , Prentice-Hall, Englewood Cliffs, NJ, 1986.

[4] E. C. Childs and N. Collis-George, "Soil geometry and soil water equilibrium," *Disc. Faraday Soc.* **3**, 78–85 (1948).

[5] H. Darcy, *Les Fountaines Publiques de la Ville de Dijon*, Dalmont, Paris, 1856.

[6] F. Garvan, *Maple V Primer: Release 4*, CRC Press, Boca Raton, FL, 1997.

[7] C. A. Hall and T. A. Porsching, *Numerical Analysis of Partial Differential Equations*, Prentice-Hall, Englewood Cliffs, NJ, 1990.

[8] R. J. Hanks, *Applied Soil Physics*, Springer-Verlag, New York, 1992.

[9] J. L. Hutson and R. J. Wagenet, *LEACHM Version 3 Documentation*, Report No. 92–3, Department of Soil, Crop and Atmospheric Sciences, Cornell University, 1992.

[10] N. V. Khusnytdinova, "The limiting moisture profile during infiltration into a homogeneous soil," *J. Appl. Math. Mech.* **31**, 783–789 (1967).

[11] D. Kirkham and W. L. Powers, *Advanced Soil Physics*, Wiley, London, 1972.

[12] F. J. Leif, J. H. Dane, and M. Th. van Genuchten, "Mathematical analysis of one-dimensional solute transport in a layer soil profile," *Soil Sci. Soc. Am. J.* **55**, 944–953 (1991).

[13] The MathWorks, Inc., *MATLAB User's Guide*, Prentice-Hall, Upper Saddle River, NJ, 1997.

[14] L. R. Petzold, "A description of DASSL: a differential/algebraic system solver," in *Scientific Computing: Applications of Mathematics and Computing to the Physical Sciences*, R. S. Stepleman et al. (Eds.), IMACS/North-Holland, Amsterdam, 1983, 65–68.

[15] Soil Science Society, *Glossary of Soil Science Terms*, Soil Science Soc. of America, Inc., Madison, WI, 1987.

[16] T. Yamaguchi, P. Moldrup, and S. Yokosi, "Using breakthrough curves for parameter estimation in the convection-dispersion model of solute transport," *Soil Sci. Soc. Am. J.* **53**, 1635–1641 (1989).

Chapter 10

Inventory Replenishment Policies and Production Strategies

P. M. Dearing
Herman Senter
Mark Fitch

Prerequisites: elementary statistics and probability, linear programming

10.1 Introduction

The models presented here derive from an analysis done for a manufacturing division of a large corporation. Although the name of the company and its actual products are not mentioned, a hypothetical manufacturing process is described which has the salient features of the real situation.

The questions addressed concern how much inventory the "Producer" should stock of a required component part, provided by a "Supplier," in order to meet daily production needs, and how the Supplier should schedule production to satisfy the demand for these parts. In the first part of the chapter, Sections 10.2 to 10.5, the manufacturing process is described and the size distribution of the product is modeled by a multinomial distribution. Three inventory reordering policies are compared in terms of their effects on inventory stock levels. The second part, starting with Section 10.6, examines production strategies for the Supplier to use in meeting the Producer's demand for parts.

10.2 *Piston Production and the Multinomial Model*

The Producer company manufactures *pistons* whose outside diameters, measured in millimeters, vary independently and randomly. Each piston must be matched with a *sleeve* whose inside diameter equals the piston's outside diameter. The sleeves are ordered from the Supplier who ships them to the Producer.

The Producer would like to have on hand an adequate supply of sleeves so that each piston can be matched as it comes from the production line. If no sleeve is on hand to match a piston (a sleeve *stockout*), the piston is temporarily stored at considerable expense until the proper size sleeve arrives.

The piston production process has the following characteristics:

- A single piston type is matched with a given sleeve type.

- Piston diameters can be divided into k distinct diameter categories based on diameter intervals of 0.1 mm (k a positive integer).

- The long-term proportion p_i of piston diameters that fall in the ith category are constrained over time; that is, the process is *stable*.

- Piston diameters are independent of one another.

- The number of pistons $n(t)$ produced on day $t \geq 1$ is known.

Based on these assumptions the distribution of piston sizes manufactured on day t can be modeled by a *multinomial distribution* [2] with

- $X_i(t)$ the random variable which counts the number of pistons in diameter category i produced on day t,

- $x_i(t)$ the number of pistons of size i produced on day t, a realization of $X_i(t)$, where $\sum_{i=1}^{k} x_i(t) = n(t)$ and $\sum_{i=1}^{k} p_i = 1$.

The joint probability distribution of piston diameters produced is then given by

$$P(x_1(t), x_2(t), \ldots, x_k(t)) = \frac{n(t)!}{x_1(t)!x_2(t)! \cdots x_k(t)!} \, p_1^{x_1(t)} p_2^{x_2(t)} \cdots p_k^{x_k(t)}.$$

In a manufacturing situation, exact values of the proportions p_i may not be known. We assume that p_i is estimated by $\widehat{p}_i$ for $i = 1, \ldots, k$. The relative frequency histogram of Figure 10.1 illustrates a hypothetical piston size distribution with $k = 30$ size categories.

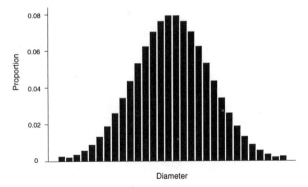

FIGURE 10.1 **Piston size distribution**

10.3 *Sleeve Inventory Safety Stocks*

In this section it is assumed that the number of pistons produced daily is constant; that is, $n(t) = n$ for all days t. The Producer orders sleeves each day and shipments from the Supplier arrive daily. Let $s_i \geq 0$ be the number of sleeves of diameter size i that the Producer would like to have in inventory at the start of each day. (Note that s_i is assumed to be the same for each day.) We shall call s_i the *safety stock level* for diameter size i and $\sum_{i=1}^{k} s_i$ the total *safe inventory* at the beginning of a day. The Producer wants to set safety stock levels s_i to minimize expected inventory costs, including the cost of stockouts. Intuitively, the Producer would like the safe inventory to be small while enjoying a low risk of a stockout in any size.

If the Producer begins day t with exactly s_i sleeves of each size i in stock, then the probability of having no stockouts that day is

$$P(\text{no stockouts} \mid s_1, \ldots, s_k) = P(X_1(t) \leq s_1, \ldots, X_k(t) \leq s_k)$$

$$= \sum_{x_1(t) \leq s_1, \ldots, x_k(t) \leq s_k} \frac{n!}{x_1(t)! \cdots x_k(t)!} \, p_1^{x_1(t)} \cdots p_k^{x_k(t)},$$

where $\sum_{i=1}^{k} x_i(t) = n$ and $\sum_{i=1}^{k} p_i = 1$.

If k is large (say $k > 30$), then accurate computation of this cumulative multinomial probability is difficult. However, $P(X_1(t) \leq s_1, \ldots, X_k(t) \leq s_k) = 1 - P(X_1(t) > s_1 \cup \cdots \cup X_k(t) > s_k)$. Using the result [4] that $P(X_1(t) > s_1 \cup \cdots \cup X_k(t) > s_k) \leq P(X_1(t) > s_1) + \cdots + P(X_k(t) > s_k)$ then gives

$$P(X_1(t) \leq s_1, \ldots, X_k(t) \leq s_k) \geq 1 - [P(X_1(t) > s_1) + \cdots + P(X_k(t) > s_k)]$$

and provides a lower bound for the probability of having no stockouts on day t, given that safety stocks s_i are on hand at the start of the day. Equivalently, $\sum_{i=1}^{k} P(X_i(t) > s_i)$ is an upper bound for the risk of at least one stockout given safety stocks s_i.

Since the marginal distribution of $X_i(t)$ is binomial with parameters (n, p_i), then $P(X_i(t) > s_i)$ is a binomial probability. For sufficiently large n (number of pistons produced daily), such as $np_i \geq 5$ and $n(1 - p_i) \geq 5$ for all i, the normal distribution provides reasonable approximations of binomial probabilities so that

$$P(X_i(t) > s_i) \approx P\left(z > \frac{s_i - np_i - 0.5}{\sqrt{np_i(1 - p_i)}}\right),$$

where z is the standard normal deviate [2].

This suggests that one possibility for setting safety stock levels s_i is to use the expected values and standard deviations of the $X_i(t)$. If safety levels are set at $s_i = E(X_i(t)) + 2\sqrt{\mathrm{Var}(X_i(t))}$ rounded to the nearest integer, then the normal approximation yields $P(X_i(t) > s_i)) \approx 0.0275$ for each i. This gives an upper bound for the probability of at least one stockout on a given day as $0.0275k$. Similarly, if $s_i = E(X_i(t)) + 3\sqrt{\mathrm{Var}(X_i(t))}$ rounded to the nearest integer, then $P(X_i(t) > s_i)) \approx 0.0013$ for each i and $P(\text{at least one stockout}) \leq 0.0013k$.

The above bounds are conditional on starting the day with exactly s_i sleeves of size i in inventory for all i. However, the number of sleeves used daily in each category i is random and incoming shipments of sleeves may not exactly replace the number used. That is, actual inventory amounts will fluctuate about some mean level depending on how sleeves are resupplied. Thus, the probability of no stockouts on day t is

$$P(X_1(t) \leq y_1(t), X_2(t) \leq y_2(t), \ldots, X_k(t) \leq y_k(t)),$$

where $y_i(t)$ is the *actual* inventory of size i sleeves at the start of day t.

The next section examines some possible reordering policies relative to their effects on actual inventory levels.

10.4   Comparison of Three Reordering Policies

The purpose of this section is to demonstrate how the multinomial model of piston production can be used to study the effects of different restocking policies on daily inventory levels of sleeves. Three reordering policies are compared. Sensitivity of these policies to production parameters and the order refill time is also examined.

Assume as before that the distribution of piston diameters can be modeled by a multinomial distribution, where p_i denotes the proportion of pistons of size i. In practice, the parameters p_i are usually estimated from historical production data. Further assume daily ordering of sleeves and daily arrivals of orders with an r-day delivery time for each order (e.g., if $r = 4$ then an order placed on day 1 arrives on day 5.) Define

- $\widehat{p}_i$ = estimated value of p_i $(i = 1, \ldots, k)$,

- $y_i(t)$ = actual stock level of size i sleeves on hand at the start of day t,

- $r_i(t)$ = number of sleeves of size i arriving during day t (by the end of day t),

- $n(t)$ = total number of pistons produced during day t (by the end of day t),

- $x_i(t)$ = number of pistons of size i produced during day t (by the end of day t),

- r = reorder period in days (a positive integer).

Assume for now a constant daily production level $n(t) = n$ and that the safety stock level $s_i(t)$ is the same for all t; i.e., $s_i(t) = s_i$. In addition, assume for simplicity of exposition that the reorder period is $r = 4$ days and that $y_i(1) = s_i$ for all i. Finally, assume the Supplier is faithful; that is, the Supplier delivers exactly the quantity ordered.

10.4.1   Order Policy 1

A *reordering policy* is a rule that specifies the number of sleeves to order at the start of day t to arrive four days later, by the end of the day $t+4$. The first order policy considered, Policy 1, computes order amounts for size i sleeves by comparing actual inventory to the target safety stock level, subtracting orders for the previous four days, and adding expected usage until the order arrives. Specifically, the amount to order of size i, at the start of day t, to arrive four days later on day $t + 4$, is given by $r_i(t + 4)$ where

$$r_i(t + 4) = [s_i - y_i(t)] - [r_i(t) + r_i(t + 1) + r_i(t + 2) + r_i(t + 3)] +$$
$$E(\widehat{X}_i(t)) + E(\widehat{X}_i(t + 1)) + E(\widehat{X}_i(t + 2)) + E(\widehat{X}_i(t + 3)) + E(\widehat{X}_i(t + 4))$$

and $E(\widehat{X}_i(t))$ is the estimated expected production on day t. We impose the initial conditions $r_i(t) = E(\widehat{X}_i(t))$ for $t = 1, \ldots, 4$ and $y_i(1) = s_i$ for all i.

Assuming a constant production level of $n(t) = n$, then $E(\widehat{X}_i(t)) = E(\widehat{X}_i) = n\widehat{p}_i$ for all t and

$$r_i(t+4) = [s_i - y_i(t)] - [r_i(t) + r_i(t+1) + r_i(t+2) + r_i(t+3)] + 5n\widehat{p}_i. \quad (10.1)$$

10.4.2   Order Policy 2

This policy always orders a constant amount, which is the estimated expected demand for that day. Thus $r_i(t) = E(\widehat{X}_i(t))$ for all t. Under the assumption that $n(t) = n$, then $E(\widehat{X}_i(t)) = E(\widehat{X}_i) = n\widehat{p}_i$ and $r_i(t) = n\widehat{p}_i$. (We assume $y_i(1) = s_i$ for all i.)

10.4.3   Order Policy 3

Under this policy the amount ordered each day is exactly what was used the previous day; i.e., $r_i(t+5) = x_i(t)$. Since $x_i(t)$ is not known until the end of day t, the order is placed on day $t+1$ to arrive 4 days later on day $t+5$. We impose the initial conditions that $r_i(t) = n\widehat{p}_i$ for $t = 1, \dots, 5$ and that $y_i(1) = s_i$ for all i.

10.4.4   Comparisons of Policies

Let $Y_i(t)$ and $R_i(t)$ be random variables denoting, respectively, the number of sleeves of size i in stock at the start of day t and the number arriving that day, with realizations $y_i(t)$ and $r_i(t)$. Rewriting equation (10.1) yields

$$y_i(t) = s_i - [r_i(t) + r_i(t+1) + r_i(t+2) + r_i(t+3) + r_i(t+4)] + 5n\widehat{p}_i. \quad (10.2)$$

However, $y_i(t+1) = s_i - [r_i(t+1) + r_i(t+2) + r_i(t+3) + r_i(t+4) + r_i(t+5)] + 5n\widehat{p}_i$. Subtracting yields $y_i(t+1) - y_i(t) = r_i(t) - r_i(t+5)$. Also, $y_i(t+1) - y_i(t) = r_i(t) - x_i(t)$, which shows that for Policy 1

$$r_i(t+5) = x_i(t). \quad (10.3)$$

That is, the daily shipment amount will vary as the actual amount of production five days before. Thus we observe that, under the assumed initial conditions (and Supplier fidelity), Policy 1 and Policy 3 are the same!

Expressing (10.3) in terms of random variables and taking expectations yields the result $E(R_i(t+5)) = E(X_i(t)) = np_i$. Thus, the average number of sleeves ordered (or arriving) each day will equal the average production of pistons for each size category.

Similarly, expression (10.2) gives

$$E(Y_i(t)) = s_i - 5np_i + 5n\widehat{p}_i = s_i - 5n(p_i - \widehat{p}_i). \qquad (10.4)$$

Also, from (10.2)

$$\begin{aligned}
\text{Var}(Y_i(t)) &= 0 + \text{Var}(R_i(t)) + \text{Var}(R_i(t+1)) + \text{Var}(R_i(t+2)) + \\
&\quad \text{Var}(R_i(t+3)) + \text{Var}(R_i(t+4)) + 0 \\
&= \text{Var}(X_i(t-5)) + \text{Var}(X_i(t-4)) + \text{Var}(X_i(t-3)) + \\
&\quad \text{Var}(X_i(t-2)) + \text{Var}(X_i(t-1)) \\
&= 5np_i(1-p_i) = 5\text{Var}(X_i). \qquad (10.5)
\end{aligned}$$

In general, the effect of Policies 1 and 3 on inventory levels with the reorder period r is to amplify the variance of production by a factor of $r+1$.

For Policy 2, the daily order amount is constant, i.e., $r_i(t) = n\widehat{p}_i$, and the change in inventory level from day t to day $t+1$ is $y_i(t+1) - y_i(t) = r_i(t) - x_i(t)$. Thus, the average daily change in inventory level is

$$E(Y_i(t+1) - Y_i(t)) = n\widehat{p}_i - np_i = n(\widehat{p}_i - p_i). \qquad (10.6)$$

Equivalently, $E(Y_i(t)) = s_i + (t-1)n(\widehat{p}_i - p_i)$.

If $\widehat{p}_i = p_i$, then the average change in actual inventory is zero so that $E(Y_i(t)) = y_i(1) = s_i$. Otherwise, $Y_i(t)$ will be driven to zero or increase without bound. For $t > 1$, $y_i(t) = y_i(1) + \sum_{j=1}^{t-1} r_i(j) - \sum_{j=1}^{t-1} x_i(j)$ and

$$\text{Var}(Y_i(t)) = \text{Var}\left(\sum_{j=1}^{t-1} X_i(j)\right) = (t-1)np_i(1-p_i). \qquad (10.7)$$

Thus, under Policy 2, the variation in actual inventory stocks grows with time.

10.4.5 Summary

If $\widehat{p}_i = p_i$ then for Policy 1 (or Policy 3) expressions (10.4) and (10.5) yield $E(Y_i(t)) = s_i$ and $\text{Var}(Y_i(t)) = 5np_i(1-p_i)$. More generally, if r is the length of the reorder period, then $\text{Var}(Y_i(t)) = (r+1)np_i(1-p_i)$. Thus, inventory variances increase with both r and n.

If $\widehat{p}_i = p_i$ then for Policy 2 the expressions (10.6) and (10.7) yield $E(Y_i(t)) = s_i$ and $\text{Var}(Y_i(t)) = (t-1)np_i(1-p_i)$, which is independent of the reorder period but unbounded over time.

If $\widehat{p}_i \neq p_i$ then for Policy 1, $E(Y_i(t)) = s_i - 5n(p_i - \widehat{p}_i)$. Thus, the average inventory level is stable with a fixed bias of $-5n(p_i - \widehat{p}_i)$ relative to s_i. If $\widehat{p}_i$ overestimates p_i, we will carry on average more than s_i stock, and if $\widehat{p}_i < p_i$, we will average less than s_i sleeves in stock. However, $\text{Var}(Y_i(t)) = 5np_i(1 - p_i)$ regardless of our estimate $\widehat{p}_i$.

If $\widehat{p}_i \neq p_i$ then for Policy 2, the average daily change in inventory amount will be $E(Y_i(t+1) - Y_i(t)) = n(\widehat{p}_i - p_i)$. If $\widehat{p}_i > p_i$, the stock of size i sleeves will continue to increase, while if $\widehat{p}_i < p_i$, the stock will be driven to zero. Also, the variance in the stock levels is unbounded as t increases. Clearly, Policy 2 is inferior to Policy 1.

The previous equations admit negative values of $y_i(t)$ which might occur if s_i is relatively small. Negative $y_i(t)$ can be interpreted as inventory deficits, corresponding to finished pistons which are temporarily stored until correctly-sized sleeves arrive.

10.5   Variable Piston Production Quantities

In practice the total number of pistons produced daily is not constant, but is a stochastic quantity which will be denoted by $N(t)$. To address this extension, Wald's equations [3] can now be used:

$$E(X_i(t)) = E(N(t))p_i$$
$$\text{Var}(X_i(t)) = E(N(t))p_i(1 - p_i) + \text{Var}(N(t))(p_i(1 - p_i))^2.$$

Under Policy 1, equations (10.4) and (10.5) become

$$E(Y_i(t)) = s_i - 5E(N(t))(p_i - \widehat{p}_i)$$
$$\text{Var}(Y_i(t)) = 5[E(N(t))p_i(1 - p_i) + \text{Var}(N(t))(p_i(1 - p_i))^2].$$

The Producer would like to set safety stocks s_i to control the risk of stockouts at an acceptable level. The probability of at least one stockout in size category i on day t is

$$P(X_i(t) > Y_i(t)) = \sum_y \left[P(Y_i(t) = y) \sum_{u>y} P(X_i(t) = u) \right].$$

That is, the chance of a stockout depends on the actual stock amount and how much it varies. The previous analysis shows that under Policy 1, the variance in inventory stocks $Y_i(t)$ is larger than that of piston production $X_i(t)$ by a factor of $r + 1$. Shortening the reorder period r would reduce safe inventory levels by decreasing the variance in daily inventory stocks. Setting safety stocks according to a rule based on variation in inventory, such as $s_i = E(X_i) + 3\sqrt{\text{Var}(Y_i)}$, incorporates effects of variation in daily piston amounts $x_i(t)$, total daily volume of production $n(t)$, and the length r of the reorder period.

Notice that under Policy 1, $r_i(5) = s_i - y_i(1) + n\widehat{p}_i$ while under Policy 3, $r_i(5) = n\widehat{p}_i$ holds by assumption. Also under Policy 1, $E(Y_i(t)) = s_i - 5n(p_i - \widehat{p}_i)$ while under Policy 3, $E(Y_i(t)) = y_i(1) - 5n(p_i - \widehat{p}_i)$. If $y_i(1) = s_i$ as we have assumed (and if the Supplier delivers exactly the quantity ordered), then Policy 1 and Policy 3 are the same. If the Supplier cannot always deliver the ordered amount, Policy 1 and Policy 3 yield different levels of actual inventory and are not equivalent policies.

10.6 The Supplier's Production Problem

To this point we have assumed the Supplier can produce and deliver sleeves of the sizes ordered. However, the Supplier's sleeve production process yields a distribution of sleeve diameters analogous to that of the piston diameters. In fact, in the real world problem, the distribution of sleeve sizes required by the Producer spans some 35 size classes. But for a fixed *target size* (nominal size), the size distribution of sleeves manufactured by the Supplier is less variable, spanning some 20 size classes. As a result, the Supplier must vary its target size over a range of values in order to ensure a full range of sleeve diameters to meet the Producer's demand. Is there a rational, perhaps optimal, way to shift the target size? What inventory levels should the Supplier maintain? We would like to find a production schedule that would yield the number of sleeves of each size required by the Producer.

10.6.1 The Supplier's Target Size Selection

Some preliminary observations follow based on simplified examples. Attention will be given to the amount of inventory the Supplier must carry to meet demand from the Producer.

Example 10.1

Assume the Producer has constant daily production of $n = 16$ pistons that fall into 7 consecutive size categories as shown in Figure 10.2. Also assume the Supplier produces 16 sleeves daily and must have at least 16 on hand at the start of each day to ship to the Producer to meet the Producer's demand.

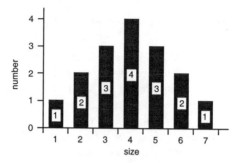

FIGURE 10.2   **Producer's distribution of piston diameters**

Case 1: If the Supplier's size distribution matches the Producer's exactly then the Supplier would need a daily inventory of exactly 16 sleeves (1 of size 1, 2 of size 2, etc.). This is an optimal situation and represents minimum inventory requirements for the Supplier.

Case 2: Assume the Supplier's control is excellent and that all 16 sleeves produced daily are exactly the target size. In this case, the target value must be shifted (repositioned) to at least seven locations.

Some possible production schedules, or cycles, by which the Supplier could meet the Producer's needs (without excess sleeves) are given below. A *shift* occurs when the Supplier must relocate the target size. A schedule with fewer shifts is preferred because it reduces expensive setups.

Schedule 1: In the following schedule, there are 16 shifts of the target value in 16 days.

Day	1	2	3	4	5	6	7	8
Target	4	3	2	5	4	6	3	1
Day	9	10	11	12	13	14	15	16
Target	4	5	2	3	4	6	5	7

The yield of sleeves over the 16-day cycle is

Category	1	2	3	4	5	6	7
Production	16	32	48	64	48	32	16

which gives precisely the required amounts of each size. An initial inventory of 8, 6, 6, 4, 14, 12, 16 sleeves in each respective size will ensure that sufficient sleeves are always available. Each day, the same total of 66 sleeves is maintained in inventory since 16 sleeves are demanded and 16 sleeves are resupplied each day.

For any feasible resupply schedule, the minimum initial inventory level can be determined by the following procedure. Assume a large starting inventory of, say, $M = 100$ sleeves in each size. Work through the production and resupply for each day following the resupply schedule. Recall that production is taken out of existing inventory, then resupply occurs for the target size. Table 10.1 illustrates this procedure for Schedule 1, showing for each size the ending daily inventory for each of the 16 days of the resupply cycle. For each size, subtract the minimum inventory level from the large inventory M and add the daily demand for that size. The resulting minimum initial inventory is shown as the last line in the table.

Table 10.1   Calculation of initial inventory for Schedule 1

	Size	1	2	3	4	5	6	7
	Large Inventory M	100	100	100	100	100	100	100
Day	Target							
1	4	99	98	97	112	97	98	99
2	3	98	96	110	108	94	96	98
3	2	97	110	107	104	91	94	97
4	5	96	108	104	100	104	92	96
5	4	95	106	101	112	101	90	95
6	6	94	104	98	108	98	104	94
7	3	93	102	111	104	95	102	93
8	1	108	100	108	100	92	100	92
9	4	107	98	105	112	89	98	91
10	5	106	96	102	108	102	96	90
11	2	105	110	99	104	99	94	89
12	3	104	108	112	100	96	92	88
13	4	103	106	109	112	93	90	87
14	6	102	104	106	108	90	104	86
15	5	101	102	103	104	103	102	85
16	7	100	100	100	100	100	100	100
	minimum	93	96	97	100	89	90	85
	$M-$ minimum	7	4	3	0	11	10	15
	demand	1	2	3	4	3	2	1
	initial inventory	8	6	6	4	14	12	16

Table 10.2 shows, for each size, the ending daily inventory for each of the 16 days in a resupply cycle, beginning with the initial inventory calculated in Table 10.1. Observe that the initial inventory appears as the ending inventory for day 16. In summary, the minimum inventory needed for Schedule 1 is $8 + 6 + 6 + 4 + 14 + 12 + 16 = 66$ sleeves. However, this schedule involves 16 shifts — one for each day of the cycle.

Table 10.2   Daily inventory by size for Schedule 1

Day	Size / Target	1	2	3	4	5	6	7
1	4	7	4	3	16	11	10	15
2	3	6	2	16	12	8	8	14
3	2	5	16	13	8	5	6	13
4	5	4	14	10	4	18	4	12
5	4	3	12	7	16	15	2	11
6	6	2	10	4	12	12	16	10
7	3	1	8	17	8	9	14	9
8	1	16	6	14	4	6	12	8
9	4	15	4	11	16	3	10	7
10	5	14	2	8	12	16	8	6
11	2	13	16	5	8	13	6	5
12	3	12	14	18	4	10	4	4
13	4	11	12	15	16	7	2	3
14	6	10	10	12	12	4	16	2
15	5	9	8	9	8	17	14	1
16	7	8	6	6	4	14	12	16

Schedule 2: In the following schedule, there are 9 shifts of the target size in 16 days.

Day	1	2	3	4	5	6	7	8
Target	7	6	6	5	5	3	3	4
Day	9	10	11	12	13	14	15	16
Target	4	4	4	2	2	5	1	3

In order to ensure sufficient sleeves are available on all days, a total stock level of 106 sleeves is found to be necessary under this schedule.

Schedule 3: In the following schedule, there are 10 shifts of the target size in 16 days.

Day	1	2	3	4	5	6	7	8
Target	7	6	6	5	5	3	3	2
Day	9	10	11	12	13	14	15	16
Target	2	1	4	4	4	5	4	3

In order to ensure sufficient sleeves are available on all days, a total stock level of 105 sleeves will be necessary.

A number of other production schedules are possible. What can be observed from the examples presented here is that the price of not shifting targets as often is higher inventory. That is, more frequent target shifts generally mean lower inventory for the Supplier.

Case 3: Suppose that the Supplier's size distribution is as given in Figure 10.3. Here the target size corresponds to the midpoint of the distribution.

FIGURE 10.3   **Supplier's distribution of sleeve diameters**

To produce the required 7 diameters, targets could be set at positions $a = 1.5$ through $f = 6.5$ as indicated below.

Location	a	b	c	d	e	f	
Diameter	1	2	3	4	5	6	7

Production over 16 days would yield the following quantities:

Size	1	2	3	4	5	6	7
2 @ a	16	16					
2 @ b		16	16				
4 @ c			32	32			
4 @ d				32	32		
2 @ e					16	16	
2 @ f						16	16
Total	16	32	48	64	48	32	16

A possible 8-day schedule (daily shifts) is shown below.

Day	1	2	3	4	5	6	7	8
Target	c	f	d	b	c	e	d	a

For this schedule, the daily inventory level is 40 sleeves. An alternative 8-day schedule that requires two fewer target shifts is as follows:

Day	1	2	3	4	5	6	7	8
Target	c	c	a	b	d	d	e	f

Under this schedule the daily inventory is 53 sleeves. A "best" schedule for the Supplier would be one that minimizes the cost of shifting target values plus inventory carrying costs (and exactly meets the Producer's demand).

Case 4: Suppose the Supplier's size distribution is given by Figure 10.4 when the target is set at t.

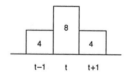

FIGURE 10.4   Supplier's distribution of sleeve diameters

Production over 16 days would then yield the following quantities:

Size	1	2	3	4	5	6	7
4 @ 2	16	32	16				
8 @ 4			32	64	32		
4 @ 6					16	32	16
Total	16	32	48	64	48	32	16

A possible 8-day schedule (two 4-day cycles) is given by the following:

Day	1	2	3	4	5	6	7	8
Target	4	2	4	6	4	2	4	6

Daily inventory under this schedule is 30 sleeves. As the Supplier's size distribution spreads out toward the Producer's, less inventory is required. However, consider yet another case.

Case 5: Suppose the Supplier's distribution of sleeve sizes, with the target value t, is that specified in Figure 10.5. Using target values of $1, 2, \ldots, 7$ (centers of the classes), the Supplier cannot meet demand without overproducing certain sizes. To produce 64 sleeves of size 4 requires 11 runs at target 4, yielding 55 sleeves of size 3 and 55 sleeves of size 5, far exceeding the 48 needed. Running at targets of 3 or 5 would only worsen the situation. In a case such as this, we may target fractional sizes (e.g., 4.5 or 4.75), that is, find selected targets $\mu_1, \mu_2, \ldots, \mu_m$ to achieve the desired distribution. Even when a range of targets is allowed, it may be impossible for the Supplier to exactly match the Producer's needs.

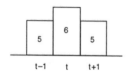

FIGURE 10.5   Supplier's distribution of sleeve diameters

10.6.2   Selection of Targets and Cycle Time

Assume the Producer's daily production distribution is $x_1, x_2, \ldots, x_k$ pistons in each of the k categories, respectively, as shown in Figure 10.6. We think of x_i as the average number of pistons of size i made daily, avoiding the more cumbersome $E(X_i)$ notation. Also suppose the Supplier's output, with target value μ_t, produces $q_i(\mu_t)$ sleeves in category i, for $i = 1, \ldots, m$; see Figure 10.7.

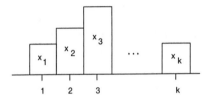

FIGURE 10.6   **Producer's distribution of pistons**

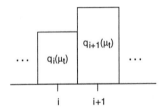

FIGURE 10.7   **Supplier's distribution of sleeve diameters**

Further assume there are T possible target values $\mu_1, \mu_2, \ldots, \mu_T$ and that $q_i(\mu_{t+j}) = q_{i-j}(\mu_t)$; that is, the shape of the Supplier's distribution of sleeve sizes is invariant with respect to target value. We want to determine nonnegative integers $n_1, n_2, \ldots, n_T$, where n_i represents the number of times that the Supplier produces sleeves set at target μ_i. The objective is to minimize the total absolute difference between what is produced and what is needed over n days. This problem can be stated as follows:

$$\text{minimize} \quad z = \sum_{i=1}^{k} \mid nx_i - (q_i(\mu_1)n_1 + \cdots + q_i(\mu_T)n_T) \mid$$

$$\text{subject to} \quad \sum_{i=1}^{T} n_i = n$$

$$n_i \geq 0 \text{ and integer.}$$

An ideal solution would have each summand in the *objective function* z equal to zero: namely,

$$nx_i - (q_i(\mu_1)n_1 + \cdots + q_i(\mu_T)n_T) = 0, \text{ for } i = 1, \ldots, k$$

so that the objective function value is zero. That is, over an n-day period the Supplier would produce exactly the number of sleeves required in each of the k size categories (by running n_j times at target μ_j, utilizing T targets). We call n the Supplier's production *cycle time*.

To formulate this problem as an integer linear programming problem [1], introduce the nonnegative variables z_i for $i = 1, \ldots, k$, replace each summand in the objective function by z_i, and add the constraints

$$\mid nx_i - (q_i(\mu_1)n_1 + \cdots + q_i(\mu_T)n_T) \mid \; \leq z_i, \text{ for } i = 1, \ldots, k.$$

This leads to the linear constraints

$$-z_i \leq nx_i - (q_i(\mu_1)n_1 + \cdots + q_i(\mu_T)n_T) \leq z_i, \text{ for } i = 1, \ldots, k.$$

The resulting linear integer programming problem is then

$$\text{minimize } \sum_{i=1}^{k} z_i$$

$$\text{subject to } -z_i \leq nx_i - (q_i(\mu_1)n_1 + \cdots + q_i(\mu_T)n_T) \leq z_i, \; i = 1, \ldots, k$$

$$\sum_{i=1}^{T} n_i = n$$

$$n_i \geq 0 \text{ and integer.}$$

The above objective function has no penalty for the production of sleeves whose diameters are outside the required size range. This problem is addressed in the next section.

Two very simple examples follow that illustrate applications of this optimization model to target selection. In both cases the obvious best solution emerges.

Example 10.2

Suppose that the Producer requires 8 sleeves daily of sizes μ_1, μ_2, μ_3 as shown in Figure 10.8 and that the Supplier's distribution matches that of the Producer. Let μ_1, μ_2, μ_3 be the candidate targets (for the center of the Supplier's distribution). In this case $k = T = m = 3$.

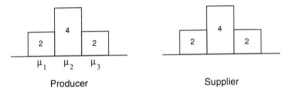

FIGURE 10.8   **Producer's and Supplier's distributions of sizes**

If we choose $n = 8$ days as the cycle time, the objective function z is

$$|16 - 4n_1 - 2n_2| + |32 - 2n_1 - 4n_2 - 2n_3| + |16 - 2n_2 - 4n_3|.$$

Ideally, one would desire $16 - 4n_1 - 2n_2 = 0$, $32 - 2n_1 - 4n_2 - 2n_3 = 0$, and $16 - 2n_2 - 4n_3 = 0$. Using $n_1 + n_2 + n_3 = 8$ and solving gives $n_1 = 0$, $n_2 = 8$, $n_3 = 0$. That is, the Supplier should run every day centered on target μ_2. Note that we had to specify the possible target values and specify the cycle time n.

Example 10.3

Suppose the Producer's and Supplier's distributions of diameters are those illustrated in Figure 10.9 and let μ_1, μ_2, μ_3 denote the candidate targets. Here $k = T = 3$ and $m = 1$.

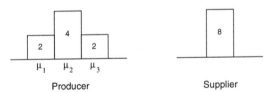

FIGURE 10.9   **Producer's and Supplier's distributions of sizes**

With $n = 8$ as the cycle time and allowable targets μ_1, μ_2, μ_3, the objective function becomes

$$|16 - 8n_1| + |32 - 8n_2| + |16 - 8n_3|.$$

Ideally, $16 - 8n_1 = 0$, $32 - 8n_2 = 0$, and $16 - 8n_3 = 0$. Solving these equations yields $n_1 = 2$, $n_2 = 4$, $n_3 = 2$. That is, in 8 days the Supplier runs twice at μ_1, four times at μ_2, and twice at μ_3.

With these examples in mind, we now turn to a more general description of this *target selection problem*, first in terms of multinomial models and later in terms of normal models for the size distributions. The target selection problem is formulated as an integer linear programming problem and then illustrated with examples.

10.7 Target Selection for Multinomial Distributions

Assume the Producer's piston diameter distribution spans k size classes of width 0.1 mm centered at class means $\mu_1, \mu_2, \ldots, \mu_k$ (e.g., $\mu_1 = 30.1$, $\mu_2 = 30.2, \ldots$) with proportions $p_1, p_2, \ldots, p_k$ of piston production in the respective classes, as illustrated in Figure 10.10. Suppose the Supplier has $m < k$ sleeve sizes with proportions $q_1, q_2, \ldots, q_m$ which are invariant with respect to location and that the average number of sleeves produced daily by the Supplier equals the Producer's average daily piston production.

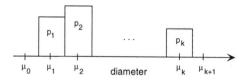

FIGURE 10.10   **Producer's piston diameter distribution**

Now assume the possible (allowable) targets for the Supplier's smallest (leftmost) size class (with proportion q_1) are at locations $\mu_0, \mu_1, \mu_2, \ldots,$ μ_{k-m+2} where $\mu_0 = \mu_1 - 0.1$. This arbitrary choice of targets allows production of sleeves at least one size below and one size above the extremes of the Producer's distribution. Let n_i be the number of times the Supplier runs at target μ_i during a production cycle, $0 \leq i \leq k - m + 2$. The problem is to find integers $n_0, n_1, \ldots, n_{k-m+2}$ to

$$\text{minimize} \quad \sum_{i=1}^{k} \mid np_i - (q_m n_{i-m+1} + \cdots + q_1 n_i) \mid + q_1 n_0 + q_m n_{k-m+2}$$

$$\text{subject to} \quad \sum_{i=0}^{k-m+2} n_i = n$$

$$n_i \geq 0 \text{ and integer.}$$

Alternatively, if we define $n_i = 0$ if $i < 0$ or $i > k - m + 2$ and $p_i = 0$ if $i < 1$ or $i > k$, the problem may be stated as

$$\text{minimize} \quad \sum_{i=0}^{k+1} \mid np_i - (q_m n_{i-m+1} + q_{m-1} n_{i-m+2} + \cdots + q_1 n_i) \mid$$

$$\text{subject to} \quad \sum_{i=0}^{k-m+2} n_i = n$$

$$n_i \geq 0 \text{ and integer.}$$

Note that the objective function now penalizes production outside the Producer's size classes (and gives equal weight to overproduction and underproduction of a size).

More generally, to allow targets to range from size l to size u, we want to find nonnegative integers $n_l, n_{l+1}, \ldots, n_u$ in order to

$$\text{minimize} \quad \sum_{i=l}^{u+m-1} \mid np_i - (q_m n_{i-m+1} + q_{m-1} n_{i-m+2} + \cdots + q_1 n_i) \mid$$

$$\text{subject to} \quad \sum_{i=l}^{u} n_i = n$$

$$n_i \geq 0 \text{ and integer}$$

where we define $n_i = 0$ if $i < l$ or $i > u$ and $p_i = 0$ if $i < 1$ or $i > k$.

Example 10.4

Suppose the Producer has $k = 35$ sizes and the Supplier has (for any fixed target) $m = 20$ sizes, so the allowable targets are from size 0 to size $35 - 20 + 2 = 17$. Then we want to find, for a specified n, nonnegative integers $n_0, n_1, \ldots, n_{17}$ (summing to n) in order to

$$\text{minimize} \quad \sum_{i=0}^{36} \mid np_i - (q_{20} n_{i-19} + \cdots + q_1 n_i) \mid,$$

where $n_i = 0$ if $i < 0$ or $i > 17$ and $p_i = 0$ if $i < 1$ or $i > 35$.

Example 10.5

To illustrate a solution of the integer linear program, we return to Example 10.1 where the Producer has seven sizes with $p_1 = p_7 = \frac{1}{16}$, $p_2 = p_6 = \frac{2}{16}$, $p_3 = p_5 = \frac{3}{16}$, $p_4 = \frac{4}{16}$. For the Supplier, let $q_1 = q_3 = \frac{4}{16}$, $q_2 = \frac{8}{16}$, corresponding to Case 4 of that example.

Using candidate target sizes of $0, 1, \ldots, 6$, we want to determine the number of times n_i to run at target i ($i = 0, 1, \ldots, 6$) during an n-day cycle in order to

$$\text{minimize} \quad \sum_{i=0}^{8} \mid np_i - (q_3 n_{i-2} + q_2 n_{i-1} + q_1 n_i) \mid,$$

where $n_i = 0$ if $i < 0$ or $i > 6$ and $p_0 = p_8 = 0$. For a 16-day cycle ($n = 16$), the integer linear program is given by

$$\text{minimize} \quad \sum_{i=0}^{8} \left| 16 p_i - \left(\frac{1}{4} n_{i-2} + \frac{1}{2} n_{i-1} + \frac{1}{4} n_i \right) \right|$$

$$\text{subject to} \quad \sum_{i=0}^{6} n_i = 16$$

$$n_i \geq 0 \text{ and integer.}$$

For convenience, multiply the objective function by the constant 4 (which does not change the optimal values of the n_i), leading to the problem of minimizing

$$\begin{aligned}
n_0 + n_6 &+ |4 - (2n_0 + n_1)| + |8 - (n_0 + 2n_1 + n_2)| \\
&+ |12 - (n_1 + 2n_2 + n_3)| + |16 - (n_2 + 2n_3 + n_4)| \\
&+ |12 - (n_3 + 2n_4 + n_5)| + |8 - (n_4 + 2n_5 + n_6)| \\
&+ |4 - (n_5 + 2n_6)|
\end{aligned}$$

subject to $\sum_{i=0}^{6} n_i = 16$, $n_i \geq 0$ and integer. Eliminating the absolute value functions gives the following integer linear programming problem.

$$\text{minimize} \quad n_0 + z_1 + z_2 + z_3 + z_4 + z_5 + z_6 + n_6$$
$$\text{subject to}$$

$$-z_1 \leq 4 - 2n_0 - n_1 \leq z_1$$
$$-z_2 \leq 8 - n_0 - 2n_1 - n_2 \leq z_2$$
$$-z_3 \leq 12 - n_1 - 2n_2 - n_3 \leq z_3$$
$$-z_4 \leq 16 - n_2 - 2n_3 - n_4 \leq z_4$$
$$-z_5 \leq 12 - n_3 - 2n_4 - n_5 \leq z_5$$
$$-z_6 \leq 8 - n_4 - 2n_5 - n_6 \leq z_6$$
$$-z_7 \leq 4 - n_5 - 2n_6 \leq z_7$$
$$n_0 + n_1 + n_2 + n_3 + n_4 + n_5 + n_6 = 16$$
$$n_i \geq 0 \text{ and integer.}$$

An optimal solution to this problem is $n_1 = 4$, $n_3 = 8$, $n_5 = 4$, and $n_i = 0$ otherwise. The objective function value is also zero, meaning that the Supplier is able to provide exactly the required number of sleeves. The solution obtained here corresponds to that presented in Section 10.6.1, Case 4, and was obtained using LINDO [5].

10.8   The Supplier's Cost Function

The analysis of the previous section requires that both the target sizes and the cycle time n be specified. Only then are the number of runs at each target determined to minimize differences between sleeve demand and sleeve production. How should candidate target sizes and the cycle time be specified? A longer cycle time generally means the Supplier must maintain a larger daily inventory, depending on the target-shifting schedule, and changing the number of targets could alter the optimum target-shifting pattern. The Supplier's ultimate goal is to minimize the cost of producing a sleeve diameter distribution which approximates the Producer's needs. To quantify these issues, let

- c_1 = cost of overproduction of a size including unusable sizes (\$/sleeve),

- c_2 = cost of underproduction of a size (\$/sleeve),

- c_3 = inventory carrying cost (\$/sleeve/day),

- c_4 = cost of shifting the target size (\$/shift),

- N = average daily sleeve production (sleeves).

Using the notation of the previous section with i denoting the target size class, define $\Delta_i = -1$ if $np_i - (q_m n_{i-m+1} + \cdots + q_1 n_i) < 0$, and 0 otherwise. That is, $\Delta_i = -1$ if size i is overproduced during a cycle and $\Delta_i = 0$ if size i is not overproduced. Let V denote the average daily inventory over the production cycle and S the number of target shifts made during a cycle. The Supplier's production cost per cycle is then

$$\sum_i(-\Delta_i c_1 + (1+\Delta_i)c_2)N[np_i - (q_m n_{i-m+1} + \cdots + q_1 n_i)] + nVc_3 + Sc_4.$$

Note that both V and S depend on the location of targets and the n_i. Determining a set of targets, a production cycle time n, and a schedule of targets shifts to minimize cost might be approached by a two-step iterative procedure. First, select a set of targets and a production cycle time n, and then calculate a set of run times n_i to minimize

$$\sum_i(-\Delta_i c_1 + (1 + \Delta_i)c_2)[np_i - (q_m n_{i-m+1} + \cdots + q_1 n_i)]$$

subject to $\sum_i n_i = n$. Second, for those run numbers n_i, determine a target-shifting order that minimizes $nVc_3 + Sc_4$. Examining individual components of cost may suggest a direction for improvement. For example, if inventory costs are relatively large, then a shorter cycle time n might produce a lower-cost solution. However, this is essentially a trial-and-error process, and there is no guarantee that an absolute or even a local minimum cost solution will be found.

10.9   Target Selection Using Normal Distributions

In real world problems, both the Producer's and the Supplier's diameter distributions are roughly symmetric and reasonably normal in shape.

Suppose that the Producer's piston diameter distribution can be modeled by a density $f(x)$ that is normal

$$f(x) \sim N(\mu, \sigma^2)$$

with k diameter categories defined by the partition $x_0, x_1, \ldots, x_k$. Also assume that the Supplier's sleeve diameter distribution for target μ_t is approximately normal

$$g_{\mu_t}(x) \sim N(\mu_t, \sigma_s^2)$$

with $\sigma_s < \sigma$ and with σ_s independent of the location μ_t.

Given a set of m locations (targets) $\mu_1 < \mu_2 < \cdots < \mu_m$ for the mean of the Supplier's distribution and a cycle time of n days, we want to determine integers $n_1, n_2, \ldots, n_m$ to

$$\text{minimize} \quad \sum_{i=1}^{k} \left| n \int_{x_{i-1}}^{x_i} f(x)\, dx - \sum_{j=1}^{m} n_j \int_{x_{i-1}}^{x_i} g_{\mu_j}(t)\, dt \right|$$

$$\text{subject to} \quad \sum_{j=1}^{m} n_j = n$$

$$n_j \geq 0 \text{ and integer.}$$

Ideally, we would like to achieve the conditions

$$n \int_{x_{i-1}}^{x_i} f(x)\, dx = \sum_{j=1}^{m} n_j \int_{x_{i-1}}^{x_i} g_{\mu_j}(t)\, dt, \quad \text{for } i = 1, \ldots, k,$$

where $\sum_{j=1}^{m} n_j = n$, with the n_j nonnegative integers.

Letting $\Phi(x)$ denote the standard normal cumulative distribution function [2], the objective function above can be expressed as

$$\sum_{i=1}^{k} \left| n \left[\Phi\left(\frac{x_i - \mu}{\sigma}\right) - \Phi\left(\frac{x_{i-1} - \mu}{\sigma}\right) \right] \right.$$

$$\left. - \sum_{j=1}^{m} n_j \left[\Phi\left(\frac{x_i - \mu_j}{\sigma_s}\right) - \Phi\left(\frac{x_{i-1} - \mu_j}{\sigma_s}\right) \right] \right|. \quad (10.8)$$

Ideally, for $i = 1, \ldots, k$ we want

$$n \left[\Phi\left(\frac{x_i - \mu}{\sigma}\right) - \Phi\left(\frac{x_{i-1} - \mu}{\sigma}\right) \right]$$

$$= \sum_{j=1}^{m} n_j \left[\Phi\left(\frac{x_i - \mu_j}{\sigma_s}\right) - \Phi\left(\frac{x_{i-1} - \mu_j}{\sigma_s}\right) \right].$$

One way to avoid excessive sleeve production outside the interval $[x_0, x_k]$ is to require $x_0 + 2\sigma_s < \mu_j < x_k - 2\sigma_s$ for all j; however, this will not be possible if $\sigma_s > (x_k - x_0)/4$. Alternatively, we can modify the objective function to minimize

$$
\sum_{i=1}^{k} \left| n \int_{x_{i-1}}^{x_i} f(x)\,dx - \sum_{j=1}^{m} n_j \int_{x_{i-1}}^{x_i} g_{\mu_j}(t)\,dt \right|
$$
$$
+ \sum_{j=1}^{m} n_j \int_{-\infty}^{x_0} g_{\mu_j}(t)\,dt + \sum_{j=1}^{m} n_j \int_{x_k}^{\infty} g_{\mu_j}(t)\,dt,
$$

or in terms of $\Phi(x)$

$$
\left(\sum_{i=1}^{k} \left| n \left[\Phi\left(\frac{x_i - \mu}{\sigma}\right) - \Phi\left(\frac{x_{i-1} - \mu}{\sigma}\right) \right] \right. \right.
$$
$$
\left. \left. - \sum_{j=1}^{m} n_j \left[\Phi\left(\frac{x_i - \mu_j}{\sigma_s}\right) - \Phi\left(\frac{x_{i-1} - \mu_j}{\sigma_s}\right) \right] \right| \right)
$$
$$
+ \sum_{j=1}^{m} n_j \left[\Phi\left(\frac{x_0 - \mu_j}{\sigma_s}\right) - \Phi\left(\frac{x_k - \mu_j}{\sigma_s}\right) \right] + n.
$$

A more general formulation is to remove the assumption that each target is uniquely specified and then find m and $\mu_1, \mu_2, \ldots, \mu_m$ so that the following function is minimized:

$$
\sum_{i=1}^{k} \left| m \int_{x_{i-1}}^{x_i} f(x)\,dx - \sum_{j=1}^{m} \int_{x_{i-1}}^{x_i} g_{\mu_j}(t)\,dt \right|
$$
$$
+ \sum_{j=1}^{m} \int_{-\infty}^{x_0} g_{\mu_j}(t)\,dt + \sum_{j=1}^{m} \int_{x_k}^{\infty} g_{\mu_j}(t)\,dt.
$$

Here m is the cycle time and μ_i is the target on day i ($n_i = 1$). We have replaced each "unknown" n_i by an "unknown" μ_i.

Example 10.6

Assume the Producer's size distribution is $N(16, 1)$, and consider size categories 13–14, 14–15, $\ldots$, 18–19. Here $k = 6$ with $x_0 = 13$, $x_1 = 14$, $\ldots$, $x_6 = 19$. Further suppose the Supplier's distribution is $N(\mu_t, 0.0625)$. Arbitrarily set $\mu_1 = 14.5$, $\mu_2 = 15.5$, $\mu_3 = 16.5$, $\mu_4 = 17.5$. Substituting these parameters into equation (10.8), the problem now involves determining nonnegative integers $n_1, \ldots, n_4$ in order to minimize

$$\left| n[\Phi(-2) - \Phi(-3)] - \{n_1[\Phi(-2) - \Phi(-6)] + n_2[\Phi(-6) - \Phi(-10)] \right.$$
$$\left. + n_3[\Phi(-10) - \Phi(-14)] + n_4[\Phi(-14) - \Phi(-18)]\} \right| + \cdots$$
$$+ \left| n[\Phi(3) - \Phi(2)] - \{n_1[\Phi(18) - \Phi(14)] + n_2[\Phi(14) - \Phi(10)] \right.$$
$$\left. + n_3[\Phi(10) - \Phi(6)] + n_4[\Phi(6) - \Phi(2)]\} \right|.$$

Since $n = n_1 + n_2 + n_3 + n_4$, this leads to the minimization of

$$| - 0.0013n_1 + 0.0215n_2 + 0.0215n_3 + 0.0215n_4|$$
$$+ | - 0.8186n_1 + 0.1131n_2 + 0.1359n_3 + 0.1359n_4|$$
$$+ |0.3186n_1 - 0.6132n_2 + 0.3185n_3 + 0.3413n_4|$$
$$+ |0.3413n_1 + 0.3186n_2 - 0.6132n_3 + 0.3185n_4|$$
$$+ |0.1360n_1 + 0.1360n_2 + 0.1133n_3 - 0.8185n_4|$$
$$+ |0.0214n_1 + 0.0214n_2 + 0.0214n_3 - 0.0013n_4|.$$

For a selected value of n, this can be solved as an integer linear program, using an optimization package such as LINDO [5].

Example 10.7

We reconsider the problem of Example 10.6, in which the producer's distribution of sizes is $N(16, 1)$, with six size categories defined by $x_0 = 13, x_1 = 14, \ldots, x_6 = 19$. The more general formulation which does not specify the number of targets or their locations is to find m and $\mu_1, \mu_2, \ldots, \mu_m$ to minimize

$$\sum_{i=1}^{k} \left| m \int_{x_{i-1}}^{x_i} f(x)\, dx - \sum_{j=1}^{m} \int_{x_{i-1}}^{x_i} g_{\mu_j}(t)\, dt \right|$$

or in this case

$$\sum_{i=1}^{6} \left| m \left[\Phi\left(\frac{x_i - 16}{1}\right) - \Phi\left(\frac{x_{i-1} - 16}{1}\right) \right] \right.$$
$$\left. - \sum_{j=1}^{m} \left[\Phi\left(\frac{x_i - \mu_j}{0.25}\right) - \Phi\left(\frac{x_{i-1} - \mu_j}{0.25}\right) \right] \right|.$$

This approach seems less promising because of the intractability of $\Phi(z)$. A possible remedy is to approximate Φ by a more tractable cumulative distribution function such as the logistic one.

10.10 Conclusion

This chapter has considered the problem of how much inventory a manufacturing facility (Producer) should carry in stock, taking into account the stochastic nature of the production process. The multinomial distribution provides a model for describing the size distribution of the product. It also provides a basis for comparing alternative reordering policies.

In order to fill the daily orders for various sizes placed by the manufacturing facility, the Supplier has various strategies available, arising from the stochastic nature of the Supplier's production process. Namely, the Supplier need to choose which target sizes to produce (the target setting problem) in order to minimize the total cost associated with production scheduling to meet or approximate the Producer's size distribution. Approaches to solving this problem based on multinomial and on normal models have been outlined. If the cycle time n and the target classes or locations $\mu_1, \mu_2, \ldots, \mu_m$ are specified, then we can find the numbers of runs n_i at each target μ_i to minimize "mismatches" using an integer linear programming formulation. However, we have not solved the more general case of finding n, the μ_i, and the n_i to minimize the Supplier's cost of production scheduling. Work on that challenging case remains an open modeling problem.

10.11   Exercises and Projects

1. Assume that the piston Producer's manufacturing process can be modeled by the multinomial distribution of Section 10.2 and that the total number of pistons produced each day is a fixed number n.

    a. Show that the marginal distribution of $X_i(t)$ is a binomial distribution with parameters n and p_i, for $i = 1, \ldots, k$. Find the mean and variance of $X_i(t)$.

    b. It can be shown that the covariance of $X_i(t)$ and $X_j(t)$ is $-np_i p_j$ for $1 \leq i, j \leq k$ and $i \neq j$. Explain why one would expect this covariance to be nonzero and negative-valued.

    c. Consider a simple case in which the Producer makes $n = 10$ pistons daily, whose diameters fall into three distinct size classes ($k = 3$) with probabilities $p_1 = \frac{1}{5}$, $p_2 = \frac{3}{5}$, $p_3 = \frac{1}{5}$. Find $P(2, 5, 3)$. Find the mean and variance of $X_2(t)$, and calculate $P(X_2(t) = 4)$. Find the covariance of $X_1(t)$ and $X_3(t)$.

2. Consider a fourth ordering policy, Policy 4, which is to order for each size i the expected daily need of sleeves plus any deviations of the actual inventory of size i sleeves from the safety stock level. For simplicity, assume the p_i are known and that the safety stock level is the

same for each day; i.e., $s_i(t) = s_i$ for all t. Using the notation of Section 10.4, Policy 4 can be stated as follows:

$$r_i(t + 4) = E(X_i(t)) + s_i - y_i(t), \text{ for } t \geq 1,$$

with the initial conditions $r_i(1) = r_i(2) = r_i(3) = r_i(4) = np_i$ and $y_i(1) = s_i$. Show that for this policy, $E(Y_i(t)) = s_i$ and $E(R_i(t)) = np_i$ for $t \geq 1$. (**Hint:** note that $E(X_i(t)) = np_i$ and then use mathematical induction.)

3. Refer to Example 10.1 of Section 10.6.1 and consider Schedule 2 for Case 2. This exercise is to verify that the minimum total number of sleeves needed at the start of any day of the schedule (to avoid stockouts) is 106 sleeves. Construct a table similar to the one shown in Table 10.1, and use it to compute the required minimum inventory.

4. This exercise is similar to Exercise 10.3 above. We again refer to Example 10.1 of Section 10.6.1 and consider the two 8-day schedules given for Case 3. The objective of this exercise is to verify that daily inventories of 40 and 53 are required, respectively, under these two schedules.

a. The first 8-day schedule is

Day	1	2	3	4	5	6	7	8
Target	c	f	d	b	c	e	d	a

Under this schedule, eight sleeves of each of two sizes will be produced daily.

Day	1	2	3	4	5	6	7	8
Sizes	3, 4	6, 7	4, 5	2, 3	3, 4	5, 6	4, 5	1, 2

Construct a table analogous to Table 10.1, and use it to verify that the minimum daily inventory needed to avoid stockouts is 40 sleeves.

b. The second 8-day schedule is

Day	1	2	3	4	5	6	7	8
Target	c	c	a	b	d	d	e	f

Working as in part (a) above, verify that the total daily inventory needed to exactly meet demand is 53 sleeves on any given day.

5. Formulate the integer linear program for Example 10.1, Case 5, using $n = 10$ and $n = 16$. Then solve using an integer linear programming package. Does the optimal solution meet the required amounts exactly?

6. Verify that the optimal solution $n_1, \ldots, n_6$ given in Example 10.5 is in fact optimal: it satisfies all the constraints of the stated integer linear programming program and minimizes the objective function z. (You will need to compute the associated values of the z_i and z variables.)

7. (Project) Formulate the integer linear program for Example 10.6 using normal distributions and solve using an integer linear programming package. Investigate for several values of n.

10.12   References

[1] V. Chvátal, *Linear Programming*, Freeman, New York, 1983.
[2] W. Feller, *An Introduction to Probability Theory and Its Applications*, Volume I, 3rd ed., Wiley, New York, 1968.
[3] S. M. Ross, *Applied Probability Models with Optimization Applications*, Holden-Day, San Francisco, 1970.
[4] S. M. Ross, *Introduction to Probability Models*, 5th ed., Academic Press, Boston, 1993.
[5] L. Schrage, *Optimization Modeling with LINDO*, Duxbury, Pacific Grove, CA, 1997; also see <http://www.lindo.com/>

Chapter 11

Modeling Nonlinear Phenomena by Dynamical Systems

Jinqiao Duan

Prerequisites: differential equations, dynamical systems, linear algebra

11.1 Introduction

Nonlinear differential equations are usually mathematical models of complex physical or engineering problems representing the change of processes in time. Dynamical systems theory describes the behavior of solutions of differential equations and is particularly useful in understanding the solutions of nonlinear differential equations. Dynamical systems theory provides a conceptual framework for investigating mathematical models that describe such nonlinear phenomena. In particular, this framework enables us to quantify nonlinear phenomena, and to devise strategies for controlling or exploiting such nonlinear phenomena. Ideas used in the study of dynamical systems range from elementary graphical and analytical mathematics to sophisticated concepts from other branches of modern mathematics, as differential and algebraic topology.

Dynamical systems theory has been used by applied mathematicians, scientists, and engineers in the observation and interpretation of dynamical patterns in laboratory and natural systems. Applying the ideas and methods of dynamical systems theory to physical and engineering problems has been a thoroughly interdisciplinary effort.

In this chapter we will use a well-known physical example, the simple pendulum, to illustrate some dynamical systems ideas and methods, and to demonstrate (together with a project assigned to students) how they can be used to understand nonlinear and chaotic motions of the simple pendulum under external time-periodic forcing. Students who are interested in the dynamics of a simple nonlinear system can easily be intimidated by the extensive reading suggested by the literature of the subject. This chapter shows it is possible to determine important behavior with a modest background, especially if the reader will accept some (in our case two) easily stated theorems without proof.

11.2   Simple Pendulum

In the absence of damping and external forcing, the motion of a simple pendulum is governed by the differential equation

$$m\frac{d^2\theta}{dt^2} = -m\frac{g}{L}\sin\theta, \tag{11.1}$$

where θ is the angle from the downward vertical, m is the mass of the pendulum, g is the acceleration due to gravity, and L is the length of the pendulum. This simple model is illustrated in Figure 11.1, in which the entire mass of the pendulum is assumed to be concentrated at its end. Equation (11.1) is easily derived from Newton's second law; see for example [2]. By redefining the dimensionless time to be $\sqrt{g/L}t$, this equation becomes

$$\ddot{\theta} + \sin\theta = 0. \tag{11.2}$$

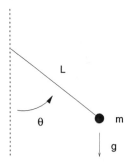

FIGURE 11.1   **A simple pendulum**

To understand the motions of the simple pendulum, it is instructive to introduce a number of dynamical systems concepts. We need two variables to determine the motion of the pendulum: angular position θ

and angular velocity $\dot{\theta}$. The *state* is thus a point in the plane, whose coordinates are the angle θ and angular velocity $\dot{\theta}$. Hence we rewrite the simple pendulum equation (11.2) as

$$\dot{\theta} = v \tag{11.3}$$
$$\dot{v} = -\sin\theta, \tag{11.4}$$

where v is the (dimensionless) angular velocity.

If we plot solutions of this system of differential equations in the θ-v plane, we can "see" the motions of the simple pendulum. The θ-v plane is the so-called *phase plane* (a concept first introduced by Poincaré about a hundred years ago); a more appropriate name should be *state plane* or *state space*. Graphs of solutions in the phase plane are called *orbits* or *trajectories* for the simple pendulum. As the pendulum swings back and forth, the state moves along an orbit in the phase plane. The phase plane together with some representative orbits is called the *phase portrait* of the simple pendulum.

Note that the system (11.3)–(11.4) is 2π-periodic in angle θ. Thus we only need to consider the system for $\theta \in [-\pi, \pi]$. Also note that $(0,0)$ and $(\pm\pi, 0)$ are equilibrium solutions or *equilibrium states*, since at these states the pendulum has zero velocity and zero acceleration and remains in these positions. The point $(0,0)$ is the rest position of the pendulum. The two points $(\pm\pi, 0)$ on the phase plane are actually for the same position of the pendulum (upside down), and they should be regarded as one point, since physically they are the same. We can also check that the Jacobian matrix of the right-hand sides of (11.3)–(11.4) at the rest state $(0,0)$ is

$$A = \begin{bmatrix} 0 & 1 \\ -1 & 0 \end{bmatrix}$$

and the Jacobian matrix at the equilibrium state $(\pm\pi, 0)$ is

$$B = \begin{bmatrix} 0 & 1 \\ 1 & 0 \end{bmatrix}.$$

The eigenvalues of these Jacobian matrices, A and B, dictate the dynamical behavior of the pendulum system (11.3)–(11.4) near $(0,0)$ and $(\pm\pi, 0)$, respectively.

Since the two eigenvalues of A are purely imaginary, the state $(0,0)$ is termed a *center*. The matrix B has one positive and one negative eigenvalue, and so $(\pm\pi, 0)$ is called a *saddle point*. We will see that these equilibrium states are important in the phase portrait of the pendulum.

In order to generate phase portraits of nonlinear dynamical systems, computer software packages are generally needed. However, for the simple pendulum system (11.3)–(11.4) it is easy to generate the phase portrait directly. We can check that

$$H = H(\theta, v) = \frac{1}{2}v^2 - \cos\theta$$

is a constant of motion, since

$$\frac{d}{dt}H = v\dot{v} + \dot{\theta}\sin\theta = \dot{\theta}(\ddot{\theta} + \sin\theta) = 0.$$

This says that $H(\theta, v)$ is constant along any orbit in the phase plane. Therefore, all orbits should lie on the level curves $H = c$, with c being an arbitrary constant.

To generate the phase portrait for the simple pendulum, we plot the level curves $H = \frac{1}{2}v^2 - \cos\theta = c$ in the θ-v phase plane for various constants c; see Figure 11.2. Due to periodicity, the phase portrait need only be generated for $\theta \in [-\pi, \pi]$.

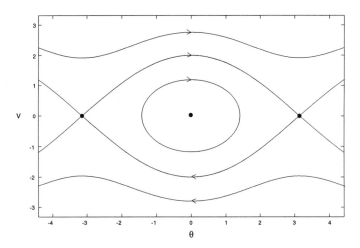

FIGURE 11.2   **Phase portrait for a simple pendulum**

A closed loop in the phase portrait in Figure 11.2 is called a *closed orbit* and it represents the oscillatory or periodic motion of the pendulum (ordinary pendulum swing). The upper and lower wavy orbits represent the motion in which the pendulum started with a high enough speed for it to continue rotating forever, so that θ increases indefinitely (where $\dot{\theta}$ is positive) on the upper orbit, and decreases on the lower orbit.

We conclude that the motions for a simple pendulum without damping and external forcing are predictable and hence are not complicated (not "chaotic").

From the analysis above, we see that geometrical structures (phase portraits) can be used to describe or understand nonlinear phenomena (nonlinear motions of a simple pendulum). Therefore there is a direct relationship between geometrical structures and nonlinear phenomena in the study of dynamical systems.

The two orbits connecting the saddle point $(\pm\pi, 0)$ to itself are called *homoclinic orbits* as they asymptotically approach the same equilibrium point $(\pm\pi, 0)$ as $t \to +\infty$ and as $t \to -\infty$. Both homoclinic orbits have $H = \frac{1}{2}\dot{\theta}^2 - \cos\theta = 1$ and we can write down an explicit solution for $\theta(t)$, $v(t)$ for the homoclinic orbit:

$$\theta(t) = \pm 2\arctan(\sinh t)$$
$$v(t) = \pm 2\operatorname{sech} t.$$

We also see that they separate different motions of the pendulum. The set of orbits emanating from a saddle point form an *unstable manifold* and the set of orbits running into a saddle point form a *stable manifold* of the saddle point. The stable manifold is invariant in the sense that any orbit starting on it will stay inside it due to the uniqueness of the solutions for the system (11.3)–(11.4). Likewise, the unstable manifold is also invariant. For a simple pendulum without damping and external forcing, the stable and unstable manifolds of the saddle point coincide (to form the homoclinic orbits); in this case, there is no chaotic motion as all motions are predictable — either oscillations or rotations.

11.3 *Periodically Forced Pendulum*

However, for a periodically forced simple pendulum, we have a different story. The periodically forced simple pendulum model is given by

$$\ddot{\theta} + \sin\theta = \epsilon\sin(\omega t)\sin\theta$$

or

$$\dot{\theta} = v \tag{11.5}$$
$$\dot{v} = -\sin\theta + \epsilon\sin(\omega t)\sin\theta, \tag{11.6}$$

where the positive constants ϵ and ω are the forcing amplitude and frequency. The external forcing is periodic with period $T = 2\pi/\omega$. This is a nonautonomous system: i.e., the right-hand side of the differential equation explicitly depends on time. If we plot solutions in the θ-v plane, they will generally intersect with each other and so it is difficult to understand pendulum motions in the phase plane.

Nevertheless, we can use Poincaré's idea to "look" at a solution or record the solutions on a copy of the θ-v plane (called the *Poincaré section*) at successive periods $T, 2T, 3T, \ldots$. In this way, continuous orbits are represented by sets of discrete points on the Poincaré section. Since solutions are defined at any point on the θ-v plane, the above way of recording the solutions defines a map P (called the *Poincaré map*) on the Poincaré section. The Poincaré map sends each point (θ_0, v_0) on the Poincaré section to another point $P(\theta_0, v_0)$, which is the location of the solution starting at (θ_0, v_0) at time $T = 2\pi/\omega$. The evolution of orbits is described by the iteration of the Poincaré map.

It can be shown that P is well-defined, invertible, and both P and P^{-1} are differentiable. See Holmes [4], Guckenheimer and Holmes [3], or Wiggins [9] for further details regarding Poincaré maps. We can study the dynamics of maps such as the Poincaré map P. For example, we can define orbits, equilibrium or fixed points, stable and unstable manifolds (which are continuous curves), and homoclinic orbits (also continuous curves) to saddle type fixed points, for the Poincaré map P. The stable and unstable manifolds are invariant. The map P captures the dynamics of the periodically forced simple pendulum.

The project for this chapter is to understand the irregular dynamical swinging of a periodically forced pendulum. We will verify that the periodically forced pendulum has chaotic motions (irregular or chaotic swinging) by studying the chaotic dynamical behavior of the associated Poincaré map P defined above. That is, the pendulum motions in this case may be very different even when they start with very close or nearby initial angular position and initial angular velocity. This is called *sensitive dependence on initial conditions*. This would then imply that the motions of the periodically forced pendulum are unpredictable. Such motions are called *chaotic motions* [1, 3, 9]. We follow the presentation in Holmes [4].

To begin we state the *Smale-Birkhoff Homoclinic Theorem*.

THEOREM 11.1

Let P be an invertible map on the plane and assume that P and P^{-1} are both differentiable. We also assume that P has a saddle type fixed point (x_0, y_0): $P(x_0, y_0) = (x_0, y_0)$, and that the stable and unstable manifolds of the saddle point intersect at a point (x^*, y^*) at a nonzero *angle: i.e., the two manifolds (curves) intersect but are not tangent at* (x^*, y^*). *Then the dynamics of P sensitively depend on some initial conditions.*

What we need to do is to apply this theorem to the Poincaré map P for the periodically forced pendulum. That is, we need to check that the Poincaré map P satisfies the assumptions in the Smale-Birkhoff Homoclinic Theorem.

To this end, consider the following general periodically forced system:

$$\dot{x} = f_1(x, y) + \epsilon g_1(x, y, t)$$
$$\dot{y} = f_2(x, y) + \epsilon g_2(x, y, t),$$

where g_1, g_2 are periodic in t with period T and $\epsilon > 0$ is small. Therefore, a Poincaré map can be defined for this system. It is assumed here that the unperturbed system ($\epsilon = 0$) has a saddle point with a homoclinic orbit $(\gamma_1(t), \gamma_2(t))$ approaching it asymptotically in both time directions. The associated Poincaré map will also have a nearby saddle point for ϵ sufficiently small. The stable and unstable manifolds of the perturbed saddle point for the Poincaré map will generally not coincide; that is, the homoclinic orbit for the case $\epsilon = 0$ will break up for ϵ sufficiently small. To use the Smale-Birkhoff Homoclinic Theorem, we still need to show that the stable and unstable manifolds intersect at a nonzero angle at some point.

The following *Melnikov function* provides an estimate of the distance between the stable and unstable manifolds at time t_0:

$$M(t_0) = \int_{-\infty}^{\infty} e^{- \int_{t_0}^{t} [f_{1x} + f_{2y}](\gamma_1(s), \gamma_2(s)) ds} \cdot [f_1(\gamma_1(t), \gamma_2(t))$$
$$g_2(\gamma_1(t), \gamma_2(t), t + t_0) - f_2(\gamma_1(t), \gamma_2(t)) g_1(\gamma_1(t), \gamma_2(t), t + t_0)] \, dt.$$

The following *Melnikov theorem* gives information about the geometrical structure of the stable and unstable manifolds for the the Poincaré map P associated with the periodically forced pendulum (11.5)–(11.6).

THEOREM 11.2

If the Melnikov function $M(t_0)$ *has a simple zero, then for ϵ sufficiently small, the stable and unstable manifolds of the perturbed saddle point for the Poincaré map intersect at a nonzero angle at some point.*

Note that the stable and unstable manifolds are invariant, and their intersection point has to be on both manifolds at each iteration of the

Poincaré map P. Thus, when the stable and unstable manifolds intersect at a nonzero angle at *one* point, they will intersect at a countably infinite number of points, and thus form a *homoclinic tangle*. In this case the phase portrait of P looks like the sketch in Figure 11.3.

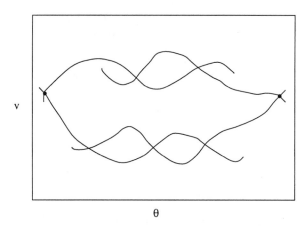

FIGURE 11.3   **Stable and unstable manifolds for map** P

11.4   Exercises and Projects

1. Find the eigenvalues of the Jacobian matrices A and B given in Section 11.2.

2. A pendulum with damping is modeled by

$$\ddot{\theta} + \alpha\dot{\theta} + \sin\theta = 0$$

or

$$\dot{\theta} = v$$
$$\dot{v} = -\alpha v - \sin\theta,$$

where $\alpha > 0$ is the damping constant.

    a. Find the Jacobian matrix at the equilibrium point $(0,0)$ and find the eigenvalues of this matrix.

    b. Verify that $(0,0)$ is a *stable node* [2] for $\alpha \geq 2$ (strong damping) by showing the eigenvalues are both negative.

    c. Verify that $(0,0)$ is a *stable focus* [2] for $\alpha < 2$ (weak damping) by showing the eigenvalues are complex but with negative real parts.

d. Verify that the equilibrium point $(\pi, 0)$ is a saddle point.

3. Generate the phase portrait for the damped pendulum system modeled in Exercise 11.2 for various damping constants $\alpha > 0$, using the ode command in MATLAB [6].

A pendulum that moves with damping or friction eventually comes to a halt, which in the phase plane means the orbit approaches the equilibrium point $(0,0)$, as should be seen in your phase portrait. The equilibrium point $(0,0)$ is called an *attractor*, as it attracts (nearby) orbits. This attractor is what the behavior of the pendulum settles down to, or is attracted to. It will also be seen that the stable and unstable manifolds of the saddle point $(\pm\pi, 0)$ do not coincide, and there are no homoclinic orbits. Therefore, for a damped simple pendulum, the stable and unstable manifolds of the saddle do not intersect, and pendulum motions are clearly predictable (no chaotic motions).

4. (Project) The objective for the project is to show that the periodically forced pendulum (11.5)–(11.6) has chaotic motions (chaotic swinging) via the Poincaré map P. The map P captures the dynamics of the periodically forced pendulum.

Our goal is to show that the Poincaré map P for the periodically forced simple pendulum satisfies the assumptions of the Smale-Birkhoff Homoclinic Theorem, for ϵ sufficiently small. Then we can conclude that the dynamics of the Poincaré map P have sensitive dependence on some initial conditions and hence P has unpredictable (chaotic) orbits. Since the Poincaré map P captures the dynamics of the periodically forced simple pendulum, this implies that the periodically forced pendulum has chaotic motions.

a. For the periodically forced pendulum (11.5)–(11.6), verify that the Poincaré map P has saddle type fixed points near $(\pm\pi, 0)$ for ϵ sufficiently small; calculate the Melnikov function $M(t_0)$ and verify that it has simple zeros. Therefore the stable and unstable manifolds for P intersect at a nonzero angle at some point, by the Melnikov theorem. (**Hint:** show that $M(t_0) = \text{constant} \cdot \cos(\omega t_0)$.)

b. Conclude that the periodically forced simple pendulum has chaotic motions, using the Smale-Birkhoff Homoclinic Theorem. Explain this result physically.

c. What does "sensitive dependence on initial conditions" mean in this context?

d. Generate the phase portrait of the Poincaré map P, using the ode command in MATLAB [6].

We remark that for the periodically forced pendulum, the stable and unstable manifolds of a saddle point intersect to form a homoclinic tangle. The breakup of the manifolds gives a mechanism for chaotic orbits or chaotic pendulum motions.

As you complete this project, you can see that dynamical systems ideas such as the geometrical relation between stable and unstable manifolds are useful in understanding the dynamical behavior of physical and engineering systems.

11.5   References

[1] K. Alligood, T. Sauer, and J. A. Yorke, *Chaos: An Introduction to Dynamical Systems*, Springer-Verlag, New York, 1997.

[2] P. Blanchard, R. L. Devaney, and G. R. Hall, *Differential Equations*, PWS Publishing, Boston, 1995.

[3] J. Guckenheimer and P. Holmes, *Nonlinear Oscillations, Dynamical Systems, and Bifurcations of Vector Fields*, Springer-Verlag, New York, 1983 (4th printing, 1993).

[4] P. Holmes, "Poincaré, celestial mechanics, dynamical systems theory and 'chaos'," *Physics Reports* **193**, 137–163 (1990).

[5] U. Kirchgraber and D. Stoffer, "Chaotic behavior in simple dynamical systems," *SIAM Review* **32**, 424–452 (1990).

[6] The MathWorks, Inc., *Using MATLAB Version 5*, The MathWorks, Inc., Natick, MA, 1997. <http://www.mathworks.com/>

[7] T. S. Parker and L. O. Chua, *Practical Numerical Algorithms for Chaotic Systems*, Springer-Verlag, New York, 1989.

[8] S. H. Strogatz, *Nonlinear Dynamics and Chaos — with Applications to Physics, Biology, Chemistry, and Engineering*, Addison-Wesley, Reading, MA, 1994.

[9] S. Wiggins, *Introduction to Applied Nonlinear Dynamical Systems and Chaos*, Springer-Verlag, New York, 1990.

Chapter 12

Modulated Poisson Process Models for Bursty Traffic Behavior

Robert E. Fennell
Christian J. Wypasek
James M. Westall

Prerequisites: probability, queueing theory, stochastic processes

12.1 Introduction

In queueing theory, the Poisson process has been a fundamental building block, primarily because of the ease of analysis that it provides. In practice, however, the assumptions of stationarity and independent increments may not hold. Computer network traffic in particular is often "bursty" in nature. That is, there are intervals of higher than normal activity mixed among intervals of lower than normal activity.

This chapter emphasizes the basic construction of *modulated* counting process models along with simulation and data analysis exercises. These models are useful, for example, in describing workstation utilization behavior within a computer network and can be applied to more general traffic analysis. Our models are based upon the premise that a workstation or traffic source can be viewed as existing in a small number of states representing characteristic levels of activity. In each state the termination of idle or busy periods is approximated by a Poisson process with a rate that is a characteristic of the state. In addition to intuitive motivation, our models can be generalized to address time-correlated behavior and provide a mechanism for Kalman-type filtering to estimate future activity. Our presentation is based upon results obtained in [14].

The chapter exercises provide a guide to the simulation of various types of stochastic processes. Simulation exercises also demonstrate that queueing delays at a GI/M/1 queue will be underestimated if it is assumed that the arrival process is Poisson when the true distribution of interarrival times has a heavier tail than the exponential distribution. In addition, the observation of self-similarity in network traffic is discussed and demonstrated in the exercises.

12.2 *Workstation Utilization Problem*

Rapid advances in microprocessor technology have equipped the scientist or engineer with unprecedented desktop computing power. The benefits of dedicating a workstation to an individual user are well established, but it is common for many elements of a network of dedicated workstations to sit idle while others are heavily loaded. Theimer and Lantz [13] found that fully one third of the 70 workstations in a research laboratory were typically idle even at the busiest times of the day. Litzkow, Livny, and Mutka [9] found that in a five-month period only 30 percent of the capacity of a collection of workstations at the University of Wisconsin was utilized. In a study of utilization conducted over a one-month period at the University of Missouri-Rolla, Clark and McMillin [4] found that workstations were idle 60% of the time. In our own study of faculty and student workstations in the Department of Computer Science at Clemson University, we found that during a 12-week period in the fall semester of 1993 CPU utilization was below 7% more than 75% of the time. Both the number and computational capacity of the workstations available in academic and research laboratories are rapidly increasing at present. Thus, the percentage of idle workstations in most systems should also continue to increase.

To better use this computing capacity, a number of load-sharing schedulers have been proposed and implemented. These include Butler [11], Condor [9], DAWGS [4], and Stealth [8]. These schedulers employ the following *fair use* principles:

- Users needing to schedule multiple work units concurrently should be able to distribute them onto workstations that would otherwise be idle.

- Each workstation is primarily dedicated to serving its local user. If the local user becomes active, distributed work units must be preempted. Possible preemption mechanisms include migrating the work unit, fully or partially quiescing it, or terminating it and restarting it on another workstation.

One important objective of a fair use scheduler is to minimize the delays and system overhead associated with the preemption of distributed work units. No delay or overhead is incurred when a work unit is placed on a workstation that remains idle longer than the amount of time required to complete the work unit. Therefore, a scheduler with the ability to estimate accurately the amount of idle time remaining on each idle workstation should have a real advantage in reducing preemption related overhead. Therefore, the objective of a solution to the workstation utilization problem is to predict how long an idle workstation will remain idle. However, the difficulty of solving the easier problem of predicting which one of a collection of idle workstations will remain idle longest can be extreme.

The similarities between the workstation utilization problem and reliability (failure-repair) problems motivates the use of Markov-type models. The assumptions implicit in the scheduling policies of the load-sharing schedulers identified above are consistent with this model view. If the lengths of idle periods of all workstations are independent and have identical exponential distributions, then random selection among idle workstations is optimal. The *random selection* (RS) policy was used in the Butler system. If the lengths of idle periods of all workstations are independent and have identical hyperexponential distributions (mixture of exponentials), then the simple strategy of selecting the workstation that has been idle longest at the time of scheduling is again optimal. We refer to this policy as *longest current idle* (LCI), and it was used in the DAWGS system.

LCI schedulers provide measurably better performance than RS schedulers when hyperexponential behavior exists. However, the hyperexponential model has some potential limitations of its own. First, the lengths of idle periods may appear to be hyperexponentially distributed according to the Kolmogorov-Smirnov test [6] when the lengths are actually correlated to the time of day. In this case, a *modulated Poisson process* is a more appropriate model. A second limitation of the hyperexponential model is that an LCI policy is optimal only in the asymptotic sense. As the maximum idle length of all idle workstations approaches zero, longest current idle assignment approaches random assignment.

These limitations motivated the development of our modulated Poisson process model. Its objectives are to detect and exploit time of day dependencies where they exist and to perform no worse than LCI or RS when idle period lengths are truly hyperexponentially or exponentially distributed. Under hyperexponential assumptions, one expects to see long and short idle period lengths occurring uniformly mixed with ratios determined by the branching probabilities.

We believe that it is most useful to model workstations as existing in a small set of *activity states*, each representing a characteristic level of activity. For example, idle periods should be expected to terminate at a high rate during intervals in which the owner of the workstation was actively developing and testing new software, but at a much lower rate during intervals in which the owner was out of the office. We assume that in any given state the idle termination process is a Poisson process having a rate that is characteristic of the state.

The following sections describe the details of how the state transition process can be characterized and how the constant rates associated with each state can be estimated. The model includes effects of both the long-term usage patterns of the workstation and its observed recent history. A Kalman-type filter for state estimation is obtained and results of its application to the workstation utilization problem are described. This filter incorporates the effects of unexpected, but observed, recent behavior into the predicted long-term behavior.

12.3   Constructing a Modulated Poisson Process

12.3.1   Intensity of a Counting Process

The key to understanding the construction of a modulated counting process is to understand its *intensity*, or expected rate of arrivals. This section provides an intuitive description of the intensity of a modulated counting process; for a rigorous treatment, see Protter [12] or Brémaud [3]. The concept of a *censored intensity* motivates the state-space representation of a Markov chain in Section 12.3.2 and the simulation technique of Section 12.4.3.

In renewal processes, if the first arrival occurs at time 0, the expected rate of the next arrival at time t is the *failure rate* or *hazard rate* of the interarrival period. Namely, if the positive continuous random variable Y has cumulative distribution function F_Y and density f_Y, then the failure rate of Y is defined to be

$$h_Y(t) = \lim_{\Delta t \to 0} \frac{1}{\Delta t} P(Y \in (t, t + \Delta t] \,|\, Y > t).$$

Thus

$$
\begin{aligned}
h_Y(t) &= \lim_{\Delta t \to 0} \frac{1}{\Delta t} \frac{P(\{Y \in (t, t + \Delta t]\} \cap \{Y > t\})}{P(Y > t)} \\
&= \lim_{\Delta t \to 0} \frac{1}{\Delta t} \frac{F_Y(t + \Delta t) - F_Y(t)}{1 - F_Y(t)}
\end{aligned}
$$

$$= \frac{f_Y(t)}{1 - F_Y(t)}$$

$$= -\frac{d}{dt} \log\left(1 - F_Y(t)\right),$$

and this implies that

$$F_Y(t) = 1 - e^{-\int_0^t h_Y(s)ds}.$$

From the definition of the failure rate, for Δt sufficiently small, the approximation $P(Y \in (t, t + \Delta t] \mid Y > t) \approx h_Y(t)\Delta t$ holds. One can easily show that the exponential random variable is the only random variable with a constant failure rate.

The definition of the intensity of a counting process can be motivated in the following way. Consider a counting process[1] $N = \{N_t : t \geq 0\}$ with $N_0 = 0$ and a "nonexplosive" regularity condition: i.e., suppose $P(N_{t+\Delta t} - N_t \geq 2) = o(\Delta t)$ and assume that the failure rate $\lambda(t)$ is a deterministic function of time. For example, the standard Poisson process satisfies these conditions. If the interval $[0, t]$ is partitioned into n segments of width $\Delta t = \frac{t}{n}$, then

$$E(N_t) = \sum_{k=1}^{n} E(N_{k\Delta t} - N_{(k-1)\Delta t}),$$

so for Δt sufficiently small,

$$E(N_t) = \sum_{k=1}^{n} 1 \times P(N_{k\Delta t} - N_{(k-1)\Delta t} = 1) + o(\Delta t)$$

$$\approx \sum_{k=1}^{n} P\big(\text{an arrival in } ((k-1)\Delta t, k\Delta t]\big)$$

$$\approx \sum_{k=1}^{n} \lambda((k-1)\Delta t)\Delta t$$

$$\rightarrow \int_0^t \lambda(s)\,ds \quad \text{as} \quad \Delta t \rightarrow 0.$$

Thus, the expected number of arrivals before time t is the integral of the failure rate. In this case, the failure rate is also called the *intensity*

---

[1]Notationally, stochastic processes as indexed families of random variables will be denoted without subscripts: i.e., $H = \{H_t : t \geq 0\}$, with H_t indicating a random variable at time t. Deterministic functions of t will be denoted $f(t)$.

of the counting process. For a Poisson process, the final result would be $E(N_t) = \lambda t$.

The reader is referred to the texts by Protter [12] and Brémaud [3] for a more complete description of the intensity of a counting process. In general, the intensity of a counting process is history dependent. The information available at time t is commonly represented by a collection $\mathcal{F}_t$ of subsets of the event space, and the intensity is defined by

$$\lambda(t) = \lim_{\Delta t \to 0} \frac{E(N_{t+\Delta t} - N_t \mid \mathcal{F}_t)}{\Delta t} = \frac{E(dN_t \mid \mathcal{F}_t)}{dt}.$$

Throughout this chapter, the notation $E(\cdot \mid \mathcal{F}_t)$ denotes the expectation conditioned upon the available information at time t. The process dN is the symbolic stochastic differential of the counting process, in the sense that $\int_0^t dN_s = N_t$ and

$$E(N_t) = E\left(\int_0^t dN_s\right) = \int_0^t E(dN_s) = \int_0^t \lambda(s)\, ds.$$

The intensity of a counting process does not need to be constant, or in fact, deterministic, but it is a predictive rate at which the counting process increments. In the case of a stochastic intensity,

$$E(N_t) = E\left(\int_0^t dN_s\right) = E\left(\int_0^t \lambda_s\, ds\right) = \int_0^t E(\lambda_s)\, ds.$$

A counting process and its integrated intensity are integral parts of what is called the *classical semi-martingale decomposition*. We can write N_t as

$$N_t = \int_0^t \lambda(s)\, ds + \left(N_t - \int_0^t \lambda(s)\, ds\right).$$

Essentially, the counting process can be decomposed into an integrable drift term and a 0-mean process

$$m_t = N_t - \int_0^t \lambda(s)\, ds$$

satisfying $E(m_t \mid \mathcal{F}_s) = m_s$ whenever $0 \le s < t$.

Of particular interest in this chapter are counting processes that are censored by a second process. For example, speeding cars may pass a certain point on the highway according to a Poisson process, but such cars can be registered only during times when a monitoring process (patrol, automated surveillance) is present. The arrivals of speeding cars that are registered form a *censored counting process*. Monitoring is performed according to an indicator process, say $Y = \{Y_t : t \geq 0\}$, a right-continuous process possessing left-hand limits which only takes on the values 0 or 1. Consider the counting process defined by

$$M_t = \int_0^t Y_{s-}\, dN_s = \sum_{0 \leq s \leq t} Y_{s-}\, \Delta N_s.$$

Only those events counted by N when $Y_{s-} = 1$ are registered by M, whereas events occurring when $Y_{s-} = 0$ are lost. The intensity of M is $Y_{s-}\lambda(s)$ and $E(M_t) = \int_0^t E(Y_{s-}\lambda(s))\, ds$. Thus, if the counting process is censored, the intensity is censored by the same process.

12.3.2   *State-Space Representation of a Markov Chain*

A continuous time Markov chain (CTMC) consists of a directed graph whose n nodes, or *states*, are visited according to a random walk. The order in which states are visited forms a discrete time Markov chain (see Chapter 13) and the random sojourn times are exponential random variables with parameter dependent on the state. For an illustration, consider a courier who runs between n different offices. The courier waits in office i until a message needs to be sent to office j, then he remains there until office j sends a message. In this CTMC, $X = \{X_t : t \geq 0\}$ represents the office that the courier is in at time t. The courier waits an exponential amount of time in each office, where the "message production rate" depends upon the office. The courier then proceeds to another office where the distribution of the destination is a multinomial random variable dependent only on the originating office. We assume that each office generates messages at a constant rate and these messages are lost if the courier is not present to receive them.

For our state-space model of a CTMC, consider the supervisor of the courier. He has a bulletin board with a place dedicated to each office; if the courier is in office i then a push pin is in place i. There is always a pin in some place and there is never more than one pin on the board. The current office of the courier represents all of the information that is available and that information is totally portrayed by the bulletin board. Let $Z = \{Z_t : t \geq 0\}$ be an $n \times 1$ probability-vector valued indicator process: i.e., for each t, Z_t is a random 0-1 vector. Take the ith component of Z_t to be 1 if $X_t = i$ and 0 otherwise, written $Z_t^i = 1_{\{X_t = i\}}$. The state indicator Z represents the supervisor's bulletin board.

In our example we said that an office produces messages at an exponential rate, even when the courier is not present. For office i, the process representing the number of messages produced by time t is a Poisson process, say $\alpha^i = \{\alpha_t^i : t \geq 0\}$. Let the process which counts messages produced by office i designated for office j be denoted $\alpha^{ij} = \{\alpha_t^{ij} : t \geq 0\}$; therefore $\alpha^i = \sum_{j \neq i} \alpha^{ij}$. Also, let the intensities of the α^{ij}, α^i processes be denoted $a^{ij}(t)$, $a^i(t)$, respectively. In a *time-stationary* CTMC, the transition rates $a^{ij}(t)$, $a^i(t)$ are constant with respect to time, but in the formulation below, time dependence is allowed for a more general approach. In the workstation utilization problem, it is critical that certain behaviors are time correlated. Furthermore, time dependence actually facilitates data fitting techniques.

Notice that the office transition time for the courier occurs at the same time as a message is produced at the current office. Following Brémaud [3], consider the following equation for Z_t^k:

$$Z_t^k = Z_0^k + \sum_{\substack{i=1 \\ i \neq k}}^{n} \sum_{0 \leq s \leq t} Z_{s-}^i \Delta \alpha_s^{ik} - \sum_{\substack{j=1 \\ j \neq k}}^{n} \sum_{0 \leq s \leq t} Z_{s-}^k \Delta \alpha_s^{kj},$$

or, in terms of Lebesgue-Stieltjes integration,

$$Z_t^k = Z_0^k + \sum_{\substack{i=1 \\ i \neq k}}^{n} \int_0^t Z_{s-}^i \, d\alpha_s^{ik} - \sum_{\substack{j=1 \\ j \neq k}}^{n} \int_0^t Z_{s-}^k \, d\alpha_s^{kj}.$$

Every counting process α^{ij} or α^i is censored by Z^i. If $Z_{s-}^i = 1$ then Z_s^i becomes 0 if α^i increments and if so, some α^{ik} must increment causing Z^k to become 1.

After compensation (adding and subtracting the integrals of the intensities),

$$Z_t^k = Z_0^k + \sum_{\substack{i=1 \\ i \neq k}}^{n} \int_0^t Z_{s-}^i a^{ik}(s) \, ds - \sum_{\substack{j=1 \\ j \neq k}}^{n} \int_0^t Z_{s-}^k a^{kj}(s) \, ds$$

$$+ \sum_{\substack{i=1 \\ i \neq k}}^{n} \int_0^t Z_{s-}^i (d\alpha_s^{ik} - a^{ik}(s) \, ds) - \sum_{\substack{j=1 \\ j \neq k}}^{n} \int_0^t Z_{s-}^k (d\alpha_s^{kj} - a^{kj}(s) \, ds)$$

$$= Z_0^k + \int_0^t \Big\{ \sum_{\substack{i=1 \\ i \neq k}}^{n} Z_{s-}^i a^{ik}(s) \, ds - Z_{s-}^k a^k(s) \, ds \Big\} + m_t,$$

where m_t is a vector valued, 0-mean process, equaling the remaining terms. In a matrix equation, we have the compact form

$$Z_t = Z_0 + \int_0^t A^T(s) Z_{s-} \, ds + m_t,$$

where the matrix valued function $A(t)$, called the *generator*, is

$$A(t) = \begin{bmatrix} -a^1(t) & a^{12}(t) & \cdots & a^{1n}(t) \\ a^{21}(t) & -a^2(t) & \cdots & a^{2n}(t) \\ \vdots & \vdots & \ddots & \vdots \\ a^{n1}(t) & a^{n2}(t) & \cdots & -a^n(t) \end{bmatrix}$$

and its transpose is denoted $A^T(t)$. The expected value of the space-space representation is the integral equation

$$
\begin{aligned}
P(t) &= E(Z_t) \\
&= E\left(Z_0 + \int_0^t A^T(s) Z_{s-}\, ds + m_t \right) \\
&= E(Z_0) + E\left(\int_0^t A^T(s) Z_{s-}\, ds \right) + E(m_t) \\
&= P(0) + \int_0^t A^T(s) P(s-)\, ds.
\end{aligned}
$$

Moreover, for each t, $P(t)$ is a probability vector since $\sum_{i=1}^n Z_t^i = 1$.

12.3.3   *Markov Modulated Poisson Processes*

A *Markov Modulated Poisson Process* (MMPP) is a counting process with a stochastic intensity that is controlled by a continuous time Markov chain. For simplicity suppose in our courier example that two types of messages are generated, say type A and type B. Type A is a message to be taken immediately to another office and type B is outside mail that the courier collects and takes to the post office at the end of the day. Further suppose that each office generates messages of type B at possibly different rates, independent of the production of messages of type A. If the courier is not present at the time of production, then type B mail is sent by different means. The courier's mailbag increments according to a counting process, say N, but the rate of arrivals depends upon the current office. Let $\lambda^i(t)$ be the rate of production of messages of type B at office i at time t. If X is our CTMC with indicator process Z, then the intensity μ of N is given by

$$\mu_t = \sum_{i=1}^n 1_{\{X_t = i\}} \lambda^i(t) = \lambda^T(t) Z_t.$$

Ultimately, at time t, if it is known that $X_t = j$, N behaves like a counting Poisson process with intensity $\lambda^j(t)$.

12.3.4   Filtering in the Workstation Utilization Problem

In the workstation utilization problem detailed in Section 12.2, one would like to estimate the future behavior of a workstation based upon past history and current conditions. Data that is readily available is the overall profile of an individual workstation, call it $Y = \{Y_t : t \geq 0\}$. The profile is the indicator of a busy workstation and can be written $Y_t = N_t^I - N_t^B$, where the processes N^I, N^B count the number of idle and busy periods that have ended, respectively. The profile can also be written as the self-censoring differential equation $dY_t = (1 - Y_{t-})dN_t^I - Y_{t-}dN_t^B$.

In order to determine the most favorable workstation on which to distribute a computational unit, our primary goal is to estimate the expected excess idle time of each workstation in the network. This can be done with an estimate $\widehat{\mu}$ of the rate at which a workstation becomes busy and then appealing to the fact that[2]

$$E(\text{excess idle time at } t \mid \mathcal{F}_t) = \int_t^\infty e^{-\int_t^\tau \widehat{\mu}_s \, ds} \, d\tau.$$

The failure rate of the excess idle times is the same as the intensity of N^I. Because of the linear nature of the intensity, $\mu_t = \lambda^T(t)Z_t$ where Z_t is the indicator process for characteristic workstation states and the components of λ are the corresponding intensities for these states. If λ is deterministic but not necessarily constant, then

$$\widehat{\mu}_t = E(\mu_t \mid \mathcal{F}_t) = \lambda^T(t)E(Z_t \mid \mathcal{F}_t) = \lambda^T(t)\widehat{Z}_t,$$

where $\widehat{Z}_t = E(Z_t \mid \mathcal{F}_t)$. If X is a single chain which modulates both activity during an idle period and activity during a busy period with conditional intensities λ^I and λ^B, the time-dependent filtering equation for $E(Z_t \mid \mathcal{F}_t)$ is given [14] by the following:

$$\widehat{Z}_t = \widehat{Z}_0 + \int_0^t A^T(s)\widehat{Z}_s \, ds +$$

$$\int_0^t \left(D(\lambda^B(s))\widehat{Z}_{s-}(\lambda^B(s)'\widehat{Z}_{s-})^\oplus - \widehat{Z}_s \right) Y_{s-}(dN_s^B - \lambda^B(s)'\widehat{Z}_{s-} \, ds) +$$

$$\int_0^t \left(D(\lambda^I(s))\widehat{Z}_{s-}(\lambda^I(s)'\widehat{Z}_{s-})^\oplus - \widehat{Z}_s \right)(1 - Y_{s-})(dN_s^I - \lambda^I(s)'\widehat{Z}_{s-} \, ds).$$

Here, $a^\oplus$ is the pseudo-inverse of a and $D(\lambda)$ denotes the diagonal matrix with diagonal elements taken from λ. One can view the filtered estimate

---

[2]The expected excess idle time calculation requires a state-space estimate into the future, but the failure rate is for the current idle period. Therefore, the observable information is taken to be $Y_s = 0$, $s \geq t$.

for Z as a sum of an integrable drift term which takes into account long run average trends and a second term that allows current behavior to override long run trends.

For purposes of discussion, assume X modulates just the activity during idle periods (the filtering equation would be the same as above, but without the $\int \cdot dN^B$ term). The matrix and vector valued parameters, A and $\lambda = \lambda^I$, need to be estimated from actual workstation data. Parameter estimation is motivated by taking X to be regenerative, in that X restarts each day. Therefore, the daily record of one workstation is sufficient to tailor parameters to that workstation. Recall that the counting process $N = N^I$ is censored according to $(1-Y_{t-})$, with general intensity $\mu_t = (1 - Y_{t-})\lambda^T(t)Z_t$. For simplicity, assume a two-state case in which λ is taken to be deterministic and constant. The censoring of N allows us to estimate μ only when $Y_{t-} = 0$, and thus the estimates for A and λ are based upon an estimate for $E(\mu_t \,|\, Y_{t-} = 0)$. Here $P(t) = E(Z_t)$ represents the probability distribution for X_t, conditioned only upon time, and

$$
\begin{aligned}
E(\mu_t \,|\, Y_{t-} = 0) &= \frac{E(\mu_t\, 1_{\{Y_{t-}=0\}})}{P(Y_{t-} = 0)} \\
&= \lambda^1 E(Z_t^1) + \lambda^2 E(Z_t^2) \\
&= \lambda^T P(t).
\end{aligned}
$$

From this it follows that

$$
P^2(t) = \frac{E(\mu_t \,|\, Y_{t-} = 0) - \lambda^1}{\lambda^2 - \lambda^1}.
$$

Therefore, given an estimate of $E(\mu_t \,|\, Y_{t-} = 0)$, our parameter estimation procedure is to take λ^1 as the minimum of $E(\mu_t \,|\, Y_{t-} = 0)$ over t and to take λ^2 as the maximum of these values. This ensures $P(t)$ is a valid probability vector. The time-dependent distribution satisfies the balance equations (see Section 12.3.2)

$$
\frac{d}{dt}P(t) = A^T(t)P(t), \qquad \text{where} \quad A(t) = \begin{bmatrix} -a^1(t) & a^1(t) \\ a^2(t) & -a^2(t) \end{bmatrix}.
$$

So, if $E(\mu_t \,|\, Y_{t-} = 0)$ can be estimated in a sufficiently smooth manner, then λ, $P(t)$, and $\frac{d}{dt}P(t)$ can be approximated. Since $A(t)$ is singular for all t, the balance equations are underdetermined and consequently the coefficients $a^1(t)$ and $a^2(t)$ have the form

$$
a^1(t) = \frac{\left(\frac{dP^1(t)}{dt}\right)^- + h(t)}{P^1(t)}, \qquad a^2(t) = \frac{\left(\frac{dP^1(t)}{dt}\right)^+ + h(t)}{P^2(t)}
$$

where $\left(\frac{dP^1(t)}{dt}\right)^+$, $\left(\frac{dP^1(t)}{dt}\right)^-$ are the positive, negative variations in $dP^1(t)$ and h is some function of t. The function h has a dramatic effect on the convergence rate to steady state and, in our numerical examples, h is taken to be 0.

To obtain the differentiability and periodicity (the system regenerates itself each day) properties necessary for estimation of $E(\mu_t \mid Y_{t-} = 0)$, the impulses of idle terminations within the sample period are first superimposed over one 24-hour period. Smoothing techniques are used, as discussed in [14], to obtain a Nelson-Allen [1] estimate of $E(\mu_t \mid Y_{t-} = 0)$.

For the purposes of job distribution, meaningful performance measures for the filtering technique revolve around enabling a job scheduler to locate a workstation with a long remaining idle period. There are too many uncertainties involved with fitting appropriate time-dependent distributions to network data to justify any theoretical approach to workstation selection policies. Indeed, if idle period lengths are actually hyperexponentially distributed, LCI is the optimal selection policy, but hyperexponentially distributed idle periods are not necessarily the case.

The idle workstation selection policies that we consider are Random Selection (RS), Longest Current Idle (LCI), Predicted Excess (PEXC), and Optimal Selection (OPT). Both RS and LCI were described in Section 12.2. Using a two-state filtering model, PEXC selects the workstation with the greatest predicted remaining or excess idle time. OPT always makes the best selection. The OPT policy can only be realized in a full information setting, such as during simulations, but is a benchmark for comparison. For each selection policy, the performance statistic considered is the time remaining on the selected workstation or the total time remaining on selected workstations for a collection of distributed jobs. Results reported here are averages over a collection of 30 trials.

Three test networks were used for this study. Data for each of the sample networks were gathered over 12 weeks. The first six weeks were used strictly for modeling. For the next six weeks the models for PEXC were updated every night based upon a fixed number of prior weeks of data. Only weekdays were used. The first sample network consisted of eight faculty workstations within the Department of Computer Science at Clemson University. To match the size of the first network, a sample network of eight workstations was taken from a collection of lab workstations within the same department. The third network was simulated using a generator based upon a Markov modulated point process model. The three networks were analyzed separately in order to establish the applicability of the RS, LCI, PEXC, and OPT selection policies to different environments. The faculty workstations, in general, are lightly utilized units dedicated to individuals with fairly regular work patterns.

Lab workstations are under higher utilizations with mixed sets of users. The simulated network portrays workstations which have high utilization with differing but regular work patterns. The OPT policy selects workstations at the beginning of very long idle periods; eventually, these same workstations will be selected by LCI but possibly at the end of their idle periods. The simulated network tests performance when the optimal selection is not expected to rest with a single workstation for more than 3 hours. A 24-workstation network (3 networks probabilistically similar to the 8-workstation network) was also simulated. Table 12.1 compares the variability in average utilizations of the sample networks.

Table 12.1 Sample network utilizations

Time Interval	Faculty Data	Lab Data	Simulated Data
Full Day	22.5%	39.4%	58.5%
8 a.m. to 8 p.m.	20.1%	42.7%	58.3%
10 a.m. to 6 p.m.	18.5%	45.0%	58.3%
12 noon to 4 p.m.	17.7%	46.0%	59.0%

The first approach used for comparison of the selection policies was simply to consider the ability of each policy to identify the workstation with the longest remaining idle time. We refer to this comparison method as the *Longest Idle Selection* test. At the start of each of the 288 five-minute intervals in each of the days of the verification set, all policies (RS, LCI, PEXC, OPT) were used to select one of the idle workstations. The remaining idle length of each of the selections was easily determined by a table look up. From the results depicted in Table 12.2, one observes that PEXC performs quite favorably in comparison with both RS and LCI.

Table 12.2 Longest idle selection:
average selection length (hours scored)

Network	OPT	PEXC	RS	LCI
8 Faculty	39.9	33.5	13.8	24.6
		84%	35%	62%
8 Lab	13.3	7.7	5.4	6.4
		58%	41%	48%
8 Simulated	10.7	7.5	5.5	2.5
		70%	51%	22%
24 Simulated	12.9	9.0	5.5	1.8
		70%	42%	14%

For a second comparison of the selection policies, the arrival and placement of jobs over the same six-week period were simulated. After all models were fit and predicted excesses calculated, for each distributed job the sample profile was adjusted according to that job's placement to indicate the targeted workstation as busy. Thus, the simulation is performed without replacement since the idle workstation is temporarily removed from the idle pool. This comparison method is called the *Distributed Load Test*. Independently, jobs were requested at times which were normally distributed, centered at 2:00 p.m. with a standard deviation of 2 hours, and uniformly distributed across the days of the sample. A total of 300 simulated runs were performed. The expected number of jobs ranged from 1 to 10, and 30 simulations were run for each case. For every run, each selection policy addressed the same distributed workloads independently on the same network profiles. Thus, for various levels of activity and for each selection policy, the performance is represented by the total time available on selected workstations from a sample of size 30. Total performance over all workloads is depicted in Table 12.3, which indicates that PEXC performs at least as well as RS and LCI in all cases and performs much better than LCI in the simulated networks.

Table 12.3 Distributed load test: total hours scored

	OPT	PEXC	RS	LCI
8 Faculty	15,415	12,295	10,764	12,350
		80%	70%	80%
8 Lab	6,183	5,288	4,391	5,098
		86%	71%	82%
8 Simulated	8,547	7,698	7,002	6,439
		90%	82%	75%
24 Simulated	15,623	12,497	8,513	5,107
		79%	54%	33%

Overall, where near hyperexponential behavior existed, LCI performed well, but where the optimal selection changed regularly as in the simulated networks, LCI performed worse than RS. The PEXC policy performed acceptably at times when LCI performed well, and excelled during times of LCI failure.

12.4   Simulation Techniques

12.4.1   Simulation of IID Random Variables

All simulations that we consider evolve as the sequential simulation of univariate random variables. Consider the M/M/1 or GI/G/1 queue. The interarrival times form an independent and identically distributed (*iid*) sequence of random variables independent of the service times which also form an *iid* sequence. A good pseudo-random number generator algorithm produces a sequence of random variables which generally appear to be *iid* uniform $(0, 1)$ random variables. Therefore, we try to produce a sequence with the desired statistical properties based on functions of a pseudo-random sequence.

The case of a continuous random variable is fairly straightforward based on the following argument. For any continuous random variable X with cumulative distribution function (cdf) F_X, and any invertible, differentiable, order-preserving function g, we have

$$F_{g(X)}(t) = P(g(X) \le t) = P(X \le g^{-1}(t)) = F_X(g^{-1}(t))$$

for the distribution of $g(X)$. If we restrict our attention to the subset of the real line where F_X is one-to-one, then trivially, $F_X(F_X^{-1}(t)) = t$ for $0 < t < 1$. It follows that $U = F_X(X)$ is a uniform $(0, 1)$ random variable. Similarly, if U is a uniform $(0, 1)$ random variable, then for $g(U) = F_X^{-1}(U)$,

$$F_{F_X^{-1}(U)}(t) = F_U(F_X(t)) = F_X(t).$$

Thus, $F_X^{-1}(U)$ is a random variable with the same distribution as X.

Example 12.1

If X is exponentially distributed with parameter λ, its cumulative distribution function is $F_X(t) = 1 - e^{-\lambda t}$ with inverse $F_X^{-1}(t) = \frac{\log(1-t)}{-\lambda}$. For simulations of X, one can use $\frac{\log(1-U)}{-\lambda}$, where U is a uniform $(0, 1)$ random variable.

While we are on the subject, what if F_X has a discontinuity? Then the jump size at time t is the probability that X equals t, namely

$$\Delta F_X(t) = F_X(t) - F_X(t-) = P(X = t).$$

For U a uniform $(0, 1)$ random variable, we have the following property:

$$P(U \in (F_X(t-), F_X(t)]) = F_X(t) - F_X(t-) = P(X = t).$$

The following algorithm will simulate a random variable X with cdf F_X that is either discrete, continuous (with piecewise continuous derivative), or is a mixture of the two types.

ALGORITHM 12.4: Simulation of random variable X

1. Simulate a uniform $(0, 1)$ random variable U.
2. If there exists a t^* where $F_X(t^*) = U$, take $F_X^{-1}(U) = t^*$.
3. If there does not exist a t^* in Step 2, then there exists a t' such that $U \in (F_X(t'-), F_X(t')]$; take $F_X^{-1}(U) = t'$.

Example 12.2

If X is a Poisson random variable with parameter λ, then

$$P(X = x) = \frac{\lambda^x e^{-\lambda}}{x!} \quad \text{for } x = 0, 1, 2, \ldots$$

Here we take $F_X^{-1}(U) = t'$, chosen so that $\sum_{x=0}^{t'-1} P(X = x) < U$ and $U \leq \sum_{x=0}^{t'} P(X = x)$.

To simulate a Bernoulli trial with parameter p, use the Boolean $(U > 1 - p) = (U < p)$.

12.4.2   Simulation of a Markov Modulated Poisson Process

In a Markov modulated Poisson process (MMPP), a Markov chain, called the modulator, controls the "state" of the counting process. With each of the n states of the Markov chain one associates a Poisson process. It is assumed that there are n independent Poisson processes and these are independent of the Markov chain. The "firings" (events) of the Poisson processes occur whether or not one physically observes them. The modulator dictates which Poisson process is being observed: i.e., if $X = j$, Poisson process j is observed and only firings of process j are recorded. All firings across all Poisson processes look identical except in their rates. Assume Poisson process j has rate λ^j for $j = 1, \ldots, n$. In terms of the observed process, when $X = j$ firings occur with rate λ^j.

Consider the two-state example in Figure 12.1. The solid line indicates the path of the Markov chain. Firings for the Poisson processes occur whether or not they are on the path of the Markov chain, but only the ones on the path are observed. Each of the Poisson processes has periods of high and low activity, but, in general, the periods of low activity in the observed process correlate to periods when the chain is in the state corresponding to the Poisson process with a lower rate. Kalman filtering for the state of the Markov chain exploits this property.

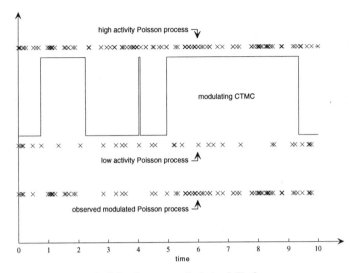

FIGURE 12.1 **A Markov modulated Poisson process**

In terms of simulation, recall that for a time-stationary continuous time Markov chain, there is a generator

$$
A = \begin{bmatrix} -a^1 & a^{12} & \ldots & a^{1n} \\ a^{22} & -a^2 & \ldots & a^{2n} \\ \vdots & \vdots & \ddots & \vdots \\ a^{n1} & a^{n2} & \ldots & -a^n \end{bmatrix},
$$

and the sojourn of the chain in state i is an exponential random variable with parameter a^i. Once the chain moves from state i, the transition probability into state j is given by $P_{ij} = a^{ij}/a^i$. Thus, in order to simulate a transition out of state i, one uses Algorithm 12.1 to determine the sojourn time in state i (an exponential random variable with rate a^i) and the next state (a discrete random variable with probabilities P_{ij}, $j \neq i$).

To simulate the two-state case of Figure 12.1, a sequence of *iid* exponentials needs to be simulated for each Poisson process and two sequences of exponentials for the sojourn corresponding to each state of the Markov chain. Moving forward in time, first determine the state of the chain, then record firings (if any) of the Poisson process associated with the current state until the next state transition. One could also exploit the memoryless property of the exponential distribution. Namely, given the current state, the next type of event is determined: either (a) a firing of the associated Poisson process, or (b) a state transition in the Markov chain. Memorylessness and time stationarity imply that this procedure can be repeated after each event.

12.4.3   Simulating a Nonconstant Failure Rate Process

Simulation of a Poisson process can be relatively fast. Event times occur at the partial sums of an infinite series of exponential random variables which can themselves be simulated by evaluating their inverse cumulative distribution function at uniform $(0, 1)$ (pseudo-) random variables. In real world problems, it is likely that the counting process is nonstationary and one can exploit the intensity to simulate nonstationarity or multivariate counting processes.

When the intensity $h(t)$ of a counting process M is nonconstant, one can discretize the interarrival times. If the increments of the discretization grid have length Δt, small enough that only one counting event can occur within $(t, t + \Delta t]$, event simulation during $(t, t + \Delta t]$ can be done with a Bernoulli random variable with parameter $h(t)\Delta t$. For slowly changing $h(t)$, larger Δt might be considered with a Poisson number of events during the increment. The looping required for such a simulation can be quite slow.

Instead, consider the simulation of a counting process with a nonconstant failure rate by *thinning* a Poisson process. This method is both fast and yields continuous time results. The remainder of this section is devoted to an intuitive derivation of this method.

Suppose $h(t)$ is the failure rate of a random variable, or the intensity of a counting process to be simulated, where $\sup_t h(t) \leq \lambda$ (constant). Let $N = \{N_t\}$ be a Poisson process with intensity λ and let $\{Y_t\}$ be an indexed family of independent Bernoulli random variables, independent of N where $p(t) = \frac{h(t)}{\lambda}$ is the parameter for Y_t. Here, Y is used to model the sampling (thinning) of the Poisson process N.

Think of N_t as being composed of M_t and M_t^c; that is, $N_t = M_t + M_t^c$, where $dM_t = Y_t \, dN_t$ and $dM_t^c = (1 - Y_t) \, dN_t$. The process M is the process of interest. Whenever an event for N occurs, a coin toss is performed to see if this firing is recorded by the M process. Specifically, if $dN_t = 0$, then Y_t has no effect; but if $dN_t \neq 0$ and $Y_t = 1$, then the M counting process increments, otherwise M^c increments. Intuitively, if $\lambda = \frac{E(dN_t)}{dt}$, then the intensity of M_t is

$$\frac{E(dM_t)}{dt} = \frac{E(Y_t \, dN_t)}{dt} = E(Y_t)\frac{E(dN_t)}{dt}$$

$$= p(t)\lambda = \frac{h(t)}{\lambda}\lambda = h(t).$$

For simulations, the parameter λ of N must be an upper bound for $h(t)$. If t is the time of a firing of N, then a simulated Bernoulli random variable with parameter $\frac{h(t)}{\lambda}$ can be used to determine if the firing is observed by M or discarded (left to its complement in N).

Through the use of a multinomial random variable, this *marked point process method* can be extended to the case of M^1, M^2, ..., M^m, M^c orthogonal (no jumps in common) counting processes. If the m processes to be simulated have respective failure rates $h^1, \ldots, h^m$, the aggregate counting process has the rate $h(t) = \sum_{i=1}^{m} h^i(t)$. Take the rate λ for the counting process N to be an upper bound for h. For each firing of N, simulate a multinomial trial with $m + 1$ outcomes where $\frac{h^i(t)}{\lambda}$ is the probability of outcome i, $i = 1, \ldots, m$, and $1 - \frac{h(t)}{\lambda}$ is the probability that the outcome is discarded. Let Y_k be the multinomial "reward" or "mark" associated with the kth firing of N. For any i, if h^i is continuous at time t and Δt is small enough to expect only one firing in $(t, t + \Delta t]$, then

$$E(M^i_{t+\Delta t} - M^i_t) = E\Big(\sum_{k \in (N_t, N_{t+\Delta t}]} 1_{\{Y_k = i\}} \Big)$$
$$\approx E(1_{\{Y_j = i\}})E(N_{t+\Delta t} - N_t)$$
$$\approx \frac{h^i(t)}{\lambda}\lambda \Delta t = h^i(t)\Delta t.$$

Returning to the case where N is a modulated counting process with modulator X, suppose X enters state j at time t and remains there for some sojourn time τ. Then, from t to $t + \tau$, N behaves like a Poisson process with rate μ_j. During this sojourn, simulation of N is easy: just iteratively simulate interarrival times using the inverse exponential distribution function on a uniform $(0,1)$ random variable until one reaches $t + \tau$. The time-nonhomogeneous multivariate counting process simulation comes into play during the simulation of the sojourn time τ.

Recall from Section 12.3.2 that during the construction of state indicator process Z, we had

$$Z^j_t = Z^j_0 + \sum_{\substack{i = 1 \\ i \neq j}}^{n} \int_0^t Z^i_{s-}\, d\alpha^{ij}_s - \sum_{\substack{l = 1 \\ l \neq j}}^{n} \int_0^t Z^j_{s-}\, d\alpha^{jl}_s.$$

If X transitions from state j at time t, then Z_t^j changes the first time

$$\sum_{\substack{l = 1 \\ l \neq j}}^{n} \int_0^t Z_{s_-}^j \, d\alpha_s^{jl} = 1.$$

Here, α^{jl} has intensity a^{jl}. We use the collection $\{a^{jl}\}$ with the marked point process method in order to determine τ.

Thus, for fixed $t \geq 0$ and $j \in \{1, \ldots, n\}$, iteratively simulate exponential random variables with rate

$$\lambda > \sup_{s > t} \sum_{\substack{l = 1 \\ l \neq j}}^{n} a^{jl}(s).$$

Let

$$a^{jc}(s) = \lambda - \sum_{\substack{l = 1 \\ l \neq j}}^{n} a^{jl}(s).$$

Note that a^{jc} is the complementary rate of events that are discarded, leaving the chain in state j. Let $V_1, V_2, \ldots$ be the generated sequence of exponentials. At potential sojourn times, $\tau_1 = V_1, \ldots, \tau_k = \sum_{l=1}^{k} V_l, \ldots$, simulate multinomial random variables with probabilities for the kth, $p^{kl} = a^{jl}(\tau_k)/\lambda$ for $l \in \{1, \ldots, n\}$, $l \neq j$, while $p^{kj} = a^{jc}(\tau_k)/\lambda$. Let τ be the smallest $\{\tau_k\}$ where the associated multinomial random variable l does not equal j. At this point the process transitions into state l.

12.5   Analysis Techniques

12.5.1   Performance in a G/M/1 Queue

It was shown in [7] that expected queue lengths by customers at a GI/M/1 queue can be significantly lower when the arrival process is a Poisson process than when the arrival process is a renewal process with a regular varying distribution (e.g., Pareto). To demonstrate this variability, let us consider a simple simulation technique [5] for the waiting time encountered by customers at the time of an arrival.

Suppose our initial arrival to an empty G/M/1 queue occurs at time $t = 0$ and let that customer be considered the 0th arrival. Let $\{S_n : n \geq 0\}$ be a sequence of exponential service times with rate γ, and let $\{I_n : n \geq 1\}$ be the sequence of interarrival times. The sequence of waiting times $\{W_n : n \geq 0\}$ that customers incur waiting for service can be generated in the following way. First, let $W_0 = 0$ so that the initial customer does not have to wait. If W_n is the waiting time of the nth customer, then $W_n + S_n$ is the total work in the system at the nth arrival. During the $(n + 1)$st interarrival period, the maximum amount of work that can be accomplished is the length of the interarrival period I_{n+1}. At the $(n + 1)$st arrival, there is still $W_n + S_n - I_{n+1}$ work to be done and this equals W_{n+1}, if this quantity is positive, of course.

Let the sequence $\{U_n : n \geq 1\}$ be defined by $U_n = S_{n-1} - I_n$, which is the contribution to waiting time brought by the nth arrival. If W_0 is taken to be 0, then the nth waiting time W_n is the maximum of $W_{n-1} + U_n$ and 0, written $W_n = \max\{W_{n-1} + U_n, 0\} = (W_{n-1} + U_n)^+$. Waiting times can also be defined as the difference between a random walk and its lowest excursion as follows. Set $U_0 = 0$ and let $\{V_n : n \geq 0\}$ be the sequence of partial sums of $\{U_n\}$; namely, $V_n = \sum_{i=0}^{n} U_i$. Define $\{\underline{V}_n : n \geq 0\}$ as the sequence of lowest excursions of V, where $\underline{V}_n = \min\{V_i : 0 \leq i \leq n\}$. This sequence can also be defined by $\underline{V}_n = \min\{\underline{V}_{n-1}, V_{n-1} + U_n\}$. It follows that $W_n = V_n - \underline{V}_n$. Whenever V_n reaches a new lowest excursion, then $W_n = 0$. Although the sequence $\{W_n\}$ is correlated, for large enough samples (simulated sequences), we can still consider $\overline{W}$ as the sample average waiting time at arrivals. In the exercises, the reader will see that deviations from Poisson assumptions can lead to considerable changes in the waiting time distribution.

12.5.2   *Notions of Self-Similarity*

A topic of recent interest is the self-similarity of network traffic. We will use the term "self-similar" in reference to a family of aggregate processes associated with a traffic model. Viewing such a process at different resolutions yields essentially the same behavior. A counting process model $N = \{N_t : t \geq 0\}$ for arrival traffic is *self-similar* if the aggregate processes $N^{\Delta t} = \{N_k^{\Delta t} \equiv N_{(k+1)\Delta t} - N_{k\Delta t} : k \geq 0\}$ exhibit similar correlation structures independent of the aggregate size Δt, over a broad range of aggregate sizes. In a strict sense, the term is reserved for cases where similarity holds for all Δt, and the term "asymptotically self-similar" is used if the similarity holds as $\Delta t \to \infty$.

If we assume that the variance of N_t has stationary increments and $\sigma_{N_t}^2 = \sigma_{N_1}^2 t^{2H}$, then the autocorrelations satisfy

$$\frac{\text{Cov}(N_{i+k}^{\Delta t}, N_i^{\Delta t})}{\sqrt{\text{Var}(N_{i+k}^{\Delta t})} \sqrt{\text{Var}(N_i^{\Delta t})}} = \frac{\text{Cov}(N_{1+k}^{\Delta t}, N_1^{\Delta t})}{\sqrt{\text{Var}(N_{1+k}^{\Delta t})} \sqrt{\text{Var}(N_1^{\Delta t})}}.$$

Furthermore, one can show [2] that the autocorrelations satisfy

$$r^{\Delta t}(1) \equiv \frac{\text{Cov}(N_2^{\Delta t}, N_1^{\Delta t})}{\sqrt{\text{Var}(N_2^{\Delta t})}\,\sqrt{\text{Var}(N_1^{\Delta t})}} = 2^{2H-1} - 1$$

and more generally

$$r^{\Delta t}(k) \equiv \frac{\text{Cov}(N_{1+k}^{\Delta t}, N_1^{\Delta t})}{\sqrt{\text{Var}(N_{1+k}^{\Delta t})}\,\sqrt{\text{Var}(N_1^{\Delta t})}} = \frac{(k+1)^{2H} + (k-1)^{2H} - 2k^{2H}}{2}.$$

Thus N_t is self-similar in the strict sense. The parameter H is called the *Hurst parameter*. It measures the degree of self-similarity, since it encapsulates both the rate of growth of the covariance function and the rate of aggregate variance growth. For a process with *iid* increments, $H = 0.5$; that is, aggregate variances climb proportionally to the aggregate size.

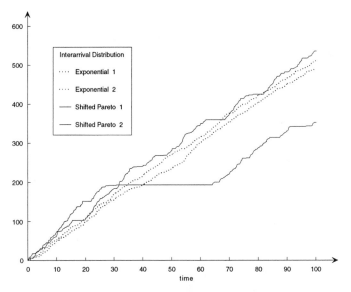

FIGURE 12.2   **Realizations of several renewal processes**

The exercises of Section 12.6 will demonstrate that Poisson processes exhibit linear variance growth and hence no interaggregate dependence, whereas shifted Pareto interrenewals for a renewal process can exhibit long range dependence. MMPPs are self-similar only in a local sense [7], for the order of aggregate variance growth is higher than linear only during its transient period. (The transient period is the time it takes for the solution of the Chapman-Kolmogorov equations to reach steady state.) Therefore, the range of self-similar behavior can be extended through adjustments to the generator. Although self-similarity may only be a local property, achievability of such behavior reaffirms the use of MMPPs for traffic models. For selected arrival data to a GI/M/1 queue (see Figure 12.2), simulated queue lengths are depicted in Figure 12.3 and aggregate variances are portrayed in Figure 12.4. MMPPs can also be combined with regular varying distributions to endow the counting process with more persistent correlated behavior [7].

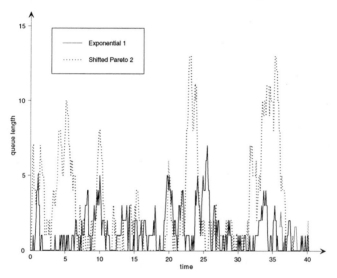

FIGURE 12.3 **Simulated queue length for selected arrival data**

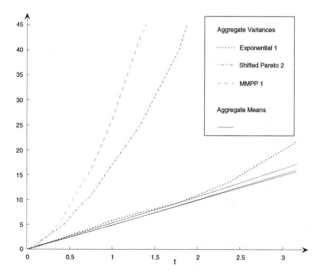

FIGURE 12.4 **Sample aggregate statistics for selected arrival data**

12.6   Exercises and Projects

Some suggestions for carrying out computations in the following exercises are given using the notation of MATLAB [10]. Additional discussion of MATLAB issues pertinent to these exercises can be found at the website <http://www.math.clemson.edu/modeling/> for this book.

1. Given a sample of data $X_1, X_2, \ldots, X_n$, the sample cumulative distribution function (cdf) is a function of sample estimates of F_X: namely, $\widehat{F}_X(t) = (\#X_i \leq t)/n$. For the following cdfs, simulate 100 realizations using the inverse cdf method. For each sample, compare the sample cdf to the actual cdf over an appropriate grid. After sampling 100 realizations, try a larger number (say, 1000) to illustrate the improved estimation of the cdf by the sample cdf.

   [In MATLAB, if x is a vector of sample data and t is a sampling grid, then the following commands will generate scdf, the sample cdf at the times of t: `for i=1:n, scdf(i)=sum(x<=t(i))/n; end, plot(t,scdf)`. A simple method that will not plot discontinuities is: `t=sort(x); plot(t,(1:n)/n)`.]

   a. Exponential: $F_X(t) = 1 - e^{-\lambda t}$ for $\lambda > 0$. Try values for λ of 0.1, 1, and 10.

   b. Weibull: $F_X(t) = 1 - e^{-(\frac{t-\nu}{\alpha})^\beta}$ for $t > \nu$ and $\alpha, \beta > 0$. Try values $\nu = 0$; $\alpha = 1, 4$; and $\beta = 2, 4$.

   c. Pareto: $F_X(t) = 1 - \left(\frac{\alpha}{t}\right)^\beta$ for $t > \alpha$ and $\alpha, \beta > 0$. In this distribution, α is called the *location* parameter and β is called the *shape* parameter. This distribution is common in self-similarity discussions because the conditional distribution of X having survived past a time T is also a Pareto distribution with the same shape parameter but a different location parameter (see Exercise 12.2). A shifted Pareto distribution has the cdf $F_X(t) = 1 - \left(\frac{\alpha}{t+\alpha}\right)^\beta$. For simulations, try a shifted Pareto with $\alpha = 0.5$ and $\beta = 1.5, 1.75, 3$.

   d. Bernoulli: $F_X(t) = 1$ if $t \geq p$ and $F_X(t) = 0$ otherwise. Use the values $p = 0.5, 0.75, 0.9$. To make the problem interesting and to demonstrate the central limit theorem, look at sample averages of Bernoulli simulations.

   [In MATLAB, for a sample size of n and a given p, you can try the following: `x=sum(rand(n,1)<p)/n.`]

   e. Binomial: $F_X(t) = \sum_{k=0}^{\lfloor t \rfloor} \frac{n!}{k!(n-k)!} p^k (1-p)^{n-k}$ for n a positive integer and $0 < p < 1$.

   [In MATLAB, for given n and p, try the following to generate the probability mass function pmf: `k=0:n; pmf=(gamma(n+1)./ (gamma(k+1).*`

gamma(n-k+1))).* p.^k.*(1-p).^(n-k); So dist=cumsum(pmf); will
generate the cdf at values of k, and sum(dist<rand(1,1)) will simulate
the binomial.]

f. Discrete Poisson: $F_X(t) = \sum_{k=0}^{\lfloor t \rfloor} \frac{\lambda^k e^{-\lambda}}{k!}$. For simulations, use a
   truncated Poisson: i.e., neglect the upper tail and use the MAT-
   LAB hint above for the binomial.

2. If X is Pareto, show that the conditional distribution of X having
survived past a time T is also a Pareto distribution with the same shape
parameter but a different location parameter. Find the values for β that
cause X to have an infinite mean and/or infinite variance.

3. Simulate counting processes with the following interarrival distributions, until $t = 100$; then graph the resulting realizations. To understand
the average behavior of each process, simulate multiple realizations and
overlay the graphs. (See Figure 12.2.) Which process has the greatest
variability of its sample paths?

   a. Poisson process, with $\lambda = 1$.

      [In MATLAB, to simulate just arrival times, a quick method is to use
      cumsum(-log(rand(n,1))) and plot(cumsum(-log(rand(n,1))),1:n).]

   b. Weibull, with $\beta = 2$, $\alpha = 1$, $\nu = 0$.

   c. Shifted Pareto, with $\beta = 1.5$, $\alpha = 0.5$.

4. Simulate a time-dependent counting process using the technique of
thinning a Poisson process described in Section 12.4.3. In particular,
simulate a counting process with intensity $h(t) = \cos(t) + 1$. Use the
graphical techniques of the preceding problem to graph and to interpret
your results.

5. The following problem involves simulating a CTMC.

   a. Simulate a two-state CTMC using the technique of Section 12.4.2,
      where transition rates are independent of time: $a^1 = 0.5$, $a^2 = 1$.
      Graph a sample path.

   b. Simulate a two-state CTMC using the technique of Section 12.4.3,
      where transition rates are time dependent: $a^1(t) = \cos(t) + 1$,
      $a^2 = \cos(t + \pi) + 1$. Graph a sample path.

Remember, one simulation will not totally describe the dynamics, so
try several. If time permits, simulate a larger number of states, or for

multiple simulations, evaluate on a grid and describe the distributions of the states at grid times.

6. Using the simulated chains from Exercise 12.5 and the techniques described in Sections 12.3.3 and 12.4.2, simulate the following MMPPs.

   a. Use $\lambda^1 = 3$ and $\lambda^2 = 9$ for Exercise 12.5(a). See Figure 12.5.

   b. Use $\lambda^1 = 1$ and $\lambda^2 = 9.5$ for Exercise 12.5(b). See Figure 12.6.

7. For the arrival processes simulated in the exercises above, perform the following steps.

   a. If the arrival processes are given in terms of the arrival times $\{\tau_n\}$, generate the sequence of interarrival times via $I_n = \tau_n - \tau_{n-1}$. Find the sample arrival rate $\overline{\lambda} = 1/\overline{\tau}$. A fair comparison between different arrival processes dictates that the average arrival rate λ should be constant across processes. Scaling the arrival process through $I_n \lambda / \overline{\lambda}$ represents a time scale change and will result in comparable average waiting times.

   b. For queue utilizations $\rho = 0.25, 0.5, 0.75$ (or others), find the appropriate service rate γ, where $\rho = \lambda / \gamma$. If there are n interarrival times, simulate n exponential service times with rate γ.

   c. Generate the sequence of waiting time contributions $\{U_n\}$, its sequence of partial sums $\{V_n\}$, and the sequence of lowest excursions $\{\underline{V}_n\}$. Find the waiting times through $W_n = V_n - \underline{V}_n$.

   [In MATLAB, consider the cumsum command to construct the $\{V_n\}$.]

   d. Plot the sequences above against the arrival times. Also consider graphs of the work in the system at time t, which is found through $W_t = \max\{W_{N_t} + S_{N_t} - (t - Z_{N_t}), 0\}$ where $Z_n = \sum_{k=1}^n I_k$. Find the average waiting time $\overline{W}_n$ for this sample of customers. Compare the average waiting time of this arrival process with the average waiting time of the other arrival processes and across different utilizations.

   e. Another interesting problem is to consider the sample distribution of the number of customers in the queue at arrivals. The distributions should have the appearance of geometric distributions.

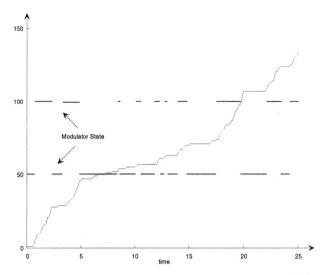

FIGURE 12.5   **Simulated MMPP for Exercises 12.5(a), 12.6(a)**

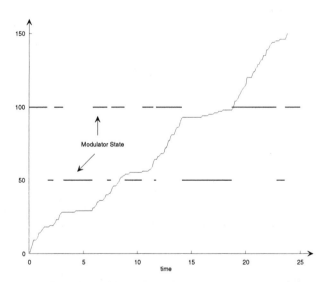

FIGURE 12.6   **Simulated MMPP for Exercises 12.5(b), 12.6(b)**

8. In the following problems, plot the sample variances versus the aggregate size.

   a. For a Poisson process, show that the aggregate variances climb linearly: $\text{Var}(N_t) = \lambda t$.

b. For a shifted Pareto interrenewal process show that the aggregate variances climb at a faster rate than linear: $\text{Var}(N_t) \approx \sigma t^{2H}$. Can you estimate the Hurst parameter? (**Hint:** try using loglog plots.)

c. For the MMPP arrival processes show that the aggregate variances climb faster than linear over a local range, then climb linearly.

d. For a two-state MMPP, the generator has the form

$$A = \begin{bmatrix} -a^1 & a^1 \\ a^2 & -a^2 \end{bmatrix}.$$

The range of local variance growth at a rate faster than linear is a function of the eigenvalues of the generator A. Consequently, the range of self-similar behavior can be extended by scaling the generator, thus scaling the eigenvalues. Reduce the generator by a factor of 10, simulate the MMPP, and perform the variance analysis. What is the effect on the variance growth?

For calculation of the variance growth curves, pick a set of aggregate sizes $\{\Delta t_1, \Delta t_2, \ldots, \Delta t_k\}$ and a desired sample size n. For graphical purposes, higher than linear growth patterns might be more readily observed using loglog plots, in which case taking $\Delta t_i = c\Delta^i$ for some constants c and Δ would be useful. For each aggregate size Δt_i, generate a sample of aggregates $N_1^{\Delta t_i}, N_2^{\Delta t_i}, \ldots, N_n^{\Delta t_i}$, where $N_j^{\Delta t_j} = N_{\tau_j + \Delta t_i} - N_{\tau_j}$. The sampling times $\tau_1, \tau_2, \ldots, \tau_n$ ideally should be chosen such that sampling intervals do not overlap and are preferably separated so as to reduce correlation. For the sample, determine the sample average $\overline{N^{\Delta t_i}}$ and the sample variance $\text{Var}(N^{\Delta t_i})$.

12.7   References

[1] P. K. Anderson, Ø. Borgan, R. D. Gill, and N. Keiding, *Statistical Models Based on Counting Processes*, Springer-Verlag, New York, 1993.

[2] J. Beran, *Statistics for Long-Memory Processes*, Chapman & Hall, New York, 1994.

[3] P. Brémaud, *Point Processes and Queues, Martingale Dynamics*, Springer-Verlag, New York, 1981.

[4] H. Clark and B. McMillin, "DAWGS — a distributed compute server utilizing idle workstations," *Journal of Parallel and Distributed Computing* **14**, 175–186 (1992).

[5] W. Feller, *An Introduction to Probability Theory and Its Applications*, Volume I, 3rd ed., Wiley, New York, 1968.

[6]   R. V. Hogg and E. A. Tanis, *Probability and Statistical Inference*, 3rd ed., Macmillan, New York, 1983.

[7]   P. C. Kiessler, C. J. Wypasek, R. E. Fennell, and J. M. Westall, "Markov renewal models for traffic exhibiting self-similar behavior," *Proceedings of the IEEE SOUTHEASTCON '96*, 76–79, April 1996.

[8]   P. Krueger and R. Chawla, "The Stealth distributed scheduler," *Proceeding of the 11th International Conference on Distributed Computing Systems*, 336–343, May 1991.

[9]   M. J. Litzkow, M. Livny, and M. W. Mutka, "Condor — a hunter of idle workstations," *Proc. IEEE 1988 Conference on Distributed Computer Systems*, 104–111, 1988.

[10] The MathWorks, Inc., *Using MATLAB Version 5*, The Math-Works, Inc., Natick, MA, 1997. <http://www.mathworks.com/>

[11] D. A. Nichols, "Using idle workstations in a shared computing environment," *ACM Operating Systems Review* **21**, 5–12 (1987).

[12] P. Protter, *Stochastic Integration and Differential Equations: A New Approach*, Springer-Verlag, Berlin, 1990.

[13] M. M. Theimer and K. A. Lantz, "Finding idle machines in a workstation-based distributed system," *Proc. IEEE 1988 Conference on Distributed Computing Systems*, 112–122, 1988.

[14] C. J. Wypasek, *Stochastic Models for Workstation Utilization*, Ph.D. Dissertation, Clemson University, Clemson, SC (1994).

Chapter 13

Graph-Theoretic Analysis of Finite Markov Chains

J. P. Jarvis
D. R. Shier

Prerequisites: abstract algebra, data structures, graph theory, probability

13.1 Introduction

Markov chains arise frequently in the modeling of physical and conceptual processes that evolve over time. For example, the diffusion of liquids across a semi-porous membrane, the spread of disease within a population, and the flow of personnel within the ranks of an organization can all be modeled using Markov chains. In each of these cases, the system can be found in any of a finite number of states, and transitions between states occur at discrete instants according to specified probabilities.

As one illustration, suppose that there are M molecules in a vessel, separated into two chambers by a membrane, across which molecules can pass. A typical configuration of the system at any instant can be described by the distribution of the M molecules between the two chambers. If there are k_1 molecules in the first chamber, then there will be $k_2 = M - k_1$ molecules in the second chamber. Transitions from the current state (k_1, k_2) can occur by the movement of a single molecule from the first chamber to the second, or from the second chamber to the first. These two new states are represented by $(k_1 - 1, k_2 + 1)$ and $(k_1 + 1, k_2 - 1)$, respectively. In one possible model of this process, the probability of transition from (k_1, k_2) to $(k_1 - 1, k_2 + 1)$ is given by k_1/M, whereas the probability of transition to $(k_1 + 1, k_2 - 1)$ is $k_2/M = 1 - k_1/M$. This quantifies the idea that if more molecules are present in (say) chamber 1, then it is more likely for some molecule to transfer next from chamber 1 to chamber 2. Using this mathematical model, one can answer questions such as: (a) under what conditions do the molecules achieve an equilibrium configuation, (b) what are the (probabilistic) characteristics of this configuration, and (c) at what rate is this equilibrium approached?

The above is an instance of a *finite-state Markov chain*, which is the topic of the present chapter. Several applications discussed in this book are based on such Markov chain models; see Chapters 4, 8, 12. Normally, this subject is presented in terms of the (finite) matrix describing the Markov chain. Our objective here is to supplement this viewpoint with a graph-theoretic approach, which provides a useful visual representation of the process. A number of important properties of the Markov chain (typically derived using matrix manipulations) can be deduced from this pictorial representation. Moreover, certain concepts from modern algebra will also be illuminated in the process of developing this approach. In addition, the graph-theoretic representation immediately suggests several computational schemes for calculating important structural characteristics of the underlying problem. This chapter indicates how appropriate data structures and algorithms enable such calculations to be carried out in an efficient manner.

13.2   State Classification

In this section the basic concepts of finite Markov chains are defined, leading to the notion of classifying states of the chain as either "recurrent" or "transient." A combination of algebraic and graph-theoretic approaches turns out to be useful in identifying such states. Algorithmic techniques for carrying out state classification are also discussed.

13.2.1   Preliminaries

Suppose that $\mathcal{M}$ is a finite-state Markov chain with states $\{1, 2, \ldots, n\}$. At every (discrete) instant t the chain $\mathcal{M}$ will be in one of these states. The quantity p_{ij} denotes the conditional probability that $\mathcal{M}$ will be in state j at time $t + 1$, given that it was observed in state i at time t. Implicitly, we are assuming that the Markov chain is *homogeneous*: namely, the values p_{ij} are independent of time t. These (one-step) state transition probabilities define the *transition probability matrix* $P = [p_{ij}]$. In general, let p_{ij}^k denote the probability that $\mathcal{M}$ proceeds from state i to state j after k transitions. Then the *k-step transition probability matrix* $P^{(k)} = [p_{ij}^k]$ is given by $P^{(k)} = P^k$, the kth matrix power of the transition probability matrix P. Notice that when $k = 0$ this produces

$P^{(0)} = P^0 = I_n$, the $n \times n$ identity matrix, agreeing with the observation that after $k = 0$ steps the Markov chain is still in its initial state. The Markov chain is called *irreducible* if, for every pair of states i and j, there exist $r, s \geq 0$ with $p_{ij}^r > 0$ and $p_{ji}^s > 0$.

Example 13.1

Figure 13.1 shows the transition probability matrix P for a five-state Markov chain, on the states $1, 2, 3, 4, 5$. Also shown is the third power P^3 of P. Accordingly, the conditional probability of being in state 4 at time 5, given that the system is observed in state 3 at time 2, is $p_{34}^3 = 0.186$. As will be seen later, this chain is not irreducible.

$$P = \begin{bmatrix} 0 & 0 & 1 & 0 & 0 \\ 0 & 0 & 0 & 0 & 1 \\ .5 & 0 & .2 & .3 & 0 \\ .4 & .1 & 0 & .2 & .3 \\ 0 & 1 & 0 & 0 & 0 \end{bmatrix}, \quad P^3 = \begin{bmatrix} .220 & .030 & .540 & .120 & .090 \\ 0 & 0 & 0 & 0 & 1 \\ .318 & .102 & .328 & .186 & .066 \\ .216 & .164 & .160 & .128 & .332 \\ 0 & 1 & 0 & 0 & 0 \end{bmatrix}$$

FIGURE 13.1   **A transition probability matrix P**

Associated with the Markov chain $\mathcal{M}$ is a *digraph* (directed graph) $G = G_{\mathcal{M}}$ having the set of nodes $N = \{1, 2, \ldots, n\}$ and the set of edges E. Each node corresponds to a state of $\mathcal{M}$, and G contains edge $(i, j) \in E$ if and only if $p_{ij} > 0$. Thus the digraph, or *state transition diagram*, G captures the structure of the possible one-step state transitions; the actual numerical values of the state transition probabilities are ignored. In graph-theoretic terms, $p_{ij}^k > 0$ means there is a directed *path* Q of *length* $l(Q) = k$ (number of edges) from node i to node j in G. If this holds for some $k \geq 0$, then node j is *accessible* from node i, written $i \to j$. Observe that $i \to i$ since we consider node i to be reachable from itself by a path of length 0. If there is no path in G from i to j, then we write $i \not\to j$. If both $i \to j$ and $j \to i$ hold, then we say that states i and j *communicate*, written $i \leftrightarrow j$. A path joining a node to itself is called a *circuit*. If this circuit contains no repeated nodes, then it is a *cycle*.

Example 13.2

Figure 13.2 shows the state transition diagram G for the Markov chain in Figure 13.1. Notice that there is a path from node 1 to node 2, but no path from node 2 to node 1; thus $1 \to 2$ whereas $2 \not\to 1$. The node sequence $[1, 3, 4, 1]$ defines a cycle in G. It is also seen that $1 \leftrightarrow 3$, $1 \leftrightarrow 4$, and $3 \leftrightarrow 4$.

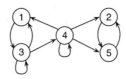

FIGURE 13.2   **The state transition diagram G**

An important concept in the analysis of Markov chains is the categorization of states as either recurrent or transient. The Markov chain, once started in a recurrent state, will return to that state with probability 1. However, for a transient state there is some positive probability that the chain, once started in that state, will never return to it. This same concept can be illuminated using graph-theoretic concepts, which do not involve the numerical values of probabilities. Namely, we define node i to be *transient* if there exists some node j for which $i \to j$ but $j \not\to i$. Otherwise, node i is called *recurrent*. (These definitions apply only to finite Markov chains.) Let $\mathcal{T}$ denote the set of all transient nodes, and let $\mathcal{R} = N - \mathcal{T}$ be the set of all recurrent nodes.

13.2.2   Algebraic and Graph-Theoretic Considerations

Certain states in a Markov chain behave in a similar fashion. The following algebraic fact underlies this type of "state classification."

LEMMA 13.1

The relation $\leftrightarrow$ is an equivalence relation: namely, (a) $i \leftrightarrow i$ for all $i \in N$; (b) if $i \leftrightarrow j$, then $j \leftrightarrow i$ for all $i, j \in N$; and (c) if $i \leftrightarrow j$ and $j \leftrightarrow k$, then $i \leftrightarrow k$ for all $i, j, k \in N$.

Consequently, the equivalence relation $\leftrightarrow$ partitions N into a number of *equivalence classes* (communicating classes): that is, disjoint sets whose union is N. Notice that $\mathcal{M}$ is irreducible precisely when there is just a single equivalence class under $\leftrightarrow$. A useful result is that nodes within a given equivalence class do indeed behave similarly.

LEMMA 13.2

If node i is recurrent and $i \leftrightarrow j$, then node j is recurrent. If node i is transient and $i \leftrightarrow j$, then node j is transient.

In the theory of directed graphs, G is called *strongly connected* if there is a path between any pair of nodes i, j in G. In other words, $i \to j$ holds for all i, j, meaning that $i \leftrightarrow j$ for all i, j. Thus an irreducible Markov chain $\mathcal{M}$ is simply one whose digraph G is strongly connected. In general, the communicating classes of $\mathcal{M}$ are just the maximal strongly connected subgraphs of G — the *strong components* of G. It is known that if nodes i and j lie on a common circuit, then they belong to the same strong component, and conversely. As a consequence, when the nodes within each strong component are combined into a new "supernode," then the *condensed* graph $\widehat{G}$ governing these supernodes can contain no cycles — $\widehat{G}$ is an *acyclic* graph.

Example 13.3

The state transition diagram in Figure 13.2 has two strong components: $K_1 = \{1, 3, 4\}$ and $K_2 = \{2, 5\}$. Thus the Markov chain $\mathcal{M}$ is not irreducible. For the larger state transition diagram shown in Figure 13.3, the four strong components are $K_1 = \{2, 5, 6\}$, $K_2 = \{3, 4\}$, $K_3 = \{1, 7\}$, and $K_4 = \{8\}$. The condensed graph $\widehat{G}$ on these components is also displayed. All nodes in components K_2 and K_4 are recurrent, whereas all nodes in components K_1 and K_3 are transient. Recall that Lemma 13.2 assures us that all nodes within a strong component have the same classification.

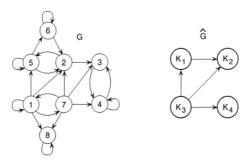

FIGURE 13.3   **A sample digraph G with its condensed graph**

This example suggests that the recurrent components are precisely those with no leaving edges in the graph $\widehat{G}$. This is true in general.

LEMMA 13.3

The recurrent nodes of graph G are those nodes whose corresponding supernodes have no leaving edges in $\widehat{G}$.

In view of Lemma 13.3, inspection of the state transition diagram G of the finite chain $\mathcal{M}$ allows one to classify each node as either recurrent or transient. In addition, these concepts are important in connection with *stationary distributions* $\pi = [\pi_1, \pi_2, \ldots, \pi_n]$ for a Markov chain having states $N = \{1, 2, \ldots, n\}$. These probabilities represent the long run proportion of time the chain $\mathcal{M}$ spends in each state. Such probabilities can be found by solving the linear system $\pi = \pi P$, $\sum_{j \in N} \pi_j = 1$. The following standard results in the theory of Markov chains are stated in terms of the state transition diagram G for $\mathcal{M}$.

THEOREM 13.1

If G is strongly connected, then there is a unique stationary distribution π for $\mathcal{M}$. Moreover, this distribution satisfies $\pi_j > 0$ for all $j \in N$.

THEOREM 13.2

If the condensed graph $\widehat{G}$ for G has a single supernode with no leaving edges, then there is a unique stationary distribution π for $\mathcal{M}$. Moreover, this distribution satisfies $\pi_j > 0$ for all $j \in \mathcal{R}$, and $\pi_j = 0$ for all $j \in \mathcal{T}$.

13.2.3   Depth-First Search Algorithm

A natural question concerns computational methods for identifying the recurrent and transient nodes of G. Fortunately, this can be done very efficiently by use of a depth-first search (or, pre-order traversal) of the digraph $G = (N, E)$. A depth-first search produces a numbering or labeling of the nodes of G, where each node is numbered in the order in which it is first encountered. This numbering can then be used in conjunction with the depth-first search to produce the strong components of G. In general, the entire procedure has time complexity $O(n + m)$ where $n = |N|$ and $m = |E|$.

To carry out a depth-first search on G, an arbitrary node $v_0 \in N$ is selected as the starting node and is labeled with $\text{lab}(v_0) = 1$. The algorithm will end up labeling all vertices that are accessible from v_0. When the algorithm terminates, there may be unlabeled nodes (not accessible from the starting node v_0). In that case, the algorithm is restarted with an unlabeled node. Continuing in this manner, eventually all nodes will be labeled. For simplicity, the algorithm is now described for a graph G in which all nodes are accessible from the initially selected node v_0.

The algorithm labels the nodes of G with the numbers $1, 2, \ldots, n$ according to the order in which they are first encountered, starting with $\text{lab}(v_0) = 1$. In addition, edges are partitioned into two sets: *tree* and *nontree* edges. As each newly labeled node v is processed, an edge (v, w) is considered. If node w has not yet been labeled, then w becomes labeled with the next available number, edge (v, w) is classified as a tree edge, and processing continues with the newly labeled node w. If w has already been labeled, (v, w) is a nontree edge and can be further classified as a *back edge, forward edge,* or *cross edge.* Back edges (v, w) have $\text{lab}(v) > \text{lab}(w)$ with v a descendant of w in the tree; forward edges (v, w) have $\text{lab}(v) < \text{lab}(w)$ with w a descendant of v; cross edges (v, w) have $\text{lab}(v) > \text{lab}(w)$ with w neither a descendant nor ancestor of v. (See Figure 13.4.)

When all edges incident from v have been considered, processing is continued using the node u that was used to label node v. (Note that there is a unique tree edge incident to v: namely, (u, v).) Assuming that all nodes are accessible from the initial node v_0, the depth-first search algorithm will terminate when all nodes have been labeled and all edges have been classified.

Example 13.4

Figure 13.4 shows a depth-first search tree rooted at node 1 for the state transition diagram given in Figure 13.3. Self-loops on nodes have been ignored, and the labels of nodes in Figure 13.4 turn out to be the original node numbers.

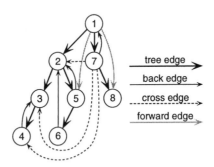

FIGURE 13.4   **Edges in a depth-first search of a digraph**

At the same time as a depth-first search is carried out on G, additional information can be collected to enable identification of the strong components of G. Our objective is not to describe the intricacies of this modification; the reader is referred to Aho et al. [1] for details. Rather we briefly describe here the overall strategy of this approach for finding strong components.

It can be verified that all nodes in the same strong component will have a common ancestor (within that component) relative to a depth-first search tree. The minimum label common ancestor (for a particular depth-first search tree) is referred to as the *root* of the strong component. Suppose G has strong components $G_1, G_2, \ldots, G_k$ and let r_i denote the root associated with G_i. The components are numbered in the order that the depth-first search from each root node is completed. The following lemma identifies the nodes in the strong components of G.

LEMMA 13.4

The nodes in strong component G_i are those nodes that are descendants of r_i but are not in G_k, for any $1 \leq k < i$.

Example 13.5

In Figure 13.4, the root nodes turn out to be $3, 2, 8, 1$. Notice that the descendants of node 3 are $\{3, 4\}$ while the descendants of node 2 are $\{2, 3, 4, 5, 6\}$, producing the strong components $G_1 = \{3, 4\}$ and $G_2 = \{2, 5, 6\}$. Similarly, $G_3 = \{8\}$ is the strong component derived from root node 8, and $G_4 = \{1, 7\}$ is the strong component derived from root node 1. These components correspond to supernodes K_2, K_1, K_4, and K_3, respectively, in Figure 13.3.

Thus the real task is to identify the roots of each strong component. The algorithm in [1] does this by executing a depth-first search and keeping auxiliary information for each node (which is updated whenever nontree edges are encountered). Overall, the strong components can be found (and states classified as recurrent or transient) very efficiently, in $O(n+m)$ time. By using appropriate data structures for representing G, the storage required by the entire algorithm is also $O(n+m)$. Recall that n is the number of states of the Markov chain and m is the number of edges in G or equivalently the number of nonzero transition probabilities in P. Accordingly, the sparsity of the transition probability matrix P can be exploited computationally by this algorithmic approach.

13.3   Periodicity

In this section, the important concept of periodicity is explored. This concept is quite useful in understanding the "limiting behavior" of a

Markov chain $\mathcal{M}$. Suppose that there is positive probability that $\mathcal{M}$, started in state i, will return to state i in a finite number of steps. That is, there is some $k > 0$ such that $p_{ii}^k > 0$. Then the *period* of state i is the largest integer d such that $p_{ii}^k = 0$ whenever k is not a positive integer multiple of d (that is, $k \neq d, 2d, 3d, \ldots$). Note that p_{ii}^k is allowed to be positive only when k is a multiple of d, but it is not necessary that all such p_{ii}^k be positive. A state with period $d = 1$ is called *aperiodic*.

Example 13.6

In the chain of Figure 13.5(a), returns to state 1 can only occur at steps $k = 3, 6, 9, 12, \ldots$, so $p_{11}^k = 0$ for k not a multiple of 3; consequently state 1 has period $d = 3$. In Figure 13.5(b), returns to state 1 can only occur at steps $k = 8, 10, 16, 18, 20$ and even $k \geq 24$; state 1 has period $d = 2$. Notice that in this case, p_{ii}^k is not positive for all even $k > 0$; namely, $p_{ii}^k = 0$ for $k = 2, 4, 6, 12, 14, 22$.

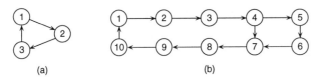

(a)   (b)

FIGURE 13.5   **Examples of periodic Markov chains**

13.3.1   *Analytical Results*

Here we state some standard facts concerning the period of a Markov chain, indicative of the importance of this concept in analyzing the long-term behavior of the chain. The remainder of this subsection focuses on setting up the machinery needed for the efficient computation of the period d. The following result shows that all states in the same communicating class have the same period.

LEMMA 13.5

If state i has period d and $i \leftrightarrow j$, then state j has period d.

When viewed in terms of the state transition diagram G, the period of strong component K_i is just the greatest common divisor (gcd) of the lengths of all directed circuits in the subgraph induced by the nodes of K_i. Equivalently, the period of K_i is the gcd of the lengths of all directed cycles in this subgraph.

Example 13.7

For the state transition diagram of Figure 13.2, all states of $K_1 = \{1, 3, 4\}$ are aperiodic while all those of $K_2 = \{2, 5\}$ have period 2.

Let $\alpha = [\alpha_1, \alpha_2, \ldots, \alpha_n]$ be the vector of *initial probabilities* of being in any of the n states at time $t = 0$. Then the vector αP^k gives the absolute probability of being in each state at time $t = k$. Under certain conditions, this probability vector approaches a vector of *limiting probabilities* as $t \to \infty$.

THEOREM 13.3

Suppose that the condensed graph $\widehat{G}$ for G has a single supernode K with no leaving edges and that all states of K are aperiodic. Then limiting probabilities exist and are independent of the initial probability vector α. Moreover, these limiting probabilities coincide with the (unique) stationary distribution for the chain.

A special but important case of Theorem 13.3 occurs when G is strongly connected and all states are aperiodic. Indeed, for the remainder of this section it will be assumed that G is strongly connected ($\mathcal{M}$ is irreducible). Lemma 13.5 then assures us that every state of $\mathcal{M}$ has the same period d. If $d = 1$ then $\mathcal{M}$ will be called *aperiodic*; if $d \geq 2$ then $\mathcal{M}$ will be called *periodic*.

The period d has been defined in terms of the gcd of the lengths of all cycles in G. Now we explore another graph-theoretic characterization of the period that will be useful in developing an efficient algorithm for computing d. The following result characterizes the period d of $\mathcal{M}$ through a decomposition of G into "cyclically moving classes."

THEOREM 13.4

$\mathcal{M}$ has period d if and only if its digraph G can be partitioned into d sets $C_0, C_1, \ldots, C_{d-1}$ such that (a) if $i \in C_k$ and $(i, j) \in E$ then $j \in C_{(k+1) \bmod d}$; and (b) d is the largest integer with this property.

Before establishing the validity of this result, we first provide an interpretation of the characterization it provides. Property (a) says that starting with any state in set C_k, the next transition must be to a state in C_{k+1}, then to a state in C_{k+2}, and so on, where the succeeding set index is taken modulo d. This situation is depicted in Figure 13.6. If $\mathcal{M}$ is aperiodic $(d = 1)$, then there is a single set C_0 containing all states.

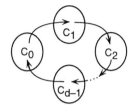

FIGURE 13.6   **State transitions in a chain with period** d

It is easy to see that if the digraph G is partitioned into d sets with properties (a) and (b), then d must be the period of $\mathcal{M}$. Since all paths in G proceed from C_k to C_{k+1} to C_{k+2}, and so forth, every path from i to i must have length 0 modulo d. In $\mathcal{M}$ this means that if $p_{ii}^r > 0$, then $r \bmod d = 0$. Hence the period of the chain must be a multiple of d. The maximality of d for the partition in G implies that the period is exactly d.

To establish the converse, it is instructive to define a relation on the nodes of the strongly connected graph G associated with $\mathcal{M}$, assumed to have period d. Namely, define $i \sim j$ if every (i,j)-path has length 0 modulo d. Note that since G is strongly connected, a path exists between every pair of nodes in G. The following result asserts that we again have an equivalence relation.

LEMMA 13.6

$\sim$ *is an equivalence relation.*

While it is conceivable that some (i,j)-paths might have length 0, while others have positive length (modulo d), this cannot occur. In fact the following general result holds concerning path lengths, modulo d.

LEMMA 13.7

For each pair of nodes i and j in G, all (i,j)-paths in G have the same length modulo d.

The equivalence classes associated with $\sim$ form the desired partition when appropriately labeled. Choose any node i_0 and a cycle containing

i_0. Since $\mathcal{M}$ has period d, this cycle must contain at least d edges. Let the first d nodes of this cycle be denoted $i_0, i_1, \ldots, i_{d-1}$. Notice that no two of these nodes can be elements of the same equivalence class because this would identify a path between those two nodes of length less than d, whereas all such paths must have length 0 modulo d. Label the equivalence class containing node i_k as C_k.

LEMMA 13.8

Let $\{C_0, C_1, \ldots, C_{d-1}\}$ be the equivalence classes induced by $\sim$ and numbered according to any (i_0, i_0) cycle. If (i, j) is an edge of G with $i \in C_k$, then $j \in C_{(k+1) \bmod d}$.

COROLLARY 13.1

Let Q_1 and Q_2 be (i, j_1) and (i, j_2)-paths in G, respectively. If $l(Q_1) \bmod d = l(Q_2) \bmod d$, then j_1 and j_2 belong to the same equivalence class.

Lemma 13.8 shows that property (a) of Theorem 13.4 holds. Property (b) follows from the maximality of the period d. Moreover, by Corollary 13.1 and the fact that there are exactly d remainders modulo d, it is seen that $\{C_0, C_1, \ldots, C_{d-1}\}$ is indeed a partition of the nodes of G. This establishes the stated characterization for the period d of an irreducible Markov chain. Consequently, determining d can be reduced to finding an appropriate partition of the nodes in the associated digraph G. Although this might seem to be a more difficult problem, it can be accomplished efficiently in conjunction with a breadth-first search of G.

13.3.2   Breadth-First Search Algorithm

In the irreducible Markov chain $\mathcal{M}$, every transition moves from a node in C_k to a node in $C_{(k+1) \bmod d}$. This property can be used to identify the sets of nodes C_k in the associated digraph G by examining the nodes of G in order of nondecreasing path length from an arbitrary starting node. A breadth-first search of G provides a systematic examination of such paths.

A breadth-first search of G from node v produces a collection of *level* sets. The kth level set contains those nodes j that can be reached from node v by a shortest path of length k; in this case define $\text{level}(j) = k$. Since all nodes within a level set share this common path length, every level set is contained in a single equivalence class (by Corollary

13.1). This property is independent of the node from which the breadth-
first search is started. The difficulty lies in determining which level sets
belong to the same equivalence class. Lemma 13.7 shows that certain
information about d can be gleaned from the knowledge of path lengths.
Sufficient information of this type can be gathered as we carry out the
breadth-first search to determine d.

A breadth-first search of the (strongly connected) digraph G visits
all nodes of G in order of nondecreasing level. In addition, this search
partitions the edges of G into *tree* and *nontree* edges. The breadth-first
search is started from an arbitrary node v, setting level$(v) = 0$. As
the search progresses, all edges incident from nodes on each level are
successively examined in order of increasing level. Suppose edge (i, j)
emanates from a node i with level$(i) = r$. Edge (i, j) is a tree edge
if node j has not yet been encountered in the traversal of G. In this
case, node j is assigned level$(j) = r + 1$. The collection of all such tree
edges induces a directed tree T (the *breadth-first search tree*) rooted from
the starting node v. If j has already been encountered, then (i, j) is a
nontree edge and j already belongs to some level set, with $s = $ level(j)
$\leq r + 1$. If j is an ancestor of i in the search tree, the edge is called a
back edge $(s < r)$; otherwise, the edge is called a *cross edge* $(s \leq r + 1)$.
See Figure 13.7. In both cases, these nontree edges provide information
regarding the period of $\mathcal{M}$.

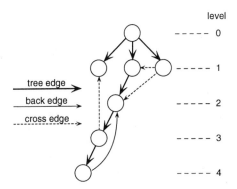

FIGURE 13.7   **Tree, back, cross edges in a breadth-first search**

If (i, j) is a back edge, then the tree edges from j to i plus the edge
(i, j) form a cycle in G. The length of the cycle, $r - s + 1$, is divisible
by the period of the Markov chain. If (i, j) is a cross edge, two distinct
paths from the root to j have been identified. One of these follows tree
edges directly to j and has length s. The other path follows tree edges
to i and then uses edge (i, j), giving a length of $r + 1$. Since (by Lemma

13.7) alternative paths between any pair of nodes have the same length modulo d, the difference in path lengths, $r - s + 1$, is again divisible by the period d of the Markov chain. When $s = r + 1$, the paths have the same length and this provides no additional information about the period. Accordingly, we define the *value* of any edge $e = (i, j)$ by val(e) = val(i, j) = level(i) − level(j) + 1 ≥ 0. Note that val(e) = 0 when $e \in T$. Also d must divide val(e) for $e \notin T$, so d must divide $g = $ gcd{val(e) > 0 : $e \notin T$}. In fact we claim that $g = d$, which will be established after first stating the implied process (Algorithm 13.1) for determining the period of a finite irreducible Markov chain $\mathcal{M}$. This approach was first developed by Denardo [3].

ALGORITHM 13.5: Finding the period d of $\mathcal{M}$

1. From an arbitrary root node, perform a breadth-first search of G producing the rooted tree T.
2. The period g is given by gcd{val(e) > 0 : $e \notin T$}.

As a practical matter, it is not necessary to keep all of the values generated by nontree edges; rather, we maintain only the current gcd g, initialized to be the first positive edge value encountered. Whenever a new val(e) > 0 is found for some edge $e \notin T$, then g is updated using $g := $ gcd{g, val(e)}. If at any step val(e) = 1 is generated (from a cross edge within the same level set), then the gcd is 1 and the Markov chain is aperiodic; the algorithm can be terminated immediately in this case.

Example 13.8

To illustrate Algorithm 13.1, consider the digraph G with six nodes shown in Figure 13.8. A breadth-first search tree for G rooted at node 1 is shown in Figure 13.9. The first nontree edge encountered is $(2, 1)$, a back edge. Then g is initialized to be level(2) − level(1)+1 = $3 - 0 + 1 = 4$. The only other nontree edge is $(6, 4)$, a cross edge. It has value level(6) − level(4) + 1 = $3 - 2 + 1 = 2$, giving the final g = gcd{4, 2} = 2. Notice that G contains no cycle of length 2, but it does have cycles of lengths 4 and 6, yielding a period of 2.

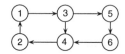

FIGURE 13.8   **A Markov chain with period 2**

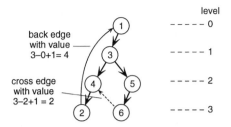

level

back edge
with value
3–0+1= 4

cross edge
with value
3–2+1 = 2

----- 0

----- 1

----- 2

----- 3

FIGURE 13.9   **Breadth-first search for a Markov chain**

PROOF   To establish the correctness of Algorithm 13.1, it suffices
to show that $g = \gcd\{\mathrm{val}(e) > 0 : e \notin T\}$ divides the length of any cycle
W in G. If so it must divide the period d, the gcd of all cycle lengths in
G. However, we have already seen that d divides g, so that $g = d$.

Relative to the breadth-first search tree T, let the edges E of G be
partitioned into sets D and R. Set D consists of *down* edges (i, j), in
which $\mathrm{level}(j) = \mathrm{level}(i) + 1$, and set R contains the *remaining* edges.
Notice that all tree edges are in D, all back edges are in R, whereas
cross edges can be in either set. Also edge e has $\mathrm{val}(e) > 0$ precisely
whenever $e \in R$. Now let $Q = [i_0, i_1, \ldots, i_k]$ be any path of length
$l(Q) = k$ in G and define $\mathrm{val}(Q) = \sum_{s=0}^{k-1} \mathrm{val}(i_s, i_{s+1})$. Since $\mathrm{val}(i_s, i_{s+1})$
$= \mathrm{level}(i_s) - \mathrm{level}(i_{s+1}) + 1$, it then follows that $\mathrm{val}(Q) = \mathrm{level}(i_0) -$
$\mathrm{level}(i_k) + l(Q)$. In particular, if W is any cycle from node i to itself in
G, then $\mathrm{val}(W) = \mathrm{level}(i) - \mathrm{level}(i) + l(W) = l(W)$. This gives $l(W)$
$= \mathrm{val}(W) = \sum \{\mathrm{val}(e) : e \in W\} = \sum \{\mathrm{val}(e) : e \in W \cap R\}$. However,
each term in the last sum is positive and divisible by g (the gcd of all
positive nontree values), so g divides $l(W)$, as required.   ∎

The computational effort associated with Algorithm 13.1 can be split
into two parts: executing the breadth-first search and then finding the
greatest common divisor associated with certain nontree edges. Let $m =$
$|E|$ denote the number of edges in the digraph G. As shown in [7],
the time complexity associated with the breadth-first search is $O(n +$
$m) = O(m)$. Also, there are $m - n + 1$ nontree edges, hence a greatest
common divisor must be found at most $m - n$ times (some nontree
edges are down edges and are thus not considered). Since $\mathrm{val}(e) =$
$\mathrm{level}(i) - \mathrm{level}(j) + 1 \leq \mathrm{level}(i) + 1 \leq n$, the complexity of finding
$m - n$ gcd values turns out to be $O(\log n + m) = O(m)$; see [2]. Hence
the overall complexity of the algorithm is $O(m)$. Using appropriate
data structures, the algorithm requires $O(n + m) = O(m)$ space. As
in the depth-first search algorithm for finding strong components, this

algorithm for determining d can exploit sparsity present in the transition probability matrix P, which often has relatively few nonzero entries.

13.4   Conclusion

The aim in this chapter is to provide a parallel development to the standard, matrix-based analysis of finite-state Markov chains [6, 8]. Our graph-theoretic interpretations not only maintain a visual representation of the model, but also reinforce a number of algebraic concepts. Specifically, the idea of an equivalence relation is quite useful in carrying out state classification and in determining periodicity. Moreover, an important step in any modeling effort is the development of efficient algorithms and data structures that "solve" the model in an effective way. This is especially important for the successful completion of large-scale modeling projects. Interestingly, certain of the Markov chain computations are most easily carried out using a depth-first search of G (state classification), while others involve a breadth-first search of G (determining the period). Overall, a mathematical sciences approach (blending discrete, algebraic, and computational elements) is illustrated here.

13.5   Exercises and Projects

1. Prove that the relation $\leftrightarrow$ is an equivalence relation (Lemma 13.1).

2. Prove that all states in a communicating class have the same character with regard to transience and recurrence (Lemma 13.2).

3. Consider the Markov chain on states $\{1, 2, \ldots, 6\}$ whose nonzero transition probabilities are indicated by '+' in the matrix below:

$$
P = \begin{bmatrix}
+ & 0 & 0 & 0 & 0 & 0 \\
0 & 0 & 0 & 0 & + & 0 \\
0 & 0 & + & + & + & 0 \\
+ & 0 & 0 & 0 & 0 & + \\
0 & + & 0 & 0 & + & 0 \\
+ & 0 & 0 & + & 0 & 0
\end{bmatrix}
$$

  a. Draw the state transition diagram G for this Markov chain and determine the communicating classes.

  b. Classify the states of this Markov chain (recurrent or transient).

  c. Does the chain have a unique stationary distribution?

   d. Draw a depth-first search tree, starting the search at node 3. Show
      the node labels and the nontree edges produced by the search.

   e. What are the root nodes for the strong components of G?

4. The *node adjacency matrix* $A = [a_{ij}]$ for a digraph $G = (N, E)$ with
$n = |N|$ is the $n \times n$ matrix with $a_{ij} = 1$ if $(i, j) \in E$, $a_{ij} = 0$ otherwise.

   a. Show that the (i, j) entry of A^k is the number of distinct (i, j)-
      paths in G having length k. (Such paths are allowed to contain
      repeated nodes and edges.)

   b. The *reachability matrix* $R = [r_{ij}]$ for G is the 0-1 matrix in which
      $r_{ij} = 1$ if $i \to j$, $r_{ij} = 0$ otherwise. Show that $R = B((I + A)^{n-1})$
      where $B(\cdot)$ is the (entrywise) Boolean function taking on the value
      1 for nonzero arguments and 0 for zero arguments.

   c. Show that node j is in the strong component of G containing node
      i if and only if $r_{ij} \cdot r_{ji}$ is nonzero. Compare the complexity of this
      approach for finding strong components with the graph-theoretic
      approach outlined in Section 13.2.3.

5. Let i be a state of a Markov chain with a positive probability of a
return to state i in a finite number of steps. Let D denote the set of all
positive integers r such that $p_{ii}^k = 0$ for all k not divisible by r.

   a. Show that $1 \in D$.

   b. Show that D is bounded above (and hence contains a largest pos-
      itive member).

6. Prove that if i and j are communicating states, then i and j have
the same period (Lemma 13.5). **Hint:** Consider paths P, Q that make
i accessible from j and j accessible from i, respectively. Then the union
of P and Q is a circuit through both i and j. Now apply the definition
of periodicity.

7. Prove that the relation $\sim$ is an equivalence relation (Lemma 13.6).

8. Prove Lemma 13.7.

9. Consider the Markov chain on states $\{1, 2, \ldots, 10\}$ whose nonzero
transition probabilities are indicated by '+' in the matrix that follows.

   a. Draw the state transition diagram for this chain. Explain why the
      chain is irreducible.

b. Show the breadth-first search tree obtained by starting with state (node) 1. Also indicate the level of each node and display the nontree edges.

c. Use the techniques described in Section 13.3.2 to determine the period d of this Markov chain.

d. Use the level sets obtained from the breadth-first search to identify the partition of states induced by periodicity (see Figure 13.6).

$$
P = \begin{bmatrix}
0 & 0 & 0 & + & 0 & 0 & 0 & 0 & 0 & 0 \\
+ & 0 & 0 & 0 & + & 0 & 0 & 0 & 0 & 0 \\
0 & 0 & 0 & 0 & 0 & 0 & + & 0 & 0 & 0 \\
0 & + & + & 0 & 0 & 0 & 0 & + & 0 & 0 \\
0 & 0 & 0 & + & 0 & 0 & 0 & 0 & 0 & 0 \\
0 & 0 & 0 & 0 & + & 0 & 0 & 0 & 0 & 0 \\
0 & 0 & 0 & + & 0 & 0 & 0 & 0 & 0 & 0 \\
0 & 0 & 0 & 0 & 0 & 0 & + & 0 & + & 0 \\
0 & 0 & 0 & 0 & 0 & 0 & 0 & 0 & 0 & + \\
0 & 0 & 0 & 0 & 0 & + & 0 & 0 & 0 & 0
\end{bmatrix}
$$

10. Suppose that an irreducible finite-state Markov chain with period d has the associated cyclic partition of states $\{C_0, C_1, \ldots, C_{d-1}\}$. Suppose that the states are numbered so that states from C_i receive smaller numbers than those from C_{i+1}, $0 \le i < d - 1$.

a. Show that the state transition probability matrix P has the block form given below.

$$
\begin{bmatrix}
0 & P_0 & 0 & 0 & \cdots & 0 \\
0 & 0 & P_1 & 0 & \cdots & 0 \\
0 & 0 & 0 & P_2 & \cdots & 0 \\
\vdots & \vdots & \vdots & \vdots & \ddots & \vdots \\
0 & 0 & 0 & 0 & \cdots & P_{d-2} \\
P_{d-1} & 0 & 0 & 0 & \cdots & 0
\end{bmatrix}
$$

b. Show that P^d is block diagonal and hence the Markov chain with one-step transition probability matrix given by P^d has d irreducible, aperiodic classes of states.

c. Show that the ith diagonal block of P^d is $(P_i \ldots P_{d-1} P_0 \ldots P_{i-1})$ for $0 \le i \le d - 1$.

d. Let π^i denote the limiting probabilities associated with the ith class of states in the Markov chain with one-step transition matrix

P^d. This distribution is unique because the classes are irreducible and aperiodic (Theorem 13.3). Show that $\pi^{i+1} = \pi^i P_i$ and that the unique stationary distribution of the original Markov chain is given by $\pi = (\pi^0, \pi^1, \ldots, \pi^{d-1})/d$.

e. By part (d), the stationary distribution for a periodic Markov chain with n states can be obtained by solving a system of equations with fewer than n equations. Which set of equations should be chosen?

11. (Project) The *Euclidean algorithm* can be used to find the greatest common divisor of two integers. Find a reference to this algorithm and then implement the method in some programming language.

13.6   References

Feller [4] presents a number of classic probability models, including finite Markov chains. A more recent treatment of Markov chains and other probability models is in similar volumes by Ross [6] and by Taylor and Karlin [8]. Traditional matrix-based computational methods for Markov chains are presented by Isaacson and Madsen [5]. Tarjan [7] develops a number of the most efficient graph algorithms and their associated data structures. An extensive treatment of topics in computational methods for discrete structures including graphs is given by Aho, Hopcroft, and Ullman [1]. The particular techniques for determining periodicity discussed in Section 13.3.2 were first presented by Denardo [3].

In addition, web-based course notes on Markov chains can be found at `<http://mscmga.ms.ic.ac.uk/jeb/or/contents.html>`. Spreadsheet add-ins that carry out a variety of Markov chain computations are available at `<http://mohican.me.utexas.edu/~jensen/addins/>`.

[1] A. V. Aho, J. E. Hopcroft, and J. D. Ullman, *The Design and Analysis of Computer Algorithms*, Addison-Wesley, Reading, MA, 1974.

[2] G. H. Bradley, "Algorithm and bound for the greatest common divisor of n integers," *Commun. ACM* **13**, 433–436 (1970).

[3] E. V. Denardo, "Periods of connected networks and powers of nonnegative matrices," *Math. Oper. Res.* **2**, 20–24 (1977).

[4] W. Feller, *An Introduction to Probability Theory and Its Applications*, Volume I, 3rd ed., Wiley, New York, 1968.

[5] D. L. Isaacson and R. W. Madsen, *Markov Chains*, Wiley, New York, 1976.

[6] S. M. Ross, *Introduction to Probability Models*, 5th ed., Academic Press, Boston, 1993.

[7] R. E. Tarjan, *Data Structures and Network Algorithms*, Society for Industrial and Applied Mathematics, Philadelphia, 1983.

[8] H. M. Taylor and S. Karlin, *An Introduction to Stochastic Modeling*, Academic Press, San Diego, 1994.

Chapter 14

Some Error-Correcting Codes and Their Applications

J. D. Key

Prerequisites: linear algebra, modern algebra

14.1 Introduction

This chapter describes three types of error-correcting codes that have been used in major applications: namely, photographs from spacecraft (first-order Reed-Muller codes), compact discs (Reed-Solomon codes), and computer memories (extended binary Hamming codes). Error-correcting codes were first developed in the 1940s following a theorem of Claude Shannon [14] that showed that almost error-free communication could be obtained over a noisy channel. The message to be communicated is first "encoded" (i.e., turned into a codeword) by adding "redundancy." The codeword is then sent through the channel and the received message is "decoded" by the receiver into a message resembling, as closely as possible, the original message. The degree of resemblance will depend on how good the code is in relation to the channel.

Such codes have been used to great effect in some important applications. This chapter describes the (linear) codes that are employed in the following three applications, showing how they can be constructed and how they can be used:

- *computer memories* [11]: the codes used are extended binary Hamming codes, the latter being perfect single-error-correcting;

- *photographs from spacecraft*: the codes initially used in this project were first-order Reed-Muller codes, which can be constructed as the orthogonal extended Hamming codes; later the binary extended Golay code was used;

- *compact discs* [7]: the codes used are Reed-Solomon codes, constructed using certain finite fields of large prime-power order.

After an introductory section on the necessary background to coding theory, including some of the effective encoding and decoding methods, we will describe how the codes can be used in each of these applications, and give a simple description how each of these classes of codes can be constructed. We will not include details of the implementation of the codes, nor of the mathematical background to the theory; the reader is encouraged to consult the papers and books in the bibliography for this. Those readers who are familiar with the elementary concepts in coding theory can pass immediately on to the applications, and refer back to Section 14.2 when necessary. The final section contains some simple exercises and some projects for further study.

14.2   Background Coding Theory

More detailed accounts of error-correcting codes can be found in Hill [6], Pless [13], MacWilliams and Sloane [10], van Lint [9], and Assmus and Key [1, Chapter 2]. See also Peterson [12] for an early article written from the engineers' point of view. Proofs of all the results quoted here can be found in any of these texts; our summary here follows [1].

The usual pictorial representation of the use of error-correcting codes to send messages over noisy channels is shown in the schematic diagram of Figure 14.1. Here a message is first given by the *source* to the *encoder* that turns the message into a *codeword*: a string of letters from some alphabet, chosen according to the code used. The encoded message is then sent through the channel, where it may be subjected to *noise* and hence altered. When this message arrives at the *decoder* belonging to the receiver, it is equated with the most likely codeword — i.e., the one (should it exist) that, in a probabilistic sense depending on the channel, was probably sent. Finally this "most likely" codeword is decoded and the message is passed on to the receiver.

Example 14.1

Suppose we use an alphabet of just two symbols, 0 and 1, and we have only two messages, for example "no" corresponding to 0, and "yes" corresponding to 1. To send a message ("yes" or "no"), we add redundancy by simply repeating the message five times. Thus we encode the message "no" as the codeword 00000. The channel might interfere with the message and could change it to, say, 10100. The decoder assesses the message and decides that of the two possible codewords, 00000 and 11111, the former is the more likely, and hence the message is decoded, correctly, as "no."

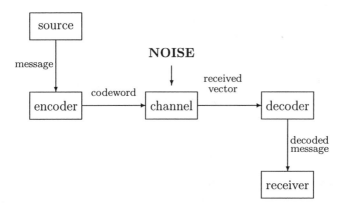

FIGURE 14.1   **A noisy communication channel**

(From E. F. Assmus, Jr. and J. D. Key, *Designs and Their Codes*, Cambridge University Press, Cambridge, 1992. Reprinted with permission of Cambridge University Press.)

We have made several assumptions in the above example. Namely, it has been assumed that the probability of an error at any position in the word is less than $\frac{1}{2}$, that each codeword is equally likely to be sent, and that the receiver is aware of the code used.

DEFINITION 14.1   *Let F be a finite set, or **alphabet**, of q elements. A **q-ary code** C is a set of finite sequences of symbols of F, called **codewords** and written $x_1 x_2 \cdots x_n$, or $(x_1, x_2, \ldots, x_n)$, where $x_i \in F$ for $i = 1, \ldots, n$. If all the sequences have the same length n, then C is a **block code** of **block length** n.*

The code used in Example 14.1 is a block code, called the *repetition code* of length 5. Such a code can be generalized to length n and to any alphabet of size q, and hence will have q codewords of the form $xx \cdots x$, where $x \in F$.

Given an alphabet F, it will be convenient, and also consistent with terminology for cartesian products of sets and for vector spaces when F is a field, to denote the set of all sequences of length n of elements from F by F^n. We call these sequences *vectors*, referring to the member of F in the ith position as the *coordinate* at i. We use either notation, $x_1 x_2 \cdots x_n$ or $(x_1, x_2, \ldots, x_n)$, for the vectors. A code over F of block length n is thus any subset of F^n.

The formal process illustrated by the reasoning in the simple case of Example 14.1 uses the concept of the "distance" between codewords.

DEFINITION 14.2   Let $v = (v_1, v_2, \ldots, v_n)$ and $w = (w_1, w_2, \ldots, w_n)$ *be two vectors in F^n. The* **Hamming distance** $d(v, w)$ *between v and w is the number of coordinate places in which they differ:*

$$d(v, w) = |\{i : v_i \neq w_i\}|.$$

We will usually refer to the Hamming distance as simply the *distance* between two vectors. It is simple to prove that the Hamming distance defines a metric on F^n:

- $d(v, w) = 0$ if and only if $v = w$;

- $d(v, w) = d(w, v)$ for all $v, w \in F^n$;

- $d(u, w) \leq d(u, v) + d(v, w)$ for all $u, v, w \in F^n$.

Nearest-neighbor decoding picks the codeword v' nearest (in terms of Hamming distance) to the received vector, should such a vector be uniquely determined. This method maximizes the decoder's likelihood of correcting errors — provided that each symbol has the same probability (less than $\frac{1}{2}$) of being received in error and each symbol of the alphabet is equally likely to occur. A channel with these two properties is called a *symmetric q-ary channel*.

DEFINITION 14.3   *The* **minimum distance** $d(C)$ *of a code C is the smallest of the distances between distinct codewords:*

$$d(C) = \min\{d(v, w) : v, w \in C, v \neq w\}.$$

If C is a code of block length n having M codewords and minimum distance d, then we say that C is an (n, M, d) *q-ary code*, where $|F| = q$. We will refer to n as the *length* of the code rather than the block length.

The following simple result is very easily proved and shows the vital importance of the distance concept for codes used in symmetric channels. Here $\lfloor z \rfloor$ denotes the floor function of z.

THEOREM 14.1

If $d(C) = d$ then C can detect up to $d - 1$ errors or correct up to $\lfloor (d - 1)/2 \rfloor$ errors.

Example 14.2

The repetition code C of length 5 given in Example 14.1 has the two codewords 00000 and 11111, so the minimum distance of C is 5. Thus code C is a $(5, 2, 5)$ 2-ary (binary) code. By Theorem 14.1 up to four errors can be detected or up to two errors can be corrected.

From the above discussion, we see that for a good (n, M, d) code C, one that detects or corrects many errors, it is desirable to have d large. However, we also prefer n to be small (for fast transmission) and M to be large (for a large number of messages). These are clearly conflicting aims, since for a q-ary code, $M \leq q^n$. In fact, there are many bounds connecting these three parameters, of which the simplest is the *Singleton bound* [1, Theorem 2.1.2]:

$$M \leq q^{n-d+1}. \tag{14.1}$$

Another bound, usually better than the Singleton bound, is the *sphere-packing bound*. To present this bound, the following definition is useful.

DEFINITION 14.4   Let F be any alphabet and suppose $u \in F^n$. For any integer $r \geq 0$, the **sphere of radius r with center u** is the set of vectors $S_r(u) = \{v \in F^n : d(u, v) \leq r\}$.

Let C be an (n, M, d) code. Then the spheres of radius $\rho = \lfloor (d-1)/2 \rfloor$ with centers at the codewords C do not overlap; they form M pairwise disjoint subsets of F^n. The integer ρ is called the *packing radius* of C. Hence we have the *sphere-packing bound*: if C is an (n, M, d) q-ary code of packing radius ρ, then

$$M \left(1 + (q - 1)n + (q - 1)^2 \binom{n}{2} + \cdots + (q - 1)^\rho \binom{n}{\rho} \right) \leq q^n. \tag{14.2}$$

The *covering radius* of a code is defined to be the smallest integer R such that spheres of radius R with their centers at the codewords cover all the words of F^n. If the covering radius R is equal to the packing radius ρ, then the code is called a *perfect ρ-error-correcting code*. Thus perfect codes are those for which equality holds in (14.2).

The binary ($|F| = 2$) repetition codes of odd length n are *trivial* perfect $(n-1)/2$-error-correcting codes; the infinite class of binary perfect 1-error-correcting codes with $n = 2^m - 1$ and $M = 2^{n-m}$ was discovered by Hamming [4] and generalized to the q-ary case by Golay [3].

A code C of length n over the finite field $F = \mathbf{F}_q$ of prime-power order q is *linear* if C is a subspace of $V = F^n$. If $\dim(C) = k$ and $d(C) = d$, then we write $[n, k, d]$ or $[n, k, d]_q$ for the q-ary code C; if the minimum distance is not specified we simply write $[n, k]$. The *information rate* is k/n and the *redundancy* is $n - k$.

Thus a q-ary linear code is any subspace of a finite-dimensional vector space over a finite field $\mathbf{F}_q$, *but with reference to a particular basis*. The standard basis for F^n, as the space of n-tuples, has a natural ordering using the numbers 1 to n, and this coincides with the spatial layout of a codeword as a sequence of alphabet letters sent over a channel. To avoid ordering the basis we may take $V = F^X$, the set of functions from X to F, where X is any set of size n. Then a linear code is any subspace of V.

For any vector $v = (v_1, v_2, \ldots, v_n) \in V = F^n$, let $S = \{i : v_i \neq 0\}$; then S is called the *support* of v, written $\mathrm{Supp}(v)$, and the *weight* of v, written $\mathrm{wt}(v)$, is $|S|$. The *minimum weight* of a code is the minimum of the weights of the nonzero codewords, and for linear codes is easily seen to be equal to $d(C)$.

For linear $[n, k, d]$ q-ary codes the Singleton bound (14.1) and the sphere-packing bound (14.2) become the following:

- *Singleton bound:* $d \leq n - k + 1$;

- *sphere-packing bound:* $\sum_{i=0}^{\rho}(q-1)^i \binom{n}{i} \leq q^{n-k}$.

A code for which equality holds in the Singleton bound is called an MDS (maximum distance separable) code. The Reed-Solomon codes (see Section 14.5.2) are MDS codes.

DEFINITION 14.5   *Two linear codes in F^n are* **equivalent** *if each can be obtained from the other by permuting the coordinate positions in F^n and multiplying each coordinate by a nonzero field element. The* *codes will be said to be* **isomorphic** *if a permutation of the coordinate positions suffices to take one to the other.*

In terms of the distinguished basis that is present when discussing codes, code equivalence is obtained by reordering the basis and multiplying each of the basis elements by a nonzero scalar. Thus the codes C and C' are equivalent if there is a linear transformation of the ambient space F^n, given by a *monomial* matrix (one nonzero entry in each row and column) in the standard basis, that carries C onto C'. When the codes are isomorphic, a *permutation* matrix can be found with this property. When $q = 2$, the two concepts are identical. Clearly, equivalent linear codes must have the same parameters $[n, k, d]$.

If C is an $[n, k]$ q-ary code, a *generator matrix* for C is a $k \times n$ array obtained from any k linearly independent vectors of C.

Via elementary row operations, a generator matrix G for C can be brought into reduced row echelon form and still generate C. Then, by permuting columns, it can be brought into a *standard form* $G' = [I_k | A]$, where G' is a generator matrix for an equivalent (in fact, isomorphic) code. Here A is a $k \times (n - k)$ matrix over F.

Now we come to another important way of describing a linear code, namely through its *orthogonal* or *dual*. For this concept we need an *inner product* defined on our space; it is the standard inner product. Specifically for $v, w \in F^n$, $v = (v_1, v_2, \ldots, v_n)$, $w = (w_1, w_2, \ldots, w_n)$, the inner product of v and w is $(v, w) = \sum_{i=1}^{n} v_i w_i$.

DEFINITION 14.6   *Let C be an $[n, k]$ q-ary code. The* **orthogonal code** *(or* **dual code***) is denoted by $C^\perp$ and is given by*

$$C^\perp = \{v \in F^n : (v, c) = 0 \text{ for all } c \in C\}.$$

Code C is called **self-orthogonal** *if $C \subseteq C^\perp$ and* **self-dual** *if $C = C^\perp$.*

From elementary linear algebra we have $\dim(C) + \dim(C^\perp) = n$ since $C^\perp$ is simply the null space of a generator matrix for C. Taking G to be a generator matrix for C, a generator matrix H for $C^\perp$ satisfies $GH^T = 0$. That is, $c \in C$ if and only if $cH^T = 0$ or, equivalently, $Hc^T = 0$. Any generator matrix H for $C^\perp$ is called a *parity-check* or *check* matrix for C. If G is written in the standard form $[I_k | A]$, then

$$H = [-A^T | I_{n-k}] \tag{14.3}$$

is a check matrix for the code with generator matrix G.

Example 14.3

The code C in Example 14.1 is a *linear* $[5, 1, 5]_2$ code, since its codewords 00000 and 11111 can be expressed as $0 \cdot 11111$ and $1 \cdot 11111$,
respectively. Thus code C has the generator matrix

$$G = [1, 1, 1, 1, 1],$$

already in standard form, giving

$$H = \begin{bmatrix} 1 & 1 & 0 & 0 & 0 \\ 1 & 0 & 1 & 0 & 0 \\ 1 & 0 & 0 & 1 & 0 \\ 1 & 0 & 0 & 0 & 1 \end{bmatrix}$$

as a check matrix.

A generator matrix in standard form simplifies encoding. Suppose
data consisting of q^k messages are to be encoded by adding redundancy
using the code C with generator matrix G. First identify the data with
the vectors in F^k, where $F = \mathbf{F}_q$. Then for $u \in F^k$, encode u by forming
uG. If $u = (u_1, u_2, \ldots, u_k)$ and G has rows $R_1, R_2, \ldots, R_k$, where each
R_i is in F^n, then $uG = \sum_i u_i R_i = (x_1, x_2, \ldots, x_k, x_{k+1}, \ldots, x_n) \in F^n$,
which is now encoded. But when G is in standard form, the encoding
takes the simpler form

$$u \mapsto (u_1, u_2, \ldots, u_k, x_{k+1}, \ldots, x_n).$$

Here the $u_1, \ldots, u_k$ are the *message* or *information* symbols, and the
last $n - k$ entries are the *check* symbols, and represent the redundancy.

In general it is not possible to say anything about the minimum weight
of $C^\perp$ knowing only the minimum weight of C but, of course, either a
generator matrix or a check matrix gives complete information about
both C and $C^\perp$. In particular, a check matrix for C determines the
minimum weight of C in a useful way.

THEOREM 14.2

*Let H be a check matrix for an $[n, k, d]$ code C. Then every choice of
$d - 1$ or fewer columns of H forms a linearly independent set. Moreover,
if every $d - 1$ or fewer columns of a check matrix for a code C are linearly
independent, then the code has minimum weight at least d.*

Notice that in terms of generator matrices, two codes C and C' with
generator matrices G and G' are equivalent if and only if there exist a

nonsingular matrix M and a monomial matrix N such that $MGN = G'$, with isomorphism if N is a permutation matrix and equality if $N = I_n$, n being the block length of the codes.

Example 14.4

The smallest nontrivial Hamming code (see Section 14.3.2) is a [7,4,3] binary code, which is a perfect single-error-correcting code. It can be given by the generator matrix G in standard form $[I_4|A]$ where

$$A = \begin{bmatrix} 1 & 1 & 1 \\ 0 & 1 & 1 \\ 1 & 0 & 1 \\ 1 & 1 & 0 \end{bmatrix}.$$

Thus a check matrix will be

$$H = \begin{bmatrix} 1 & 0 & 1 & 1 & 1 & 0 & 0 \\ 1 & 1 & 0 & 1 & 0 & 1 & 0 \\ 1 & 1 & 1 & 0 & 0 & 0 & 1 \end{bmatrix}.$$

Taking $\{a, b, c, d\}$ as the information symbols, and $\{b', c', d'\}$ as the check symbols, the diagram in Figure 14.2 (suggested by McEliece [11]) can be used to correct a single error, for any vector received, if at most a single error has occurred. The rule is that the sum of the coordinates in any of the three circles must be 0 (modulo 2); these constitute the "parity checks" as seen from the matrix H above. For example, if the vector 1011111 is received, checking the parity in the three circles R_1, R_2, R_3 (corresponding to b', c', d', respectively) shows that an error occurred in circles R_2 and R_3 but not R_1. Since the set $R_2 \cap R_3 \cap \overline{R}_1$ contains b, we conclude that the (single) error occurred at the information symbol b; this error is corrected, yielding the vector 1111111.

FIGURE 14.2   **The Hamming code $\mathcal{H}_3$**

A general method of decoding for linear codes is a method, due to Slepian [15], that uses nearest-neighbor decoding and is called *standard-*

array decoding. The *error vector* is defined to be $e = w - v$, where v is the codeword sent and w is the received vector. Given the received vector, we wish to determine the error vector. We look for that coset of the subgroup C in F^n that contains w and observe that the possible error vectors are just the members of this coset. The strategy is thus to look for a vector e of minimum weight in the coset $w + C$, and decode as $v = w - e$. A vector of minimum weight in a coset is called a *coset leader*; of course it might not be unique, but it will be in the event that its weight is at most ρ, where ρ is the packing radius, and this will always happen when at most ρ errors occurred during transmission. It should be noted that there may be a unique coset leader even when the weight of that leader is greater than ρ and thus a complete analysis of the probability of success of nearest-neighbor decoding will involve analyzing the weight distribution of all the cosets of C; in the engineering literature this is known as "decoding beyond the minimum distance." Use of a parity-check matrix H for C to calculate the *syndrome*

$$\text{synd}(w) = wH^T$$

of the received vector w assists this decoding method, the syndrome being constant over a coset and equal to the zero vector when a codeword has been received.

DEFINITION 14.7   Let C be an $[n, k, d]$ q-ary code. Define the **extended** code $\widehat{C}$ to be the code of length $n + 1$ in F^{n+1} of all vectors $\hat{c}$ for $c \in C$, where if $c = (c_1, \ldots, c_n)$ then

$$\hat{c} = \left(c_1, \ldots, c_n, -\textstyle\sum_{i=1}^{n} c_i\right).$$

This is called *adding an overall parity check*, for we see that if $v = (v_1, \ldots, v_{n+1})$, then $v \in \widehat{C}$ satisfies $\sum v_i = 0$. If C has generator matrix G and check matrix H, then $\widehat{C}$ has generator matrix $\widehat{G}$ and check matrix $\widehat{H}$. The matrix $\widehat{G}$ is obtained from G by adding an $(n + 1)$st column such that the sum of the columns of $\widehat{G}$ is the zero column; $\widehat{H}$ is obtained from H by attaching an $(n - k + 1)$st row and $(n+1)$st column to H, the row being all 1s and the column being $(0, 0, \ldots, 0, 1)^T$. If C is binary with d odd, then $\widehat{C}$ will be an $[n + 1, k, d + 1]$ code.

Example 14.5

Extending the $[7, 4, 3]$ binary Hamming code in Example 14.4 gives an $[8, 4, 4]$ binary code. This extended code is self-dual; notice that all rows of $\widehat{G}$ are mutually orthogonal and each has even weight.

An inverse process to extending an $[n, k, d]$ code is that of *puncturing*, which is achieved by simply deleting a coordinate, thus producing a linear code of length $n - 1$. The dimension will be k or $k - 1$, but in the great majority of cases the dimension will remain k. The minimum weight may change in either way, but unless the minimum weight of the original code is 1, the minimum weight of the punctured code will be either $d - 1$ or d and in the great majority of cases $d - 1$.

Another way to obtain new codes is by *shortening*. Given an $[n, k, d]$ q-ary code C, for any integer $r \leq k$, we take the subspace C' of all codewords having 0 in a fixed set of r coordinate positions, and then remove those coordinate positions to obtain a code of length $n - r$. For example, if G is a generator matrix for C in standard form, shortening by the first coordinate will clearly produce an $[n-1, k-1, d']$ code, where $d' \geq d$. In this way we obtain $[n - r, k - r, d']$ codes. This technique is used in Section 14.5.

14.3   Computer Memories and Hamming Codes

This section describes briefly how memory chips are designed, how codes may be used, and why they are needed. A fuller account of the use of error-correcting codes in computer memories may be found in [11].

14.3.1   Application

The memories of computers are built from silicon chips. Although any one of these chips is reliable, when many thousands are combined in a memory, some might fail. The use of an error-correcting code can mean that a failed chip will be detected and the error corrected. The errors in a chip might occur in the following way. A memory chip is a square array of data-storage cells, for example a $64K$ chip, where $K = 2^{10}$. The $64K$ chip stores $64K = 2^{16} = 65,536$ bits (binary digits) of data. Alternatively, there are $256K = 2^{18}$ and one-megabit (2^{20} bits) chips. In the $64K$ chip the data-storage cells are arranged in a $2^8 \times 2^8$ array, where each cell stores a 0 or a 1. Each cell can be accessed individually, and has an "address" corresponding to its row and column coordinates, usually numbered from 0 to 255. The largest address is in position $(255, 255) = (2^8 - 1, 2^8 - 1)$; the binary representation of 255 is 11111111, a sequence of eight bits, and thus the row and column addresses for a $64K$ chip require eight bits each, for a total of 16 bits. A $256K$ chip has $2^9 = 512$ rows and columns and thus requires 18 bits to address a cell, and a one-megabit chip requires 20.

The 0s or 1s stored in a memory chip are represented by the presence or absence of negative electric charges at sites in the silicon crystal whose electrical properties make them potential wells for negative charge. When a 0 is to be stored in a cell, the potential well at the site is filled with electrons; when a 1 is to be stored, the well is emptied. The cell is read by measuring its negative charge; if the charge is higher than a certain value it is read to be a 0, otherwise a 1.

Clearly, if a potential well lost its charge it would be read incorrectly. Errors do in fact occur. *Hard errors* occur when the chip itself is damaged. *Soft errors* occur when the chip itself is not damaged but alpha particle bombardment occurs and changes the charge in a cell. The latter is a common cause of error and cannot be avoided.

An error-correcting code is used to correct such errors in the following way. Suppose we have one megabyte of memory consisting of 128 64K chips. In such a memory the chips are arranged in four rows of 32 chips each; each chip contains 2^{16} memory cells, so the memory has a total of 2^{23} cells. The data are divided into words of 32 bits each, and each word consists of the contents of one memory cell in each of the 32 chips in one row. In order to correct errors, a further seven chips are added to each of the four rows, making 156 chips. Each row has now 39 chips, the seven extra bits for each word being the parity bits, and these are reserved for error correction. The code actually employed is the binary extended Hamming code of length 64, a $[64, 57, 4]$ binary code. Such a code will actually protect 57 bits of data, but designers of the codes use only 32 bits.

We now describe how the (binary) Hamming codes are defined.

14.3.2   Hamming Codes

These codes were first fully described by Golay [3] and Hamming [4, 5] although the $[7, 4]$ binary code had already appeared in Shannon's fundamental paper [14]. They provide an infinite class of perfect codes. We need here only the binary case, which was the one considered by Hamming. Consider a binary code C with check matrix H; the transposed syndrome Hy^T of a received vector y is, in the binary case, simply the sum of those columns of H where the errors occurred. To design a single-error-correcting code we want H not to have any zero columns, since errors in that position would not be detected; similarly, we want H not to have any equal columns, for then errors in those positions would be indistinguishable. If such an H has r rows, n columns and is of rank r, then it will be a check matrix for a single-error-correcting $[n, n - r]$ code. Thus, to maximize the dimension of the code, n should be chosen as large as possible. The number of distinct nonzero r-tuples available for columns is $2^r - 1$. We take for H the $r \times (2^r - 1)$ binary matrix whose columns are all the distinct nonzero r-tuples.

DEFINITION 14.8   *The binary Hamming code of length $2^r - 1$ is the code $\mathcal{H}_r$ that has for its check matrix the $r \times (2^r - 1)$ matrix H of all nonzero r-tuples over $\mathbf{F}_2$.*

THEOREM 14.3

The binary code $\mathcal{H}_r$ is a $[2^r-1, 2^r-1-r, 3]$ perfect single-error-correcting code for all $r \geq 2$.

Example 14.6

If $r = 2$ then the check matrix is

$$H = \begin{bmatrix} 1 & 0 & 1 \\ 0 & 1 & 1 \end{bmatrix}$$

and $\mathcal{H}_2$ is a $[3, 1, 3]$ code. This is simply the binary repetition code of length 3.

If $r = 3$ then $\mathcal{H}_3$ is a $[7, 4, 3]$ binary code, with check matrix

$$H = \begin{bmatrix} 1 & 0 & 1 & 0 & 1 & 0 & 1 \\ 0 & 1 & 1 & 0 & 0 & 1 & 1 \\ 0 & 0 & 0 & 1 & 1 & 1 & 1 \end{bmatrix}.$$

Decoding a binary Hamming code is very easy. We first arrange the columns of H, a check matrix for $\mathcal{H}_r$, so that column j represents the binary representation (transposed) of the integer j. Now we decode as follows. Suppose the vector y is received. We first compute the syndrome $\operatorname{synd}(y) = yH^T$. If $\operatorname{synd}(y) = 0$, then $y \in \mathcal{H}_r$ and we decode as y. If $\operatorname{synd}(y) \neq 0$, then, assuming one error has occurred, it must have occurred at position j, where the vector $(\operatorname{synd}(y))^T$ is the binary representation of the integer j. Thus we decode y as $y + e_j$, where e_j is the vector of length $n = 2^r - 1$ with 0 in every position except position j, where it has a 1. The examples given here use this ordering, but notice that the ordering of the entries in the transposed m-tuple representing a number is read from left to right. Thus the (transposed) row vector $[1011]$ represents $1+0+4+8 = 13$ rather than the customary $1 + 2 + 0 + 8 = 11$.

If we form the extended binary Hamming code $\widehat{\mathcal{H}_r}$, then we obtain a $[2^r, 2^r - 1 - r, 4]$ code which is still single-error-correcting, but which is capable of simultaneously detecting two errors. This code is useful for *incomplete decoding*; see Hill [6].

Example 14.7

Let $C = \widehat{\mathcal{H}_3}$ be the extended code obtained from $\mathcal{H}_3$. As described in the discussion following Definition 14.7, the check matrix is

$$\widehat{H} = \begin{bmatrix} 1 & 0 & 1 & 0 & 1 & 0 & 1 & 0 \\ 0 & 1 & 1 & 0 & 0 & 1 & 1 & 0 \\ 0 & 0 & 0 & 1 & 1 & 1 & 1 & 0 \\ 1 & 1 & 1 & 1 & 1 & 1 & 1 & 1 \end{bmatrix}$$

and the generator matrix is

$$\widehat{G} = \begin{bmatrix} 1 & 1 & 1 & 0 & 0 & 0 & 0 & 1 \\ 1 & 0 & 0 & 1 & 1 & 0 & 0 & 1 \\ 0 & 1 & 0 & 1 & 0 & 1 & 0 & 1 \\ 1 & 1 & 0 & 1 & 0 & 0 & 1 & 0 \end{bmatrix}.$$

The data $(1,0,1,0)$ is encoded as $(1,0,1,0)\widehat{G} = (1,0,1,1,0,1,0,0)$. If a single error occurs at position i, then the received vector y will have $(\text{synd}(y))^T$ as column i of $\widehat{H}$ and decoding can be performed. However, if two errors occur, at positions i and j, then $(\text{synd}(y))^T$ will be the sum of columns i and j of $\widehat{H}$, and thus will have 0 as the last entry. As a result, decoding will not take place.

DEFINITION 14.9   *The orthogonal code $\mathcal{H}_r^{\perp}$ of $\mathcal{H}_r$ is called the binary **simplex** code and is denoted by $\mathcal{S}_r$.*

The simplex code $\mathcal{S}_r$ clearly has length $2^r - 1$ and dimension r. The generator matrix is H. It follows that $\mathcal{S}_r$ consists of the zero vector and $2^r - 1$ vectors of weight 2^{r-1}, so that it is a $[2^r - 1, r, 2^{r-1}]$ binary code. Any two codewords are at Hamming distance 2^{r-1} and, if the codewords are placed at the vertices of a unit cube in $2^r - 1$ dimensions, they form a simplex. From the discussion of Section 14.2, we know that a check matrix for $\widehat{\mathcal{H}_r}$ is H enlarged by a column of all zeros and a row of all ones. The code $\widehat{\mathcal{H}_r}^{\perp}$ spanned by this matrix is a $[2^r, r+1, 2^{r-1}]$ binary code, and is also a first-order Reed-Muller code, denoted by $\mathcal{R}(1, r)$. It can correct $2^{r-2} - 1$ errors. The utility of this type of code will be illustrated in Section 14.4.2.

Finally now, looking back at the application to computer memories, the code used is $\widehat{\mathcal{H}_6}$, a $[64, 57, 4]$ binary code. A check matrix can easily be constructed in the manner described above. The code will now correct any single error that occurs in the way described above, and will *simultaneously* detect any two errors.

14.4   Photographs in Space and Reed-Muller Codes

In this section, an application concerning the transmission of photographs from deep space will be discussed. Two types of codes (Reed-Muller and binary Golay codes) turn out to be useful in ensuring the integrity of these photographs.

14.4.1   Application

Photographs of the planet Mars were first taken by the Mariner series of spacecraft in the 1960s and early 1970s, and the first-order Reed-Muller code of length 32 was used to obtain good quality photographs. The original black and white photographs taken by the earlier Mariners were broken down into 200×200 picture elements. Each element was assigned a binary 6-tuple representing one of 64 possible brightness levels from, say, 000000 for white to 111111 for black. Later this division was made finer by using 700×832 elements, and the quality was increased by encoding the 6-tuples using the $[32, 6, 16]$ binary 7-error-correcting Reed-Muller code $\mathcal{R}(1, 5)$ in the way described in Section 14.3.2. When color photographs were taken, the same code was used simply by viewing the same photograph through different colored filters. In the Voyager series of spacecraft after the late 1970s, the binary extended Golay code $\mathcal{G}_{24}$, a $[24, 12, 8]$ code, was used in the color photography.

Later on in the Voyager series of spacecraft, different types of codes, convolutional Reed-Solomon codes, were used; see [17]. The Reed-Solomon codes are described in Section 14.5.2.

14.4.2   First-Order Reed-Muller Codes

A full account of the Reed-Muller codes, which are all binary codes, can be found in [10] or [1, Chapter 5]. We describe here simply a way to construct the $[32, 6, 16]$ code used for space photography, although in fact we can be more general since we already have seen its construction from the extended Hamming codes. Specifically, the first-order Reed-Muller code $\mathcal{R}(1, m)$ of length 2^m is the code $\widehat{\mathcal{H}_m}^{\perp}$: i.e., the code orthogonal to the extended binary Hamming code. It is, as discussed previously, a $[2^m, m + 1, 2^{m-1}]$ binary $(2^{m-2} - 1)$-error-correcting code. A 6×32

generator matrix G for $\mathcal{R}(1,5)$ may thus be constructed as follows. First form a 5×31 matrix G_0 whose columns are the binary representations of the numbers between 1 and 31; then adjoin a column with all entries zero; and finally add a sixth row with all entries equal to 1. Thus

$$
G_0 = \begin{bmatrix}
1 & 0 & 1 & \cdots & 0 & 1 \\
0 & 1 & 1 & \cdots & 1 & 1 \\
0 & 0 & 0 & \cdots & 1 & 1 \\
0 & 0 & 0 & \cdots & 1 & 1 \\
0 & 0 & 0 & \cdots & 1 & 1
\end{bmatrix}
$$

and

$$
G = \begin{bmatrix}
 & & & 0 \\
 & G_0 & & \vdots \\
 & & & 0 \\
1 & 1 & 1 & \cdots & 1
\end{bmatrix}.
$$

14.4.3   The Binary Golay Codes

There are many ways to arrive at the perfect Golay $[23, 12, 7]$ binary 3-error-correcting code $\mathcal{G}_{23}$ and its extension, the $[24, 12, 8]$ binary Golay code $\mathcal{G}_{24}$ that was used in the Voyager spacecraft. We will give a generator matrix, as Golay originally did in [3]. That the code generated by this matrix has the specified properties can be verified very easily, or by consulting the references given in Section 14.2; for example, see [6, Chapter 9].

A generator matrix over $\mathbf{F}_2$ for $\mathcal{G}_{24}$ is $G = [I_{12}|B]$, where I_{12} is the 12×12 identity matrix and B is a 12×12 matrix given by

$$
B = \begin{bmatrix}
0 & 1 & \cdots & 1 \\
1 & & & \\
\vdots & & A & \\
1 & & &
\end{bmatrix}.
$$

Here A is an 11×11 matrix of 0s and 1s defined in the following way. Consider the finite field $\mathbf{F}_{11}$ of order 11: i.e., the eleven remainders modulo 11. The first row of A is labeled by these eleven remainders in order, starting with 0, and placing an entry 1 in position i if i is a square modulo 11, and a 0 otherwise. The squares modulo 11 are $\{0, 1, 3, 4, 5, 9\}$, and thus the first row of A is $\begin{bmatrix} 1 & 1 & 0 & 1 & 1 & 1 & 0 & 0 & 0 & 1 & 0 \end{bmatrix}$. For the remaining rows of A simply cyclically rotate this row to the left ten times, to obtain

$$A = \begin{bmatrix} 1 & 1 & 0 & 1 & 1 & 1 & 0 & 0 & 0 & 1 & 0 \\ 1 & 0 & 1 & 1 & 1 & 0 & 0 & 0 & 1 & 0 & 1 \\ 0 & 1 & 1 & 1 & 0 & 0 & 0 & 1 & 0 & 1 & 1 \\ 1 & 1 & 1 & 0 & 0 & 0 & 1 & 0 & 1 & 1 & 0 \\ 1 & 1 & 0 & 0 & 0 & 1 & 0 & 1 & 1 & 0 & 1 \\ 1 & 0 & 0 & 0 & 1 & 0 & 1 & 1 & 0 & 1 & 1 \\ 0 & 0 & 0 & 1 & 0 & 1 & 1 & 0 & 1 & 1 & 1 \\ 0 & 0 & 1 & 0 & 1 & 1 & 0 & 1 & 1 & 1 & 0 \\ 0 & 1 & 0 & 1 & 1 & 0 & 1 & 1 & 1 & 0 & 0 \\ 1 & 0 & 1 & 1 & 0 & 1 & 1 & 1 & 0 & 0 & 0 \\ 0 & 1 & 1 & 0 & 1 & 1 & 1 & 0 & 0 & 0 & 1 \end{bmatrix}. \tag{14.4}$$

(This construction is in fact quite general, and it leads to a class of Hadamard matrices and also to the quadratic residue codes; see [1, Chapters 2, 7] for further details.) An effective decoding algorithm for $\mathcal{G}_{24}$ is given in [16, Chapter 4].

The perfect binary Golay code $\mathcal{G}_{23}$ may be obtained from $\mathcal{G}_{24}$ by deleting any coordinate.

14.5   Compact Discs and Reed-Solomon Codes

This section discusses the application of coding theory to the design of compact discs. In particular, Reed-Solomon codes turn out to be useful in correcting, in real time, multiple errors that may be present in the media. A full account of the use of Reed-Solomon codes for error-correction in compact discs is given in [7], [16, Chapter 7], or [17].

14.5.1   Application

Sound is stored on a compact disc by dividing the media up into small parts and representing these parts by binary data, just as pictures are divided up, as described in Section 14.4. A compact disc is made by sampling sound waves 44,100 times per second, with the amplitude measured and assigned a value between 1 and $2^{16} - 1$, given as a binary 16-tuple. In fact, two samples are taken, one for the left and one for the right channel. Each binary 16-tuple is taken to represent two field elements from the Galois field $\mathbf{F}_{2^8}$ of 2^8 elements, and thus each sample produces four symbols from $\mathbf{F}_{2^8}$.

For error-correction the information is broken into segments called *frames*, where each frame holds 24 data symbols. The code used for error-correction is a *Cross Interleaved Reed-Solomon code* (CIRC) obtained by a process called "cross-interleaving" of two shortened Reed-Solomon codes, as described below. The 24 symbols from $\mathbf{F}_{2^8}$ from six samples are used as information symbols in a (shortened) Reed-Solomon $[28, 24, 5]$ code C_1 over $\mathbf{F}_{2^8}$. Another shortened Reed-Solomon $[32, 28, 5]$ code C_2 also over $\mathbf{F}_{2^8}$ is then used in the interleaving process, which has four additional parity-check symbols. See [7, 16, 17] for a detailed description of this process.

As a result of this interleaving process of error-correction, flaws such as scratches on a disc, producing a train of errors called an "error burst," can be corrected.

We describe now the basic Reed-Solomon class of codes.

14.5.2   Reed-Solomon Codes

The Reed-Solomon codes are a class of *cyclic* q-ary codes of length n dividing $q - 1$ that satisfy the Singleton bound (14.1); such codes are therefore also MDS codes (see Section 14.2). It would take too long to describe general cyclic codes, but we will nevertheless define these codes as being cyclic and illustrate immediately how a generator matrix and a check matrix may be found.

Let $F = \mathbf{F}_q$ and let $n | (q - 1)$. Then F has elements of order n, and we let β be such an element. Pick any number δ such that $2 \leq \delta \leq n$, and take any number a such that $0 \leq a \leq q - 2$. Then the polynomial

$$g(X) = (X - \beta^{1+a})(X - \beta^{2+a})(X - \beta^{3+a}) \ldots (X - \beta^{\delta-1+a}) \quad (14.5)$$

is the generator polynomial of an $[n, n - \delta + 1, \delta]$ code over F, a *Reed-Solomon* code. Notice that the minimum weight of such a code is thus $d = \delta$. In the special case when $n = q - 1$, the code is a *primitive* Reed-Solomon code. We obtain a generator polynomial for the code by first expanding the polynomial $g(X)$ in Equation (14.5) to obtain

$$g(X) = g_0 + g_1 X + \cdots + g_{d-2} X^{d-2} + X^{d-1}$$

where $g_i \in F$ for each i. A generator matrix can then be shown to be

$$G = \begin{bmatrix} g_0 & g_1 & g_2 & \cdots & g_{d-1} & 1 & 0 & \cdots & 0 \\ 0 & g_0 & g_1 & \cdots & g_{d-2} & g_{d-1} & 1 & \cdots & 0 \\ \vdots & \vdots & \ddots & \vdots & \vdots & \vdots & \vdots & \ddots & \vdots \\ 0 & \cdots\cdots & g_0 & \cdots & & \cdots & g_{d-2} & g_{d-1} & 1 \end{bmatrix}. \quad (14.6)$$

The general theory of BCH codes (see [1, Chapter 2]) immediately gives
a check matrix

$$H = \begin{bmatrix} 1 & \beta & \beta^2 & \ldots & \beta^{(n-1)} \\ 1 & \beta^2 & \beta^4 & \ldots & \beta^{2(n-1)} \\ \vdots & \vdots & \vdots & \ddots & \vdots \\ 1 & \beta^{d-1} & \beta^{2(d-1)} & \ldots & \beta^{(n-1)(d-1)} \end{bmatrix}, \qquad (14.7)$$

taking here $a = 0$. This is a convenient generator matrix for the orthogonal code (which is also a Reed-Solomon code) with a rather simple encoding rule: namely, the data set $(a_1, a_2, \ldots, a_{d-1})$ is encoded as
$(a(1), a(\beta), a(\beta^2), \ldots, a(\beta^{n-1}))$ where

$$a(X) = a_1 X + a_2 X^2 + \cdots + a_{d-1} X^{d-1}.$$

The theory of BCH codes [1, Chapter 2] also tells us that the reciprocal
polynomial of $h(X) = (X^n - 1)/g(X)$, given by $\bar{h}(X) = X^{n-d+1} h(X^{-1})$,
is a generator polynomial for the orthogonal code, so another check
matrix may be obtained for this in the same way as we did for the code
with $g(X)$ as generator polynomial. A more convenient method for the
purposes here is to reduce G to standard form and then simply use the
formula shown in Equation (14.3).

Example 14.8

Let $q = 11$ and $n = 5$. Then we can take $\beta = 4$ as this has order
precisely 5, and let $a = 0$ and $\delta = 3$. Then

$$g(X) = (X - \beta)(X - \beta^2) = (X - 4)(X - 5) = 9 + 2X + X^2$$

so that by Equation (14.6)

$$G = \begin{bmatrix} 9 & 2 & 1 & 0 & 0 \\ 0 & 9 & 2 & 1 & 0 \\ 0 & 0 & 9 & 2 & 1 \end{bmatrix}.$$

Since $(X^5 - 1) = (X - 1)(X - 4)(X - 5)(X - 9)(X - 3)$, we have
$h(X) = (X^5 - 1)/g(X) = (X - 1)(X - 3)(X - 9)$, and the orthogonal
code has generator polynomial

$$\bar{h}(X) = X^3(X^{-1} - 1)(X^{-1} - 3)(X^{-1} - 9)$$
$$= 1 + 9X + 6X^2 + 6X^3.$$

Thus a check matrix is

$$H = \begin{bmatrix} 1 & 9 & 6 & 6 & 0 \\ 0 & 1 & 9 & 6 & 6 \end{bmatrix}.$$

Alternatively, the check matrix can be obtained from Equation (14.7), giving

$$H' = \begin{bmatrix} 1 & 4 & 5 & 9 & 3 \\ 1 & 5 & 3 & 4 & 9 \end{bmatrix},$$

which is seen to be row-equivalent to H.

The codes used are actually *shortened* Reed-Solomon codes (see Section 14.2). The easiest way to describe these codes is to have a generator matrix of the original Reed-Solomon code C in standard form, which is possible without even having to take an isomorphic code in this case, due to the Reed-Solomon codes having the property of being MDS codes. Thus a generator matrix can be row reduced to standard form, without the need of column operations. (Sometimes it is more convenient to use the standard form for the orthogonal code, and thus have the generator matrix in the form $[A|I_k]$.) If now C, an $[n, k, d]$ q-ary code with $n - k = d - 1$, is to be shortened in the first r places to obtain a code C' of length $n - r$, we obtain a generator matrix G' for C' also in standard form by simply deleting the first r rows and columns of G. If H is the check matrix in standard form for G, then the check matrix in standard form for G' is H', obtained from H by deleting the first r columns. It is easy to see that the MDS properties of the original code show that C' is also MDS and is an $[n - r, k - r, d]$ q-ary code with $n - r = k - r + d - 1$.

The finite field used in the codes developed for compact discs is the field $F = \mathbf{F}_{2^8}$ of order 2^8. This can be constructed as the set of all polynomials over the field $\mathbf{F}_2$ of degree at most 7. That is,

$$F = \{a_0 + a_1 X + a_2 X^2 + a_3 X^3 + a_4 X^4 + a_5 X^5 + a_6 X^6 + a_7 X^7 : a_i \in \mathbf{F}_2\}.$$

Addition of these polynomials is just the standard addition, as is multiplication, except that multiplication is carried out modulo the polynomial

$$1 + X^2 + X^3 + X^4 + X^8.$$

Thus $X^8 = 1 + X^2 + X^3 + X^4$ and all other powers of X are reduced to be at most 7, using this rule. The elements of $F = \mathbf{F}_{2^8}$ are thus effectively 8-tuples of binary digits, and this is how they are treated for the application.

The length of the Reed-Solomon codes over F are divisors of $2^8 - 1 = 255 = 3 \times 5 \times 17$. The code actually used is the shortened primitive one, i.e., the shortened $[255, 251, 5]$ code over F, with certain information symbols set to zero. Then if ω is a primitive element for the field (namely, an element of multiplicative order 255), we take $\beta = \omega$. Since we want minimum distance 5, we take $k = n - 4 = 251$, and shorten to length 28 for C_1 and to length 32 for C_2. The original Reed-Solomon code of length 255 would be the same for the two codes, and would have generator polynomial

$$g(X) = (X - \omega)(X - \omega^2)(X - \omega^3)(X - \omega^4)$$
$$= \omega^{10} + \omega^{81}X + \omega^{251}X^2 + \omega^{76}X^3 + X^4.$$

The powers of ω are then given as binary 8-tuples; for example,

$$\omega^{10} = \omega^2 + \omega^4 + \omega^5 + \omega^6 = (0,0,1,0,1,1,1,0),$$

and

$$\omega^{251} = \omega^3 + \omega^4 + \omega^6 + \omega^7 = (0,0,0,1,1,0,1,1),$$

as computed using, for example, the computer package Magma [2].

For further details of the properties of finite fields, the reader might consult [8].

14.6 Conclusion

We have been brief in this outline of the use of well-known mathematical constructions in important practical examples, and have not included full details of the implementation of the codes. The reader is urged to consult the bibliography included here for a full and detailed description of the usage. We hope merely to have given some idea as to the nature of the codes, and where they are applied.

The exercises included in the next section should be accessible using the definitions described in this chapter. They are mostly quite elementary. The projects described are open problems whose solution might prove to have rather important applications; some computational results might first be done to make the questions more precise, in particular for the case of Project 2. Magma [2] is a good computational tool to use for this type of problem. Online information about this system is available at the website <http://www.maths.usyd.edu.au:8000/u/magma/>. A tutorial on using Magma for error-correcting codes can be found at <http://www.it.uq.edu.au/~magma/VERS/htmlhelp/text674.html>.

14.7 Exercises and Projects

1. Prove Theorem 14.1.

2. If C is a code of packing radius ρ, show that the spheres with radius ρ and centers at the codewords of C do not intersect. Hence deduce the sphere-packing bound, Equation (14.2).

3. Prove Theorem 14.2.

4. Let C be a binary code with generator matrix

$$G = \begin{bmatrix} 1 & 0 & 1 & 1 & 0 \\ 1 & 1 & 0 & 1 & 1 \end{bmatrix}.$$

Show that $d(C) = 3$ and construct a check matrix for C. If a vector $y = (0, 1, 1, 1, 1)$ were received, what would you correct it to?

5. State the Singleton bound and the sphere-packing bound for an (n, M, d) q-ary code. Compare these to find the best bound for M for a $(10, M, 5)$ binary code.

6. Let C be a binary code. Show that

a. if C is self-orthogonal, then every codeword has even weight;

b. if every codeword of C has weight divisible by 4, then C is self-orthogonal;

c. if C is self-orthogonal and generated by a set of vectors of weight divisible by 4, then every codeword has weight divisible by 4 (thus C is *doubly-even*).

7. Let C be a linear code with check matrix H and covering radius r. Show that r is the maximum weight of a coset leader for C, and that r is the smallest number s such that every syndrome is a linear combination of s or fewer columns of H.

8. Prove from Definition 14.8 that $\mathcal{H}_r$ is a $[2^r - 1, 2^r - 1 - r, 3]$ perfect binary code.

9. Use the diagram shown in Figure 14.2 to decode the received vectors 1101101 and 1001011.

10. Produce a check matrix for $\mathcal{H}_4$, writing the columns so that column j is the binary representation of the number j. Use the method described after Example 14.6 to decode $\sum_{i=1}^{10} e_i$ and $\sum_{j=1}^{5} e_{2j-1}$, where e_i is the vector of length 15 with an entry 1 in position i, and zeros elsewhere.

11. Show that a binary $[23, 12, 7]$ code is perfect and 3-error-correcting.

12. In the matrix A of Equation (14.4) show that the sum of the first row with any other row has weight 6, and hence that the sum of any distinct two rows of A has weight 6. What can then be said for the sum of any two rows of B?

13. If C is an $[n, k, d]$ code with generator matrix G in standard form, show that shortening of C by the first r coordinate positions, where

$r \leq k$, produces an $[n-r, k-r, d']$ code C' where $d' \geq d$. (**Hint:** look at the check matrix for C' and use Theorem 14.2.) Show that it is possible to have $d' > d$ by constructing an example.

14. Construct a primitive $[6, 4, 3]$ Reed-Solomon code C over $\mathbf{F}_7$ using the primitive element 3. Form a shortened code of length 4 and determine its minimum distance.

15. Extend the code C in Exercise 14.14 to $\widehat{C}$ (see Definition 14.7). Is $\widehat{C}$ also MDS? Give generator matrices for $\widehat{C}$ and $(\widehat{C})^{\perp}$.

16. Construct the Reed-Solomon code of length 16 and minimum distance 5 by giving its generator polynomial and a check matrix.

17. (Project) Let C be a Hamming code $\mathcal{H}_r$. Is it possible to construct a generator matrix G for C such that every row of G has *exactly* three nonzero entries, and every column of G has *at most* three nonzero entries?

18. (Project) Is it possible, in general, to construct a basis of *minimum weight* vectors for other Hamming and Reed-Muller codes?

14.8   References

[1] E. F. Assmus, Jr. and J. D. Key, *Designs and Their Codes*, Cambridge University Press, Cambridge, 1992 (Second printing with corrections, 1993).

[2] W. Bosma and J. Cannon, *Handbook of Magma Functions*, Department of Mathematics, University of Sydney, November 1994.

[3] M. J. E. Golay, "Notes on digital coding," *Proc. IRE* **37**, 657 (1949).

[4] R. W. Hamming, "Error detecting and error correcting codes," *Bell System Tech. J.* **29**, 147–160 (1950).

[5] R. W. Hamming, *Coding and Information Theory*, Prentice-Hall, Englewood Cliffs, NJ, 1980.

[6] R. Hill, *A First Course in Coding Theory*, Oxford University Press, Oxford, 1986.

[7] H. Hoeve, J. Timmermans, and L. B. Vries, "Error correction and concealment in the compact disc system," *Philips Tech. Rev.* **40**, 166–172 (1982).

[8] R. Lidl and H. Niederreiter, *Introduction to Finite Fields and their Applications*, Cambridge University Press, Cambridge, 1986.

[9] J. H. van Lint, *Introduction to Coding Theory*, Springer-Verlag, New York, 1982.

[10] F. J. MacWilliams and N. J. A. Sloane, *The Theory of Error-Correcting Codes*, North-Holland, Amsterdam, 1983.

[11] R. J. McEliece, "The reliability of computer memories," *Scientific American* **252**(1), 88–95 (1985).

[12] W. W. Peterson, "Error-correcting codes," *Scientific American* **206**(2), 96–108 (1962).

[13] V. Pless, *The Theory of Error Correcting Codes*, 2nd ed., Wiley, New York, 1989.

[14] C. E. Shannon, "A mathematical theory of communication," *Bell System Tech. J.* **27**, 379–423, 623–656 (1948).

[15] D. Slepian, "Some further theory of group codes," *Bell System Tech. J.* **39**, 1219–1252 (1960).

[16] S. A. Vanstone and P. C. van Oorschot, *An Introduction to Error Correcting Codes with Applications*, Kluwer, Boston, 1992.

[17] S. B. Wicker and V. K. Bhargava (Eds.), *Reed-Solomon Codes and Their Applications*, IEEE Press, Piscataway, NJ, 1994.

Chapter 15

Broadcasting and Gossiping in Communication Networks

R. Laskar
J. A. Knisely

Prerequisites: graph theory

15.1 Introduction

In this chapter we discuss various types of problems of information dissemination encountered by a group of individuals connected by a communication network. *Gossiping* refers to the information dissemination problem that occurs when each member of a set A of n individuals knows a unique piece of information and must transmit it to every other person.

In the world of multi-processor computing, gossiping provides a useful model for the exchange of information. For example, a given computational problem might be divided into subproblems, each allocated to a separate processor. Some computations occur, and then each processor might need to be notified of the results of all the other processors before any further computation can be done. In this case, gossiping is required in order to continue. Communication is inherently slower than processor computing; hence, the cost involved with the gossiping must be kept low. As a result, efficient gossiping schemes are paramount for implementing parallel algorithms.

Gossiping also occurs when conducting a multilocation telephone call. In this instance, there is most likely no direct link between the sites, but instead a series of links that must be made between sites. However, the underlying problem is the same. Each person's messages, expressions, or drawings need to be sent to every other location. Thus, an all-to-all information dissemination process must occur.

Broadcasting refers to the process of message dissemination in a communication network whereby a message, originated by one individual, is transmitted to all members of the network. Broadcasting occurs when messages are sent from a system or network administrator about the status of the network, such as a message about the network going down or the availability of a new feature. In these cases, all people are going to be recipients but only one is the sender. Therefore, broadcasting, not gossiping, occurs.

Another instance of broadcasting can be seen in the spread of a disease through a previously uninfected population. In this case, the information or message is actually a virus. The message (virus) is spread via the person's daily interaction with others. Thus, as the carrier comes into contact with uninfected people, the virus is exchanged. In this environment, a variation of the model is needed, since frequently a person is only a carrier for a fixed length of time. A gossiping and broadcasting scheme called *periodic* incorporates this additional constraint.

15.2　Standard Gossiping and Broadcasting

15.2.1　Standard Assumptions

The original "gossip problem" or the "telephone problem," attributed to Boyd [4], is the following problem. "There are n ladies, and each of them knows an item of scandal which is not known to any of the others. They communicate by telephone, and whenever two ladies make a call, they pass on to each other as much scandal as they know at the time. How many calls are needed before all ladies know all the scandal?" A major variant of the gossip problem is "broadcasting." In broadcasting, there is only one message originator, who transmits the information to every one else in the communication network.

The gossiping problem is solved by producing a sequence of unordered pairs (i, j), $i, j \in A$ (where A is the set of individuals involved), each of which represents a phone call made between a pair of individuals. During each such call the two people involved exchange all the information they know at that time. Also, at the end of the sequence of calls, everybody knows everything. Such a calling sequence, which spreads the gossip to everyone, is called *complete*. The broadcasting problem is solved by a similar sequence of unordered pairs, except now the terminating criterion is that all people know the original message.

Example 15.1

Table 15.1 displays an example of a complete gossiping sequence for eight persons. It shows at each step the knowledge currently known by each person. For instance, the row containing the pair (4,5) corresponds to the exchange of information previously known by person 4 with the information known by person 5. (The information known by each is listed in the previous row.) In this particular gossiping sequence, twelve exchanges are needed to complete gossiping.

Table 15.2 indicates an example of a complete broadcasting sequence, where the message originates from the first person. Each row of this table shows which individuals currently know the original message. In this case, everyone knows the message after seven exchanges.

Table 15.1 A complete gossiping sequence for eight persons

Pair	Messages known by each person							
	1	2	3	4	5	6	7	8
(1,2)	1,2	1,2	3	4	5	6	7	8
(3,4)	1,2	1,2	3,4	3,4	5	6	7	8
(5,6)	1,2	1,2	3,4	3,4	5,6	5,6	7	8
(7,8)	1,2	1,2	3,4	3,4	5,6	5,6	7,8	7,8
(2,3)	1,2	1,2,3,4	1,2,3,4	3,4	5,6	5,6	7,8	7,8
(4,5)	1,2	1,2,3,4	1,2,3,4	3,4,5,6	3,4,5,6	5,6	7,8	7,8
(6,7)	1,2	1,2,3,4	1,2,3,4	3,4,5,6	3,4,5,6	5,6,7,8	5,6,7,8	7,8
(8,1)	1,2,7,8	1,2,3,4	1,2,3,4	3,4,5,6	3,4,5,6	5,6,7,8	5,6,7,8	1,2,7,8
(1,5)	all	1,2,3,4	1,2,3,4	3,4,5,6	all	5,6,7,8	5,6,7,8	1,2,7,8
(2,6)	all	all	1,2,3,4	3,4,5,6	all	all	5,6,7,8	1,2,7,8
(3,7)	all	all	all	3,4,5,6	all	all	all	1,2,7,8
(4,8)	all	all	all	all	all	all	all	all

15.2.2 Graph-Theoretical Definitions

The construction of mathematical models may take many forms and may involve many areas of mathematics. One area of mathematics particularly well suited for problems arising in communication networks is graph theory. Throughout this chapter, $G = (V, E)$ will denote an undirected graph in which the *vertex set* is $V = V(G)$ and the *edge set* is $E = E(G)$.

An edge e is said to be *incident* with a vertex v if v is in the pair e. Two vertices u and v are *adjacent* if $(u, v) \in E$ while two edges e and f

Table 15.2   A complete 8-person
broadcasting sequence with the
message originating from the first

	Persons knowing the message							
Pair	1	2	3	4	5	6	7	8
(1,2)	*	*						
(2,3)	*	*	*					
(1,4)	*	*	*	*				
(1,5)	*	*	*	*	*			
(2,6)	*	*	*	*	*	*		
(3,7)	*	*	*	*	*	*	*	
(4,8)	*	*	*	*	*	*	*	*

are adjacent if there exists a vertex u which is incident with both e and
f. The *degree* of a vertex u is the number of vertices that are adjacent
with it. The minimum and maximum degrees in a graph G are denoted
by $\delta(G)$ and $\Delta(G)$, respectively.

A sequence of vertices $v_0, v_1, v_2, \ldots, v_k$ such that $v_{i-1}v_i \in E$ for all
$i = 1, 2, \ldots, k$ is called a *walk* from v_0 to v_k. A *path* is a walk in which
all the vertices are distinct. A walk is *closed* if $v_0 = v_k$. A closed walk
is called a *circuit*, while a closed path is called a *cycle*. The path having
n edges is denoted P_n while the cycle having n edges is denoted C_n.

A pair of vertices u and v are *connected* if there exists a path from u
to v. The *distance* from u to v is the minimum number of edges in any
path from u to v. A graph is *connected* if every pair of distinct vertices
are connected. The *diameter* of a connected graph is the maximum
distance between any two vertices in the graph.

The *complete* graph on n vertices, denoted by K_n, is the graph in
which every pair of distinct vertices are adjacent. Figure 15.1 shows the
graph K_5. A *tree* is a connected graph with no cycles.

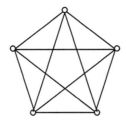

FIGURE 15.1   **A complete graph on five vertices**

Two distinct vertices or edges are *independent* if they are not adjacent in G. A set of pairwise independent edges of G is called a *matching* in G. An assignment of k colors to the edges of G so that adjacent edges are colored differently is called a *proper k-edge-coloring* of G.

Example 15.2

Figure 15.2 gives an example of a tree on eight vertices. The edges of this tree are partitioned into three matchings, as indicated by their labels. These matchings also provide a 3-edge coloring of the tree using the colors $\{1, 2, 3\}$.

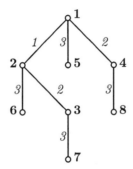

FIGURE 15.2   **Edge partitioning of a tree into three matchings**

15.2.3   *Communication Over a Network*

Standard gossiping was introduced to model communication where each individual is allowed to call anyone else. A graph can be used to illustrate which calls can occur and which may not. If the graph is complete, then no restriction exists on who can call whom. For example, in Boyd's initial problem, K_n is the underlying graph.

One reason for working with graphs and restricted communication is that it allows the model to be more accurate. One such restriction of communication occurs in an organization. A president makes a decision; it is then passed on to his subordinates. They pass it down to those for whom they are responsible, and so on. Here, the network would be a tree and broadcasting is the form of communication occurring. Since a broadcast does not benefit from having the same person receive the information from different sources, cycles are unnecessary; thus, the edges used in a broadcast form a tree. Several papers on the subject of communication over trees include [9] on gossiping and [17] on broadcasting.

15.2.4   Communication Described by Matchings

Communication can also be described by a sequence $E_1, E_2, \ldots, E_k$ of matchings where each edge in matching E_i generates a call during a given time step. The ability to have several calls occurring during the same time unit allows communication to complete more rapidly. In Tables 15.1 and 15.2, the number of time units used are 12 and 7, respectively. However, if independent calls are allowed to occur simultaneously (that is, the calls are grouped into matchings), then the time needed to complete communication is reduced.

Example 15.3

Tables 15.3 and 15.4 illustrate gossiping and broadcasting sequences of matchings that allow for the respective communications to occur in just 3 time units each. The set of pairs comprising each matching are shown grouped together (delimited by heavy horizontal lines). Also, Figures 15.3 and 15.2, respectively, are graphs showing the matchings used for communication.

Table 15.3   A complete gossiping sequence for eight persons

Pair	\multicolumn{8}{c}{Messages known at vertex}
	1	2	3	4	5	6	7	8
(1,2)	1,2	1,2	3	4	5	6	7	8
(3,4)	1,2	1,2	3,4	3,4	5	6	7	8
(5,6)	1,2	1,2	3,4	3,4	5,6	5,6	7	8
(7,8)	1,2	1,2	3,4	3,4	5,6	5,6	7,8	7,8
(2,3)	1,2	1,2,3,4	1,2,3,4	3,4	5,6	5,6	7,8	7,8
(4,5)	1,2	1,2,3,4	1,2,3,4	3,4,5,6	3,4,5,6	5,6	7,8	7,8
(6,7)	1,2	1,2,3,4	1,2,3,4	3,4,5,6	3,4,5,6	5,6,7,8	5,6,7,8	7,8
(8,1)	1,2,7,8	1,2,3,4	1,2,3,4	3,4,5,6	3,4,5,6	5,6,7,8	5,6,7,8	1,2,7,8
(1,5)	all	1,2,3,4	1,2,3,4	3,4,5,6	all	5,6,7,8	5,6,7,8	1,2,7,8
(2,6)	all	all	1,2,3,4	3,4,5,6	all	all	5,6,7,8	1,2,7,8
(3,7)	all	all	all	3,4,5,6	all	all	all	1,2,7,8
(4,8)	all	all	all	all	all	all	all	all

Generally, two types of problems arise in communication:

- determining the minimum number of calls needed to complete communication, and

- determining the minimum number of time units required to complete communication.

Table 15.4 A complete 8-person
broadcasting sequence with the
message originating from the first

Pair	Vertices knowing the message							
	1	2	3	4	5	6	7	8
(1,2)	*	*						
(2,3)	*	*	*					
(1,4)	*	*	*	*				
(1,5)	*	*	*	*	*			
(2,6)	*	*	*	*	*	*		
(3,7)	*	*	*	*	*	*	*	
(4,8)	*	*	*	*	*	*	*	*

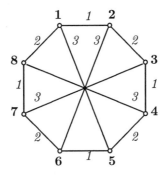

FIGURE 15.3 **Matchings used in gossiping**

15.3 Examples of Communication

15.3.1 A Broadcasting Example

One class of graphs on which information transfer has been studied is grid graphs. Figure 15.4 illustrates a grid graph with 20 vertices. Farley and Hedetniemi [2] studied these graphs because they provide a good model for city streets and also for parallel array processors. A set of city streets are modeled by placing a vertex at each intersection and using edges for roads that join two intersections. A configuration of parallel array processors can be modeled by using a vertex for each processor and by making edges for all connections between processors.

These models create two different types of grid graphs. One corresponds to the grids modeling city streets, so they have a path in each

row and column. The other type of grid models the connection of parallel array processors, so there are cycles in each row and column. This
distinction can also be produced by embedding the latter grid on a torus
so that there are no corners, as opposed to placing it in the plane where
corners are present.

Example 15.4

Suppose a message that the city was under attack needs to be broadcast
from the watchtower in one corner of a city having the street structure
in Figure 15.4. Assume the tower is located at vertex 1 so that the
message originates from there. Figure 15.5 is a broadcast tree illustrating the flow of information from vertex 1 to all vertices. Each level in
the tree corresponds to a matching or round of communication. Thus,
it takes 7 units of time for the original message to reach vertex 20.

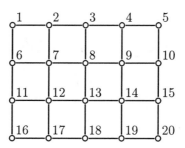

FIGURE 15.4   **A grid graph**

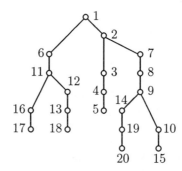

FIGURE 15.5   **A broadcast tree**

15.3.2   A Gossiping Example

Distributed processing machines such as the hypercube architecture structure are well suited for communication between processors. In this architecture, processors correspond to the vertices of an n-dimensional cube,
and they can communicate with those processors with whom they are
adjacent in the cube. Algorithms frequently have communication stages,
during which information is exchanged between some or all of the processors. One such algorithm is parallel cyclic reduction for solving tridiagonal systems of equations. Here a given large system of equations is
decomposed so that each processor gets a subset of the original equations. Each processor solves its piece, and then all partial solutions are
combined to solve the original system.

Since all the partial solutions need to be combined, a complete gossiping sequence must occur. The matchings used to gossip are formed
by using all the edges of the cube that are parallel to some edge. Thus,
the edges in each direction comprise the set of constituent matchings.
For example, in a three-dimensional cube (8 processors), the matchings
would be the edges parallel with the x-axis, the y-axis, and the z-axis.
The communication that occurs is then as efficient as possible. The
number of processors in an n-dimensional hypercube is 2^n and yet the
number of matchings used is only n.

Example 15.5

Figure 15.6 shows the partition of the edges used in the complete gossiping sequence for a hypercube computer with eight processors. Here
only 3 time units would be required to exchange all information.

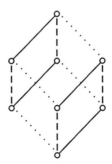

FIGURE 15.6   **Matchings in an 8-processor hypercube**

15.3.3   Restrictive Variations of Communication

The usage of networks as the framework over which communication occurs allows the model to be applicable in many situations. However, sometimes additional constraints are needed to allow the model to be more accurate. Several of these constraints are discussed in this section.

One constraint incorporates the idea that information transfer is not bi-directional. Several occasions where this form of communication would occur are communication via letter or electronic mail. The underlying network in this case would be directed; that is, the pairs of edges in the network would be ordered pairs where the information flows from the first coordinate to the second. Some of the first people to write on this form of communication were Harary and Schwenk [5, 6]. Their analysis dealt with the fewest number of calls required to complete gossiping.

A second form of restriction concentrates on the structure of the matchings. Usually, the same edge can be used several times in succession, and several edges may not be used at all. In some instances, it is desirable to use an edge only once, or to insist that each edge or link has a rest or down time following each transmission. In standard gossiping, there is no reason to use the same edge twice in succession since that pair of vertices have already exchanged everything that they know. Therefore, unless a down time of greater than one is stipulated, no additional restriction is placed on the gossiping sequence.

One type of gossiping that imposes restrictions on the matchings is *cyclic* or *periodic* gossiping. In this form of gossiping, the set of edges is partitioned into matchings $E_1, E_2, \ldots, E_k$. The edge sets E_i are used in sequence to perform the gossiping. During time t, the edge set E_i is used where $t \equiv i \pmod{k}$. If all the sets are cycled through, then all the edges are used; also, each edge is down or not being used during $k - 1$ consecutive time steps. Liestman and Richards [14] introduced this form of gossiping as a means of structuring gossiping so that it could be easily implemented in hardware.

Periodic broadcasting is useful in modeling the spread of disease in a population. Periodicity allows for the representation of two common ideas. First, people tend to interact with the same groups of people in a repetitive fashion. For example, an individual may have lunch every Tuesday with the same group of friends. Also, most infected people are infectious only for a specific period of time. Thus, the infectious period could be the period used in the communication scheme. However, one person might interact with the same person multiple times during the time that he is infected, so the restriction that the matchings form a k-coloring of the edges is violated.

A more general form of periodic gossiping drops the restriction that the matchings $E_1, E_2, \ldots, E_k$ be disjoint. Each edge should be in at least one set but an edge could appear in more than one set if so desired. This variant is sometimes called *set-periodic* gossiping to signify that a set is mapped to each edge instead of just one number. Set-periodic gossiping allows for the repetitive nature of periodic gossiping but with reduced communication cost for some structures. For example, a tree can benefit from using set-periodic gossiping instead of periodic gossiping, as the following example illustrates.

Example 15.6

Figures 15.7 and 15.8 show a tree with optimal labelings for periodic and set-periodic gossiping, respectively. In the first case, the edge set is partitioned into 3 matchings and the time required to complete periodic gossiping is 11 units. In the second case, there are 7 matchings (not all disjoint) and the time required to complete set-periodic gossiping is 9 units.

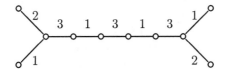

FIGURE 15.7   **Periodic gossiping**

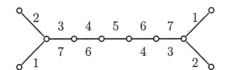

FIGURE 15.8   **Set-periodic gossiping**

The need for communication over secure channels may require that each channel be used at most once. Gossiping that uses each edge at most once is sometimes called "burning your bridges" gossiping. In cases like this, gossiping can still occur if the underlying network possesses a property called being *label-connected*. One paper on label-connected graphs is Göbel et al. [3]. This paper gives characterizations of label-connected graphs and shows how the labeling that identifies a graph as label-connected also gives rise to a gossiping technique that uses each edge exactly once. (Clearly, if gossiping completes before the last edge is used, then using any remaining edges is unnecessary.)

Gossiping and broadcasting have also been studied in certain instances where the duplication of information is to be kept to a minimum. Two main forms of gossiping result from these studies. They are called NOHO for "no one hears his own message" and NODUP for "no duplication." In NODUP, an exchange between two vertices occurs only if the two vertices contain no duplicate information.

Note that NODUP is even more restricted than "burning your bridges" gossiping because in NODUP each edge is used at most once, but the use of that edge is further restricted. West wrote extensively on the NOHO topic in [18, 19, 20, 21]. He also mentions NODUP in some of these papers, while Seres [15, 16] and Lenstra [13] also studied NODUP.

15.3.4   Random Communication

The framework of random gossiping and broadcasting can be posed in several ways. First, one can lift the restriction that all information exchanges are errorless and then assign a probability that the transmission will occur properly. In this case, one might be concerned with finding the most reliable or accurate means of communicating information. This idea of reliability can also be phrased in terms of the edges of the network remaining intact.

One means of protecting the gossiping sequence against edge or vertex failure is to have the information passed through two or more vertex-disjoint paths. Thus, if one vertex or edge fails and the information travels along two independent paths, then the remaining vertices will still receive the information. One necessity for protection against failure by k vertices is that each pair of vertices must be joined by $k+1$ vertex-disjoint paths. For example, a tree could not support duplication of information via disjoint paths since only one path joins any two vertices. One paper that examines the case when $k = 1$ is [8].

Another form of random broadcasting or gossiping reflects the fact that many information exchanges are random or occur as much by chance as by any set purpose. When a person wishes to share some gossip, one does not always call people in a set way, but frequently passes that information to the first person with whom he comes in contact. Of course, issues of a well-mixed population and concerns about the correct description of the information flow give this area a plethora of different models from which to choose. For example, Landau [11] and Landau and

Rapoport [12] assume that the exchange of information can be modeled using the same ideas used to model the spread of infectious diseases. On the other hand, Daley and Kendall [1] considered the possibility that a person might stop spreading the rumor. In this case, assuming that each person who knows the information continues to spread it would not be a correct model. However, in diseases, there are infected people who are infectious, those who are not infectious but infected, and individuals who have never been infected. Hence, even under the scenario of Daley and Kendall, an epidemic model would apply to some extent.

The variations of gossiping that exist are as many and varied as the problems they model and attempt to solve. The variations are interesting as exercises in mathematical endeavors, but their utility gives them greater worth. However, if these models are unmanageable, then we are no further ahead. Consequently, we now examine in more detail what analytical results can be obtained.

15.4   Results from Selected Gossiping Problems

15.4.1   Communication Results

The first set of results concern the number of calls needed to complete communication in a network with n vertices. For broadcasting, the number of calls needed is always $n - 1$ since the edges used form a tree. In addition, the number of calls needed for gossiping to be completed is also easy to determine.

LEMMA 15.1

The minimum number of calls required to complete gossiping is $2n - 4$ if the graph contains a four-cycle and $2n - 3$ otherwise.

PROOF   The cycle graph C_4 requires four calls to complete gossiping. If any other graph contains a C_4, then it too can complete gossiping in $2n - 4$ calls. The generation of a calling sequence that only uses $2n - 4$ calls to gossip is not hard to produce. One such method is to have each vertex call along a minimum length path to some vertex on the four-cycle. Once all the information reaches the four-cycle, the information can be exchanged in four calls. Then the total information can be dispensed out to the remaining vertices by reversing the order of the calling sequence, thus spreading the information back to the other vertices. If a four-cycle does not exist, carry out the same steps except use a single edge in the middle of the graph instead of a four-cycle; thus $2n - 3$ calls are used. ∎

A lower bound on the number of time steps needed to gossip, as well as broadcast, is given in Lemma 15.2. Here $\lceil x \rceil$ denotes the smallest integer greater than or equal to x.

LEMMA 15.2

The minimum length of time required for gossiping to complete is $\lceil \log_2 n \rceil$ for even n and $\lceil \log_2 n \rceil + 1$ for odd n.

This bound is a direct result of the fact that, during any given time step, a person can communicate with at most one other person. Thus, during each time step, the number of new people learning a piece of information can at most double. Hence, the logarithmic bound follows immediately. For graphs with an odd number of vertices, it is impossible for the information known by every vertex to double since at least one vertex is not in a matching. So, one additional time step is required.

For some classes of graphs (such as the grid graphs mentioned previously), the logarithmic bound is not tight. Instead another lower bound is more applicable. It is given in Lemma 15.3.

LEMMA 15.3

The minimum length of time required for gossiping to complete is bounded below by the diameter of the graph.

For broadcasting, the maximum length of time required for a vertex in a graph to complete broadcasting is called its *broadcast number*. Also, the broadcast number of a graph is the maximum broadcast number of any of its vertices. The set of vertices whose broadcast number equals the graph's broadcast number is called the *broadcast center* of the graph.

The determination of the length of time required to complete gossiping is much more difficult than just finding the minimum number of calls needed. The complexity of determining if the broadcast number of an arbitrary vertex exceeds some constant k is NP-complete. (NP-complete means, among other things, for all known algorithms to solve this problem there exist some graphs for which the number of steps required to solve the problem is not bounded above by a polynomial in the number of vertices in the graph.)

LEMMA 15.4

For a vertex v in a graph G, the determination of whether the broadcast number of v exceeds a fixed k is NP-complete for $k \geq 4$.

Clearly, the broadcast number of a graph is another lower bound on the length of time required to complete gossiping on that graph. The difficulty in finding the broadcast number of a graph is mirrored by the fact that it is also hard to compute the minimum number of time units needed to complete gossiping in a general graph.

15.4.2   Communication Algorithms

In general, finding matchings that form a complete gossiping sequence in minimal time is difficult. Thus, most algorithms are heuristics that produce an optimal solution for some classes of graphs and perform well for other similar types of graphs. For certain types of trees, there is an algorithm to produce an optimal labeling of the edges.

Some terms are needed to describe the algorithm. An *end vertex* is a vertex of degree one; thus, it is adjacent to only one vertex. A *caterpillar* is a tree having the property that all of its non-end vertices form a path.

In [8], a procedure is presented for properly coloring (labeling) the edges of a caterpillar to achieve an optimal periodic (cyclic) gossiping sequence as defined in Liestman and Richards [14]. Let T be a caterpillar with maximum degree $k = \Delta(T)$, and let P be the path containing all of the non-end vertices of T. Thus, P has two terminals — end vertices of the path.

ALGORITHM 15.6: Optimal periodic gossiping in a caterpillar

1. Label the edges of P alternately with the labels 1 and k starting with k.
2. Let u be a terminal of P whose edge is labeled k. Label all the edges of the end vertices that are incident with u using labels $\{1, 2, \ldots, \text{degree}(u) - 1\}$.
3. Let v be the other terminal of P, with d the label of its edge on P. Label all edges of the end vertices incident with v using labels $\{1, 2, \ldots, \text{degree}(v)\} - \{d\}$.
4. Label all remaining edges using colors $\{1, 2, \ldots, k\}$ so that the coloring remains proper.

Example 15.7

Figure 15.9 shows the application of Algorithm 15.1 to a caterpillar. Here $k = \Delta(T) = 5$ so that the (upper) path of T has its edges alternately labeled with 5 and 1. The leftmost vertex of P has its remaining incident edges labeled using $\{1, 2, 3\}$, while the rightmost vertex of P has its remaining incident edges labeled using $\{2, 3\}$. Applying Step 4 results in a periodic gossiping sequence having five sets E_i.

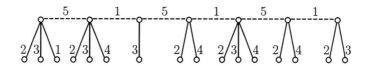

FIGURE 15.9   **Applying Algorithm 15.1 to a caterpillar**

The labeling procedure above can be extended to trees other than caterpillars. Another algorithm for coloring some trees to achieve a minimal gossip time is given in [10]. There is also a detection algorithm for finding the broadcast center of a tree (see [17]). But for most cases, no good algorithm is known for producing optimal communication schemes.

15.5   Conclusion

The field of broadcasting and gossiping is a rich area. In a survey article published in 1988, Hedetniemi, Hedetniemi, and Liestman [7] cite 135 articles that had been written on the subject prior to 1987. Some of the variations listed in Section 15.3 have either just recently been examined or are still awaiting investigation at this time. The applicability of these problems to different types of communication makes their study rewarding.

15.6   Exercises and Projects

1. Verify that the labelings in Figures 15.7 and 15.8 complete gossiping in 11 and 9 time steps, respectively.

2. Prove that the length of time required to complete gossiping in a graph of order n, for odd n, is $1 + \lceil \log_2 n \rceil$.

3. Prove that gossiping in an n-dimensional cube can be accomplished in n time steps.

4. Prove that a graph of order 2^n must contain at least $n2^{n-1}$ edges if it completes gossiping in n time steps.

5. Prove that the time required to complete gossiping in a grid graph is equal to its diameter, except for the 3×3 grid.

6. Determine the time required to complete gossiping in the caterpillar of Figure 15.9, using the indicated edge coloring.

7. (Project) Write a computer program that reads in a set of edges and determines if it forms a complete gossiping sequence; that is, after all the edges have been used for communication, every vertex in the graph knows the information from every other vertex.

8. (Project) Simulate random broadcasting. Write a program that reads in a graph and an initial vertex. From the set of vertices that currently know the piece of information from the initial vertex, randomly choose one and randomly choose an adjacent vertex with which to communicate. Repeat this process until the message has been broadcast throughout the entire graph. What is the number of calls used? How many of the calls could have been made concurrently? This case occurs when an edge used is not adjacent with the previous edge so that two (or more) calls could have been made concurrently.

15.7 References

[1] D. J. Daley and D. G. Kendall, "Stochastic rumours," *J. Inst. Maths. Applics.* **1**, 42–55 (1965).

[2] A. Farley and S. Hedetniemi, "Broadcasting in grid graphs," *Proc. Ninth SE Conf. on Combinatorics, Graph Theory, and Computing* Utilitas Mathematica, Winnipeg, 275–288 (1978).

[3] F. Göbel, J. Orestes Cerdeira, and H. J. Veldman, "Label-connected graphs and the gossip problem," *Discrete Math.* **87**, 29–40 (1991).

[4] A. Hajnal, E. C. Milner, and E. Szemeredi, "A cure for the telephone disease," *Canad. Math Bulletin* **15**, 447–450 (1972).

[5] F. Harary and A. J. Schwenk, "The communication problem on graphs and digraphs," *J. Franklin Inst.* **297**, 491–495 (1974).

[6] F. Harary and A. J. Schwenk, "Efficiency of dissemination of information in one-way and two-way communication networks," *Behavioral Sci.* **19**, 133–135 (1974).

[7] S. M. Hedetniemi, S. T. Hedetniemi, and A. L. Liestman, "A survey of gossiping and broadcasting in communication networks," *Networks* **18**, 319–349 (1988).

[8] J. Knisely, *The Study of Cyclic Gossiping in Graphs*, Ph.D. Dissertation, Clemson University, Clemson, SC (1993).

[9] R. Labahn, "The telephone problem for trees," *Elektron. Informationsverarb. u. Kybernet.* **22**, 475–485 (1986).

[10] R. Labahn, S. T. Hedetniemi, and R. Laskar "Periodic gossiping in trees," *Discrete Appl. Math.* **53**, 235–245 (1994).

[11] H. G. Landau, "The distribution of completion times for random communication in a task-oriented group," *Bull. Math. Biophys.* **16**, 187–201 (1954).

[12] H. G. Landau and A. Rapoport, "Contribution to the mathematical theory of contagion and spread of information I: spread through a thoroughly mixed population," *Bull. Math. Biophys.* **15**, 173–183 (1953).

[13] H. W. Lenstra, Jr., private communication, August 1976.

[14] A. L. Liestman and D. Richards, "Network communication in edge-colored graphs: gossiping," *IEEE Transactions on Parallel and Distributed Systems* **4**, 438–445 (1993).

[15] A. Seres, "Gossiping old ladies," *Discrete Math.* **46**, 75–81 (1983).

[16] A. Seres, "Quick gossiping without duplicate transmissions," *Graphs and Combinatorics* **2**, 363–383 (1986).

[17] P. J. Slater, E. Cockayne, and S. Hedetniemi, "Information dissemination in trees," *SIAM J. Comput.* **10**, 692–701 (1981).

[18] D. B. West, "A class of solutions to the gossip problem, part I," *Discrete Math.* **39**, 307–326 (1982).

[19] D. B. West, "A class of solutions to the gossip problem, part II," *Discrete Math.* **40**, 87–113 (1982).

[20] D. B. West, "A class of solutions to the gossip problem, part III," *Discrete Math.* **40**, 285–310 (1982).

[21] D. B. West, "Gossiping without duplicate transmissions," *SIAM J. Alg. Disc. Meth.* **3**, 418–419 (1982).

Chapter 16

Modeling the Impact of Environmental Regulations on Hydroelectric Revenues

R. Lougee-Heimer
W. Adams

Prerequisites: linear programming

16.1 Introduction

Hydroelectric power is considered to be one of the cleanest and safest energy alternatives. In contrast to fossil fuel plants, hydroelectric facilities (*hydros*) emit minimal pollution and use a renewable natural resource, water. Hydros produce electricity from the inherent energy in stream flow by redirecting the water through hydroelectric turbines. Typically, the stream flow is diverted into a reservoir for storage and then released at strategic times through the turbines. The releases are planned based on reservoir levels, turbine efficiencies, anticipated weather conditions, consumer demands for electricity, and the changing value of electricity, to make the most efficient and profitable use of the natural resource.

While damming the stream flow improves the productivity of hydroelectric generation, it leads to a conflict between hydro operators and environmental advocates. The hydro managers believe they must exploit the time-dependent value of electricity and the operating efficiencies that storing water affords, if their operations are to be economically viable. Environmentalists, on the other hand, believe that the reduction in downstream flow caused by storing water adversely impacts the environment, threatening fish and wildlife populations, and lowering water quality. To address their concerns, environmentalists have advocated legislation imposing minimum stream flow regulations, in effect requiring a minimum emission from the hydros at all times. The hydro operators oppose the proposed regulations, contending that the cost in lost revenues will be detrimental to their facilities' profitability. Thus arises the conflict.

In this chapter, we focus on a question underpinning a resolution to this conflict: how sensitive are the hydroelectric revenues to minimum stream flow regulations? The challenge is to develop a mathematical model which can be used to measure the revenue variations caused by enforcing minimum emissions, and thereby to quantify the impact of environmental regulations on hydroelectric revenues.

16.2   Preliminaries

As the basic disagreement between energy and environmental advocates centers on the regulation of stream flow, it is important to understand the benefits enjoyed by the hydros when stream flow regulations are not enforced. The hydros regulate stream flow for two main reasons: (a) to improve the yield of power production and (b) to increase the time-dependent revenues from hydroelectric sales.

The amount of power $\mathcal{P}$ in mega-watts (MW) that is produced by a given flow of water through a turbine is a function of three factors: the flow, the turbine efficiency at that flow, and the height of the head water level over the turbine intake that is directly available to the turbines, called the *net head*. Formally, this relationship is given by the formula $\mathcal{P} = kQeh$ where k is a conversion constant, Q is the flow in cubic feet per second (cfs), e is the flow-dependent turbine efficiency, and h is the net head [3]. From the formula above, it is easy to see that by improving any of these factors (while not degrading the others), hydroelectric operators can increase the amount of power produced. Due to the hydro's operating characteristics, in particular the turbine specifications, production improvements can be achieved by storing water.

A turbine has operating specifications which define e as a discontinuous function of Q. More plainly put, not all flow results in the generation of electricity. To produce electricity, a minimum amount of flow is needed to turn the blades of the turbine. Any flow below this minimum threshold produces no electricity. Similarly, a turbine has an upper limit on flow. Flow at the upper limit generates the upper bound on electricity; flow in excess of the maximum produces no additional electricity. Regardless of the cause, we will refer to the flow through a hydro facility which does not produce electricity as *spill*. On the other hand, the flow which produces electricity, necessarily lying between the turbine lower and upper specifications, will be referred to as *release*. The words *spill* and *release* will be reserved for these meanings throughout the remainder of the chapter. For our purposes, we will assume that a turbine's efficiency e is constant for any release between the lower and upper specifications, and 0 elsewhere.

The maximum and minimum turbine specifications make storing water attractive. To illustrate, suppose a hydro is experiencing relatively dry conditions, so the natural stream flow is insufficient to satisfy the turbine's lower specifications. If storing water is prohibited — that is, the hydro is a *run-of-river* facility — then no electricity can be generated. However, if storing water is permitted, the operator can dam the inflow into the reservoir until the capacity is sufficient to generate electricity. On the other extreme, suppose heavy rains occur. If the hydro is a run-of-river facility, the inflow is likely to violate the turbine's upper limit. However, if storing water is permitted, the inflow can be dammed so that the flow in excess of the upper specification is not spilled, but instead reserved for future productive use. In either case, storing water enables hydros to operate at a more favorable turbine efficiency, thereby increasing the energy production from the available water resources.

The second main reason for storing water is to increase the time-dependent electric revenues. Hydro management would like to regulate stream flow to produce electricity when it is most needed and, consequently, when it is most valued. The demand for electricity is stochastic (time-dependent), varying throughout the day and along a cyclical pattern over the year. There are many different methods to value electricity. One common method is a three-tier classification scheme. The value of electricity at a given time is classified as either on-peak, off-peak, or weekend, depending on the historic demand load faced by the electric utility. The electricity produced during these three periods is assigned successively decreasing values. By storing water and strategically timing releases, hydro operators can increase hydroelectric revenues by selecting the periods during which their electricity is produced.

Although storing water enables electricity to be generated more efficiently when it is most needed and most valuable, it unavoidably alters the volume of downstream flow. A critical question which therefore arises is the following: how sensitive are hydroelectric revenues to minimum stream flow regulations? This is the question we address, for not one hydro, but for a system of small-capacity hydros, located in series, operating cooperatively to maximize the system's revenue.

In a series configuration, each hydro along the river must decide how much water to spill, release, and store. Here, the operating possibilities of one hydro are inextricably linked to the operating decisions made upstream from it. The amount of water entering a hydro depends on how much flow is spilled and/or released at the facility immediately upstream and when the flow is emitted. The length of time it takes for water to travel from a hydro to the next facility downstream is known as the *time lag* of the emitting hydro. Another water source for a hydro is the *exogenous inflow* from the hydro's local *drainage area*. Every hydro in the system has its own turbine specifications, turbine efficiencies, reservoir capacity, net head, exogenous inflow, and drainage area. All facilities, except the furthest downstream, will have a time lag.

For such a system scenario, we use mathematical programming to quantify the effects of environmental regulations on hydroelectric revenues. We first develop a mathematical model to calculate the system's maximum revenue for a given minimum stream flow regulation. The sensitivity of the hydroelectric revenues to the proposed regulations is then analyzed by varying the minimum stream flow regulations and examining the resulting changes in the system's maximum revenue. In the next section, we develop a time-dynamic, fixed-charge network flow formulation of the problem. The time-dynamic nature of the model is dictated by the varying flows, the time-dependent electric revenues, and the time lags. In Section 16.4, we consider the amenability of this large-scale mixed 0-1 linear program to solution procedures and motivate an alternative formulation. We give computational results in Section 16.5, using historical data for a system of six hydros on the Saluda River, and compare the merits of the two formulations. Finally, in the last section we cite exercises to encourage further study of water resources problems and related discrete programming issues.

16.3   Model Formulation

This section develops a mathematical model to determine the maximum revenue a system of small-capacity hydros located in series, operating under minimum stream flow regulations, can achieve. As discussed in the previous section, operators can maximize their revenues by strategically timing the generation of electricity to take advantage of both the turbine specifications and the time-dependent values of electricity. Each hydro operator controls the quantity of available water to release, spill, and store, subject to the fundamental law of flow conservation (over time the total flow into a hydro must equal the total flow out of the hydro), the turbine lower and upper specifications, and the reservoir's capacity.

The physical flow of water in the hydro system lends itself naturally to a network flow model (for a discussion of network models, see e.g. [1]). In such a model, the hydros correspond to nodes and the flows to arcs. Figure 16.1 shows a network flow representation for a system with six hydros. Each of the two distinct emissions, release and spill, is represented as a separate arc originating at the emitting hydro. The arcs describe the progress of flow along the river and terminate at the next hydro downstream. Within this paradigm, an exogenous inflow into a hydro can be viewed as a supply at the corresponding node. For simplicity, these supplies are not explicitly depicted.

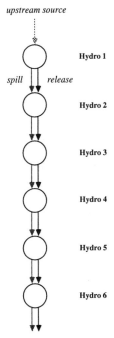

FIGURE 16.1 **Configuration of six hydros in series along a river**

To capture the time-dependent characteristics of the hydro operations, we need a time-dynamic network structure. The time-dynamic structure must accommodate the varying value of electricity, the different time lags, the changing flows and exogenous inflows, and the ability to store water. Such a model can be constructed by discretizing the (finite) planning horizon into some T time periods of equal length, and essentially reproducing the single period model of Figure 16.1, once for each period. The length of each period can be taken as any common divisor of the time lags and the number of consecutive hours that the electricity is valued at on-peak, off-peak, and weekend rates, under the assumption that each flow into or out of a hydro occurs uniformly over the period. If this uniformity assumption is not accurate, the time intervals can be accordingly subdivided into periods of shorter duration. To keep the size of the network manageable for the available computing resources, we recommend selecting the largest such common divisor and the smallest meaningful time horizon. With this time decomposition, the quantity of water to spill, release, and store must now be decided at each hydro for each period. These decisions can be represented by three distinct arcs originating at the node associated with the hydro-period pair. Water stored at a hydro from one period to the next is represented by an arc linking nodes corresponding to the same hydro in consecutive periods. Spills and releases, which take time to travel to the next facility downstream, are assigned destinations based on the originating hydro's time lag.

Example 16.1

Figure 16.2 illustrates the time-dynamic network for the six hydro example of Figure 16.1. The time lags of the hydros, starting with the furthest upstream facility are 5, 1, 0, 3, and 4 periods, respectively. As before, the exogenous inflows are envisioned as node supplies and are suppressed for simplicity.

With this time-dynamic structure as a basis, we now focus on developing a mathematical formulation of the problem. In a network model of, say, $\mathcal{H}$ hydros and $\mathcal{T}$ periods, each arc corresponds to a decision variable and each node to a constraint. Let r_{it}, s_{it}, and d_{it} denote the amount of flow in cubic feet per period that is released, spilled, and stored (i.e., $d$ammed) at hydro i in period t for all $i = 1, \ldots, \mathcal{H}$ and $t = 1, \ldots, \mathcal{T}$. We will assume for the remainder of the chapter that i iterates from 1 to $\mathcal{H}$ and that t iterates from 1 to $\mathcal{T}$ unless specifically stated otherwise.

The constraints on these decision variables include the law of flow conservation, the imposed minimum stream flow regulations, the upper and lower turbine specifications, and the reservoir capacities. Within our time-dynamic context, the law of flow conservation states that for each hydro (node), the total supply of water in the reservoir at the end of a given period must equal the supply at the beginning of that period plus the difference between the flow that enters and leaves the hydro during that period. Denoting for each (i, t) the exogenous inflow at

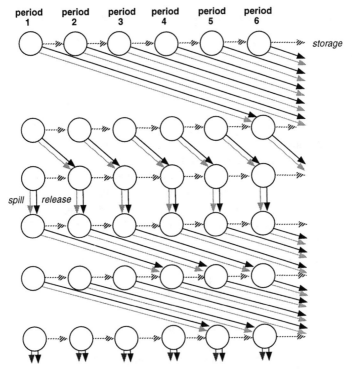

FIGURE 16.2   **Discretization of the planning horizon**

hydro i during period t as $\mathcal{I}_{it}$, and the time lag from hydro i to hydro $i + 1$ as $lag(i)$ for $i \neq \mathcal{H}$, where $lag(0) = 0$, flow conservation is enforced by the equations

$$r_{it} + s_{it} + d_{it} - \left(r_{i-1,t-lag(i-1)} + s_{i-1,t-lag(i-1)} + d_{i,t-1} \right) = \mathcal{I}_{it} \ \forall (i, t)$$

with the following notational conveniences. Since there are no hydros upstream from hydro 1, we define $r_{0t} = s_{0t} = 0$ for all t. Since time lags preclude "early" hydros from receiving upstream releases or spills, we define $r_{pq} = s_{pq} = 0$ for all (p, q), $p \neq 0$, $q \leq 0$, and note that adjustments to the exogenous inflows of the corresponding hydro-period nodes can be made to offset the flows lost from upstream sources. Due to the finite planning horizon, we consider the initial quantity of water d_{i0} for all i to be a given input parameter.

If desired, a final "global sink" flow conservation equation incorporating all water flow throughout the system can be derived. All releases and spills from hydro $\mathcal{H}$, all water in storage at the end of period $\mathcal{T}$, and all releases and spills r_{it} and s_{it}, $i \neq \mathcal{H}$, such that $t + lag(i) > \mathcal{T}$ can be envisioned as terminating at a single node, whose demand equals the total system water supply. The associated flow conservation equation is given below.

$$\sum_t (r_{\mathcal{H}t} + s_{\mathcal{H}t}) + \sum_i d_{i\mathcal{T}} + \sum_{\substack{(i,t)\,:\,i \neq \mathcal{H} \\ t + lag(i) > \mathcal{T}}} (r_{it} + s_{it}) = \sum_{(i,t)} \mathcal{I}_{it} + \sum_i d_{i0}$$

Since this equation is implied by the other flow conservation restrictions, we do not need to explicitly state it within our formulation. (Recall that since the flow conservation equations of a network contain precisely one 1 and one −1 in each column, with all other entries 0, any one such equation can be eliminated without affecting the set of solutions.) For notational convenience we will suppress this equation throughout the remainder of the chapter, but note its potential utility in preserving a partial network structure.

The reservoir capacities provide upper bounds on the storage arcs. Let $\mathcal{D}_i$ denote the storage capacity of hydro i in cubic feet. Then for each i, the quantity of water stored during each period t cannot exceed $\mathcal{D}_i$ or, notationally,

$$0 \leq d_{it} \leq \mathcal{D}_i \ \forall(i,t).$$

There are several types of constraints on the releases: the lower turbine specifications, the upper turbine specifications, and the minimum stream flow regulations. The lower turbine specifications stipulate that an emission from hydro i in period t generates electricity if and only if it is at least as great as the minimum per-period turbine specification, which we will denote by l_i. The upper turbine specifications dictate that the release cannot exceed the maximum per-period turbine specification, denoted by $\mathcal{R}_i$. A common way to capture this dichotomous ("electricity is generated or it is not") nature is through the use of binary variables [6]. For each (i,t), we let y_{it} be a 0-1 variable which has value 1 if the turbine at hydro i in period t is engaged and has value 0 otherwise. If some $y_{it} = 1$, then the release at hydro i during period t must lie between the turbine lower and upper per-period specifications. Notationally, $l_i \leq r_{it} \leq \mathcal{R}_i$. If some $y_{it} = 0$, then there cannot be a release from hydro i in period t, and we want $r_{it} = 0$.

The minimum stream flow regulations require that each hydro maintain a minimum stream flow at all times. Let m_i denote the minimum per-period amount of water that must be emitted from hydro i. As the downstream flow from hydro i in period t is the sum of the release and spill, we must enforce that $r_{it} + s_{it} \geq m_i$. Observe that, since no revenue is generated by spill, the choice will always be made to satisfy the minimum stream flow regulation via a release, if possible. With this intuitive observation, we can succinctly capture the desired restrictions in our model with the following constraints.

$$\max\{m_i, l_i\}y_{it} \le r_{it} \le \mathcal{R}_i y_{it} \ \forall(i,t)$$

$$m_i(1 - y_{it}) \le s_{it} \ \forall(i,t)$$

$$y_{it} \ \text{binary} \ \forall(i,t)$$

Note that these restrictions enforce that if some $y_{it} = 1$, then the associated release satisfies the turbine restrictions while if $y_{it} = 0$, all outflow must be treated as spill. In either case, the minimum flow regulations dictating that $r_{it} + s_{it} \ge m_i$ hold true. Implicit in the above inequalities is the assumption that the minimum flow regulations can always be met by the given system's inflows, since otherwise no feasible flows exist. Also implicit is that the problem input data satisfies $m_i \le \mathcal{R}_i$ for each i. If the minimum stream flow regulation were to exceed the upper turbine capacity of some hydro i, then that turbine would remain continuously engaged at its upper capacity, and the decision variables y_{it} for all t could be eliminated from the formulation. In fact, the turbine would remain continuously engaged if $m_i \ge l_i$, and the same reduction of variables could occur. (The proof is left as an exercise for the reader.)

Operating under the constraints we have discussed, the hydros regulate flow to maximize the system revenue. The revenue for the system is the sum of the individual hydro revenues. At each hydro, the revenue earned in a period is a product of two factors: (a) the value of electricity and (b) the amount of electricity generated. Because of our selection of period lengths, the electricity generated within a period is valued at exactly one of the three rates: on-peak, off-peak, or weekend. The amount of electricity produced, as explained in Section 16.2, is in general a cubic polynomial dependent on the release, the turbine efficiencies, and the net head. We make two simplifying assumptions that effectively render power a linear function of the release. As stated earlier, we assume that the turbine efficiencies are constant within the minimum and maximum specifications. Second, because we are concerned only with small-capacity hydros whose net head variation is slight, we make the reasonable assumption that the net head for each hydro is constant over the planning horizon. Let cr_{it} denote the revenue in dollars per cubic foot released from hydro i in period t. The total energy revenue can then be expressed as $\sum_i \sum_t cr_{it} r_{it}$, since spills do not contribute to the revenue.

There is another type of "revenue" which must be considered. Because the planning horizon is finite, we must place a value on the water remaining in storage at the end of the final period. Otherwise, with no

incentive to save, the model will drain all reservoirs in the last period
to generate as much energy revenue as possible. Let cd_i denote the revenue in dollars per cubic foot of water in storage at hydro i at the end
of period $\mathcal{T}$, so that the total value of water in storage at the end of
this period is $\sum_i cd_i d_{i\mathcal{T}}$. No other revenues exist. The following *mixed
integer* programming formulation maximizes the total revenue subject
to the above restrictions.

$$\text{MI1} : \text{maximize} \sum_i \sum_t cr_{it} r_{it} + \sum_i cd_i d_{i\mathcal{T}}$$

subject to

$$r_{it} + s_{it} + d_{it} - (r_{i-1,t-lag(i-1)} + s_{i-1,t-lag(i-1)} + d_{i,t-1}) = \mathcal{I}_{it} \; \forall(i,t)$$
(16.1)

$$\max\{m_i, l_i\} y_{it} \leq r_{it} \leq \mathcal{R}_i y_{it} \; \forall(i,t)$$
(16.2)

$$m_i(1 - y_{it}) \leq s_{it} \; \forall(i,t)$$
(16.3)

$$0 \leq d_{it} \leq \mathcal{D}_i \; \forall(i,t)$$
(16.4)

$$y_{it} \text{ binary } \forall(i,t)$$
(16.5)

16.4   Model Development

Problem MI1 is a mixed 0-1 linear program; it has both binary and continuous variables, and the objective function and constraints are linear
in the decision variables. This class of optimization problems is well-
recognized as being very difficult to solve. The difficulty stems from
the combinatorial explosion of possible solutions for the binary decision variables. Theoretically, Problem MI1 can be solved by enumerating over each possible realization of the binary variables, solving the
resultant linear programs over the continuous variables, and selecting
that feasible solution with the maximum revenue. Indeed, the special
structure of MI1 yields, for any fixed set of values for the variables y_{it},
a bounded-variable network flow problem [1] over the continuous flow
variables. Even so, this strategy is computationally prohibitive for all
but the smallest of problem instances, since it requires solving up to 2^n
such network flow problems, where $n = \mathcal{H}\mathcal{T}$ is the number of binary
variables present within MI1.

A standard approach for solving a mixed 0-1 linear program is to
construct a *linear programming relaxation* of the problem, and to then
use the relaxation to compute upper bounds on the optimal objective

function value to the discrete problem. The linear programming relaxation of a mixed 0-1 linear program is also referred to as the *continuous relaxation*. It is defined to be that linear program obtained by replacing the binary restrictions with the relaxed conditions that the variables lie in the interval [0,1]. The continuous relaxation is embedded within an implicit enumeration strategy over the binary variables, and used to generate bounds for fathoming nonoptimal and/or infeasible solutions. The idea is to explicitly enumerate as few of the 2^n binary realizations as possible by implicitly enumerating over the remainder. In this scheme, the tightness of the computed bounds plays an instrumental role in the success of the enumerative algorithm. As a rule of thumb, the tighter the bounds, the fewer the number of binary solutions that must be explicitly considered.

Given a discrete optimization problem, there typically is more than one way to formulate it as a mixed 0-1 linear program. While each of these formulations is by design equivalent when the binary restrictions are imposed, the strength of their continuous relaxations can vary greatly. For a specific problem instance, the issues of what constitutes an attractive formulation and how to construct such a formulation have been extensively studied by researchers. The general consensus is that formulations whose continuous relaxations provide "tight" representations of the feasible region of the discrete problem tend to produce better computational results than "weaker" representations. Simply put, the tighter the representation, the closer the linear programming bound will lie to the true optimal objective function value, and the fewer the number of binary solutions that will have to be explicitly considered. In the extreme case, if the continuous relaxation defines the *convex hull* of feasible solutions to the mixed 0-1 linear program itself, then the discrete problem can be solved as a linear program. (The convex hull of a given set S is defined to be the smallest convex set containing S.)

Returning to the formulation MI1, recall that our intent is to repeatedly solve this problem for a variety of input parameters in order to determine the sensitivity of the hydroelectric revenues to minimum stream flow regulations. Consequently, we are concerned with the amenability of this formulation to solution procedures. In light of the above discussion, we are specifically concerned with the strength of MI1's continuous relaxation, which we will refer to as Problem CMI1. Below, we consider two different operating scenarios, one in which CMI1 produces the true discrete optimal objective function value to MI1 and a second in which CMI1 is a poor approximation. These examples provide insight into the strengths and weaknesses of this formulation, and lead us to construct an alternative, tighter representation of the problem.

Example 16.2

Consider the simple example in which, for each hydro facility i, the minimum stream flow regulation is at least as great as the flow required to engage the turbines ($l_i \leq m_i$) and no storage is allowed ($\mathcal{D}_i = 0$). In this case, to produce the maximum energy revenue, every optimal solution will keep the turbines constantly engaged. Here, a spill will occur only when the total flow through a hydro in a given period exceeds the turbine's upper specification. For such a scenario, the linear programming relaxation is indeed tight since it can be verified that an optimal solution to CMI1 will have $y_{it} = 1 \; \forall (i,t)$.

Problem CMI1, however, is not always a tight relaxation of MI1, as seen by Example 16.3.

Example 16.3

Consider the case where for each hydro i, the positive minimum stream flow regulation is less than the minimum turbine specification ($0 < m_i < l_i$), the exogenous inflow into the system lies between these two values ($m_i \leq \mathcal{I}_{it} < l_i$), and there is no reservoir capacity ($\mathcal{D}_i = 0$). The plants are unable to generate electricity. However, the relaxation permits the production of electricity at each plant during each time period. It can be verified that the unique optimal solution to Problem CMI1 under this scenario is given by $y_{it} = (\mathcal{I}_{it} - m_i)/(l_i - m_i)$, $d_{it} = 0$, $r_{it} = l_i y_{it}$, and $s_{it} = m_i(1 - y_{it})$ [4]. Consequently, CMI1 turns out to be a weak relaxation for this scenario.

Example 16.3 exposes a weakness of Problem CMI1 and, in so doing, suggests means for devising alternate formulations with tighter continuous relaxations. By permitting the y_{it} variables to realize fractional values, CMI1 in effect allows the turbines to be partially engaged when the minimum specifications are not satisfied. Given a hydro i and a period t for which the total flow emitted from the hydro is less than the lower turbine specification, we would ideally like to enforce that all flow be treated as spill, and that the associated variable y_{it} takes the value 0. This enforcement is in contrast to the relaxed conditions afforded by (16.2) and (16.3). It is our intent to construct an alternate formulation of the water management problem that restricts, beyond that of (16.2) and (16.3), the set of feasible solutions for fractional values of y.

The formulation we pose is based on the construction of auxiliary variables reflecting the alternative paths that flow can take. By redefining our problem in a higher-dimensional space, we will be able to place restrictions on the newly-defined "path-variables" to further tighten the continuous relaxation. To illustrate our approach, let us reconsider the time-dynamic representation of Figure 16.2, and recall that in constructing MI1, a decision variable was defined for each arc in the network. We now define, for each *pair* of adjacent arcs, a variable to represent the flow of water that passes through the first arc *and* subsequently through the second arc: based on their origins, these new variables are referred to as *2-path variables*. Consider any node in the network, and observe that there are (at most) four types of arcs entering that node: inflow, storage, release, and spill, and three types of arcs exiting that node: release, spill, and storage. We construct all (twelve) possible pairwise combinations of these arcs: inflow-release, inflow-spill, inflow-storage, storage-release, storage-spill, storage-storage, release-release, release-spill, release-storage, spill-release, spill-spill, and spill-storage. For each (i, t), we define the variables er_{it}, es_{it}, and ed_{it} to be the exogenous inflow to hydro i in period t that is released, spilled, and stored, respectively. For each $(i, t), t \neq \mathcal{T}$, we define the variables dr_{it}, ds_{it}, and dd_{it} to be the amounts of water stored at hydro i in period t and then released, spilled, and stored, respectively, in the next period. Here, we need not consider $t = \mathcal{T}$ since there are only $\mathcal{T}$ periods. All other 2-path variables correspond to flows between consecutive hydros. We let the variables rr_{it}, rs_{it}, and rd_{it} denote the amounts of water released at hydro i in period t and subsequently released, spilled, and stored at hydro $i + 1$, and the variables sr_{it}, ss_{it}, and sd_{it} denote the amounts of water spilled at hydro i in period t and subsequently released, spilled, and stored at hydro $i + 1$. Since hydro $\mathcal{H}$ has no downstream facility and since flow emitted from a given hydro i requires $lag(i)$ periods to travel to the next facility, these six families of variables are defined for all (i, t) such that $i \neq \mathcal{H}$ and $t \leq \mathcal{T} - lag(i)$.

In effect, we are disaggregating each flow of Figure 16.2 into its component parts. For example, given any hydro $i \neq \mathcal{H}$ and any period $t \leq \mathcal{T} - lag(i)$, the release from hydro i at period t must be subsequently released, spilled, or stored in the next period: notationally, $r_{it} = rr_{it} + rs_{it} + rd_{it}$. Moreover, for any (i, t), the release for hydro i at period t must have originated as inflow, release, spill, or storage: notationally, $r_{it} = er_{it} + rr_{i-1,t-lag(i-1)} + sr_{i-1,t-lag(i-1)} + dr_{i,t-1}$. In essence, the sum of the flows through all 2-path variables with the same first arc must equal the flow along that first common arc, while the sum of the flows through all 2-path variables with the same second arc must equal the flow along that common second arc. We generically label the set of all such constraints as the *flow-partitioning constraints*.

Now, to construct our 2-path formulation, we replace restrictions (16.1) of MI1 with seven sets of flow-partitioning restrictions as provided in (16.6) through (16.12) below, and enforce nonnegativity of the

2-path variables as in (16.13) through (16.15). Our result is Problem MI2 below. For notational convenience, we do not explicitly exclude from consideration any 2-path variables with nonpositive subscripts, but realize that such variables do not exist within the formulation.

$$\text{MI2}: \text{maximize} \sum_i \sum_t cr_{it} r_{it} + \sum_i cd_i d_{i\mathcal{T}}$$

subject to   (16.2), (16.3), (16.4), (16.5)

$$rr_{it} + rs_{it} + rd_{it} = r_{it} \ \forall (i,t), \ i \neq \mathcal{H}, \ t \leq \mathcal{T} - lag(i) \qquad (16.6)$$

$$er_{it} + rr_{i-1,t-lag(i-1)} + sr_{i-1,t-lag(i-1)} + dr_{i,t-1} = r_{it} \ \forall (i,t) \quad (16.7)$$

$$sr_{it} + ss_{it} + sd_{it} = s_{it} \ \forall (i,t), \ i \neq \mathcal{H}, \ t \leq \mathcal{T} - lag(i) \qquad (16.8)$$

$$es_{it} + rs_{i-1,t-lag(i-1)} + ss_{i-1,t-lag(i-1)} + ds_{i,t-1} = s_{it} \ \forall (i,t) \quad (16.9)$$

$$dr_{it} + ds_{it} + dd_{it} = d_{it} \ \forall (i,t), \ t \neq \mathcal{T} \qquad (16.10)$$

$$ed_{it} + rd_{i-1,t-lag(i-1)} + sd_{i-1,t-lag(i-1)} + dd_{i,t-1} = d_{it} \ \forall (i,t) \quad (16.11)$$

$$er_{it} + es_{it} + ed_{it} = \mathcal{I}_{it} \ \forall (i,t) \qquad (16.12)$$

$$er_{it}, es_{it}, ed_{it} \geq 0 \ \forall (i,t) \qquad (16.13)$$

$$dr_{it}, ds_{it}, dd_{it} \geq 0 \ \forall (i,t), \ t \neq \mathcal{T} \qquad (16.14)$$

$$rr_{it}, rs_{it}, rd_{it}, sr_{it}, ss_{it}, sd_{it} \geq 0 \ \forall (i,t), \ i \neq \mathcal{H}, \ t \leq \mathcal{T} - lag(i) \quad (16.15)$$

We now address the relative strengths of CMI1 and the continuous relaxation of the above alternative, denoted CMI2. As may be suspected from our motivation of the flow-partitioning equations, the feasible region to Problem CMI1 is the projection of the feasible region to CMI2 onto the (r, s, d, y) variable space. This result is stated more formally in the theorem below. A sketch of the proof is provided, with the details left as exercises for the reader. To facilitate discussion, we let τ denote the set of all 2-path variables.

THEOREM 16.1

A point (r, s, d, y) is feasible to Problem CMI1 if and only if it is part of a feasible solution (r, s, d, y, τ) to CMI2.

PROOF   Let (r, s, d, y, τ) be any feasible solution to CMI2. To prove that (r, s, d, y) is feasible to CMI1, it suffices to show that equations (16.1) can be expressed as linear combinations of (16.6) through (16.12). The computations of the precise multipliers are left as an exercise.

Now, let (r, s, d, y) be feasible to CMI1. The objective is to compute a set of 2-path variables τ which, together with (r, s, d, y), constitute a feasible solution to CMI2. We accomplish this by defining the 2-path variables proportional to the r, s, and d flow values. For each (u, v) pair with $u \in \{r, s\}$ and $v \in \{r, s, d\}$, and for every (i, t) such that $i \neq \mathcal{H}$ and $t \leq \mathcal{T} - lag(i)$, define $uv_{it} = 0$ if $v_{i+1,t+lag(i)} = 0$ and $uv_{it} = (u_{it}v_{i+1,t+lag(i)})(r_{i+1,t+lag(i)} + s_{i+1,t+lag(i)} + d_{i+1,t+lag(i)})^{-1}$, otherwise. Also, for each $v \in \{r, s, d\}$ and for every (i, t) such that $t \neq \mathcal{T}$, define $dv_{it} = 0$ if $v_{i,t+1} = 0$ and $dv_{it} = (d_{it}v_{i,t+1})(r_{i,t+1} + s_{i,t+1} + d_{i,t+1})^{-1}$, otherwise. Finally, for each $v \in \{r, s, d\}$ and for every (i, t), define $ev_{it} = 0$ if $v_{it} = 0$ and $ev_{it} = (\mathcal{I}_{it}v_{it})(r_{it} + s_{it} + d_{it})^{-1}$, otherwise. These 2-path variables are clearly nonnegative, and (r, s, d, y, τ) can be shown to satisfy (16.6) through (16.12). The details of showing feasibility of (r, s, d, y, τ) to (16.6) through (16.12) are left as an exercise.   ∎

It follows directly from the above theorem and proof that Problems MI1 and MI2 are equivalent in the sense that a feasible solution to either problem identifies a feasible solution to the other problem with the same objective function value. As a result, Problem MI2 is a valid model of the energy problem. The formal equivalence is stated below as a corollary to the theorem.

COROLLARY 16.1

A point (r, s, d, y) is feasible to Problem MI1 if and only if it is part of a feasible solution (r, s, d, y, τ) to MI2.

Theorem 16.1 in effect demonstrates that the additional variables and constraints found in Problem CMI1 and not present in CMI2 do not help strengthen the continuous relaxation; in fact, given any instance of the water resource management problem, the two problems will yield the same objective function value. However, the higher-variable space found in CMI2 does promote the construction of additional inequalities, called *cutting planes*, that can serve to eliminate fractional y solutions. Below, we present nine such families of cutting planes. Note that depending on the particular problem of concern, any collection of these cuts can be selected for implementation.

We begin with any release-release variable, rr_{it}. Observe that if either turbine i at period t or turbine $i + 1$ at period $t + lag(i)$ is not engaged, then the 2-path variable rr_{it} must take value 0. If, however, both such turbines are engaged, the quantity of flow rr_{it} cannot exceed the minimum of these turbines' upper specifications. As a result, the following inequality must hold for all feasible solutions to Problem MI2.

$$rr_{it} \leq \min\{\mathcal{R}_i y_{it}, \mathcal{R}_{i+1} y_{i+1,t+lag(i)}, \mathcal{R}_{i+1} y_{it}, \mathcal{R}_i y_{i+1,t+lag(i)}\}$$

Note that the flow-partitioning equations, the turbine upper specifications, and the variable nonnegativity enforce this restriction when the minimum is achieved by either of the first two terms for all feasible solutions to CMI1, as shown below.

$$rr_{it} \leq rr_{it} + rs_{it} + rd_{it} = r_{it} \leq \mathcal{R}_i y_{it}$$
$$rr_{it} \leq er_{i+1,t+lag(i)} + rr_{it} + sr_{it} + dr_{i+1,t+lag(i)-1}$$
$$= r_{i+1,t+lag(i)} \leq \mathcal{R}_{i+1} y_{i+1,t+lag(i)}$$

However, the constraints of CMI1 do not imply that $rr_{it} \leq \mathcal{R}_{i+1} y_{it}$ when $\mathcal{R}_i > \mathcal{R}_{i+1}$ or that $rr_{it} \leq \mathcal{R}_i y_{i+1,t+lag(i)}$ when $\mathcal{R}_i < \mathcal{R}_{i+1}$. These inequalities, stated below, comprise our first two sets of cutting planes.

$$rr_{it} \leq \mathcal{R}_{i+1} y_{it} \; \forall (i,t), \; i \neq \mathcal{H}, \; \mathcal{R}_i > \mathcal{R}_{i+1}, \; t \leq \mathcal{T} - lag(i) \qquad (16.16)$$

$$rr_{it} \leq \mathcal{R}_i y_{i+1,t+lag(i)} \; \forall (i,t), \; i \neq \mathcal{H}, \; \mathcal{R}_i < \mathcal{R}_{i+1}, \; t \leq \mathcal{T} - lag(i) \qquad (16.17)$$

We demonstrate the potential utility of cuts (16.16) in Example 16.4.

Example 16.4

Figure 16.3 illustrates two hydros over a single period, with the objective being to maximize the sum of the releases ($cr_{11} = cr_{21} = 1$). No storage capacities or time lags exist, both minimum stream flow regulations are 1, the local exogenous inflow is only into hydro 1, and the lower and upper turbine capacities, l_i and $\mathcal{R}_i$, are as shown.

The optimal energy revenue for Problems MI1 and MI2 in this case is 2, with $y_{11} = 0$, $y_{21} = 1$, $r_{11} = s_{21} = 0$, and $r_{21} = s_{11} = 2$. The optimal objective function value to Problem CMI1 (and hence to CMI2 by Theorem 16.1) is $\frac{28}{9}$, with $y_{11} = \frac{1}{9}$, $y_{21} = 1$, $r_{11} = \frac{10}{9}$, $r_{21} = 2$, $s_{11} = \frac{8}{9}$, and $s_{21} = 0$. However, the inequality $rr_{11} \leq 2y_{11}$ of (16.16) strengthens CMI2 so that the objective function value improves to $\frac{20}{9}$ when this cut is active.

Two additional families of cutting planes can be computed by applying similar logic to the release-storage and storage-release variables rd_{it} and dr_{it}, respectively. The quantity of water released then stored, or stored then released, cannot exceed the associated storage capacity if the turbine is engaged, and must be 0 if the turbine is not engaged. This observation leads to the following inequalities.

$$rd_{it} \leq \mathcal{D}_{i+1} y_{it} \; \forall (i,t), \; i \neq \mathcal{H}, \; t \leq \mathcal{T} - lag(i), \; \mathcal{D}_{i+1} < \mathcal{R}_i \qquad (16.18)$$

$$dr_{it} \leq \mathcal{D}_i y_{i,t+1} \ \forall(i,t), \ t \neq \mathcal{T}, \ \mathcal{D}_i < \mathcal{R}_i \qquad (16.19)$$

We do not consider in (16.18) those indices i having $\mathcal{D}_{i+1} \geq \mathcal{R}_i$ since the restrictions $rd_{it} \leq \mathcal{R}_i y_{it} \ \forall(i,t), \ i \neq \mathcal{H}, \ t \leq \mathcal{T} - lag(i)$, are implied by the constraints of CMI1. Similarly, we do not consider in (16.19) those indices i having $\mathcal{D}_i \geq \mathcal{R}_i$ since the inequalities $dr_{it} \leq \mathcal{R}_i y_{i,t+1} \ \forall(i,t), \ t \neq \mathcal{T}$, are implied by the constraints of CMI1.

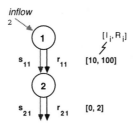

FIGURE 16.3   **Example illustrating the utility of cuts (16.16)**

This same type logic can once again be applied relative to the inflow-release variables er_{it}. Observe that for any hydro i and period t, the exogenous inflow that is released cannot exceed the total inflow if the turbine is engaged, and must be 0 if the turbine is not engaged. Notationally, we have the following inequalities.

$$er_{it} \leq \mathcal{I}_{it} y_{it} \ \forall(i,t), \ \mathcal{I}_{it} < \mathcal{R}_i \qquad (16.20)$$

In (16.20), we do not consider those indices i having $\mathcal{I}_{it} \geq \mathcal{R}_i$ since the restrictions $er_{it} \leq \mathcal{R}_i y_{it} \ \forall(i,t)$ are implied by the constraints of CMI1. Inequalities (16.20) represent our fifth family of cutting planes.

The release-release variables give rise to two additional families of cutting planes. Given any such variable rr_{it}, if both associated turbines are either engaged or disengaged, then this variable value is bounded above by each of the individual releases. On the other hand, if precisely one turbine is engaged, then the flow cannot exceed the release through the active turbine, less its lower bound. Consequently, the following two families of inequalities are valid.

$$rr_{it} \leq r_{it} + l_i(y_{i+1,t+lag(i)} - y_{it}) \ \forall(i,t), \ i \neq \mathcal{H}, \ t \leq \mathcal{T} - lag(i) \quad (16.21)$$

$$rr_{it} \leq r_{i+1,t+lag(i)} + l_{i+1}(y_{it} - y_{i+1,t+lag(i)}) \ \forall(i,t), \ i \neq \mathcal{H}, \ t \leq \mathcal{T} - lag(i) \quad (16.22)$$

Our final two families of cutting planes deal with the spill-release and release-spill variables. Observe that for any two consecutive hydros, the quantity of flow either spilled then released, or released then spilled, cannot exceed the individual spill and cannot exceed the individual spill less the associated minimum flow regulations if neither hydro is engaged. These restrictions can be stated notationally as follows.

$$sr_{it} \leq s_{it} + m_i(y_{it} + y_{i+1,t+lag(i)} - 1) \ \forall(i,t), \ i \neq \mathcal{H}, \ t \leq \mathcal{T} - lag(i) \tag{16.23}$$

$$rs_{it} \leq s_{i+1,t+lag(i)} + m_{i+1}(y_{it} + y_{i+1,t+lag(i)} - 1) \ \forall(i,t), \ i \neq \mathcal{H}, \ t \leq \mathcal{T} - lag(i) \tag{16.24}$$

Our 2-path formulation is now constructed by appending inequalities (16.16) through (16.24) to Problem MI2. The resulting formulation, denoted Problem 2-P, is the following.

$$\text{2-P : maximize} \sum_i \sum_t cr_{it} r_{it} + \sum_i cd_i d_{i\mathcal{T}}$$

subject to   (16.2)–(16.24)

We mention here that while Problem 2-P was developed explicitly for the water resource management problem, the constructs can be generally applied to other network flow problems. Indeed, the thought process can be extended to 3-paths, or more generally p-paths, where restrictions on flows along paths can be enforced.

16.5   Case Study

In this section, we focus attention on a specific hydro system as a case study for our models. We consider this system with two objectives in mind: (a) to quantify the effects of minimum stream flow regulations on hydroelectric revenues and (b) to evaluate the computational merits of our two alternative mathematical programming formulations.

The test case used was recommended by the South Carolina Water Resources Commission (SCWRC). It consists of a collection of six small-capacity hydros located in series on the Saluda River. Beginning with the furthest upstream and progressing downstream, the facilities' names are Saluda, Piedmont, Upper Pelzer, Lower Pelzer, Hollidays Bridge, and Ware Shoals. A diagram of the system is shown in Figure 16.4. This diagram is essentially a duplicate of Figure 16.1 where the individual nodes now represent specific hydro facilities and the time lags are shown in hours.

16.5.1   Input Data

In order to construct the formulations MI1 and 2-P for the Saluda River case study, we need three sets of input data: (a) the electric rate schedule,

(b) the operating characteristics of the hydros, and (c) the exogenous inflows into the system. This data was supplied to us by Duke Power, the SCWRC, and the United States Geological Survey (USGS) [7], respectively.

The rate schedule for electricity is based on a seven-day week. For the purposes of our study, on-peak hours occur from 0:00 to 16:00 hours on Monday through Friday. Off-peak hours comprise the remainder of the weekdays, and weekend hours run all day Saturday and Sunday. The on-peak, off-peak, and weekend rates are $25, $20, and $15 per megawatt-hour, respectively.

Based on this rate schedule and the time lags of the hydros given in Figure 16.4, we chose the period duration to be 2 hours, the largest common divisor of the factors. All model parameters were suitably calculated from the input data to match this time period selection. For example, the on-peak rates for electricity were valued at $50 per megawatt-period. Note that with this period duration, Figure 16.2 accurately reflects the time lags in the time-dynamic representation of the Saluda River system. Our test runs were conducted for a one-week planning horizon beginning at 0:00 hours on Monday, the shortest horizon that included the entire electric rate schedule. These conservative choices yielded fairly large and challenging instances of formulations MI1 and 2-P, each having $T = 84$ periods and 504 binary variables y_{it}.

The objective function coefficients were calculated in the following manner. Consistent with the explanation of Section 16.3, the power generated at a hydro i in period t is defined as a linear function of the release as $kehr_{it}$, where k is a conversion constant, e is the turbine efficiency, h is the net head, and r_{it} is the release. The power is computed in units of megawatt-periods where, in keeping with formulations MI1 and 2-P, we assume the release is made uniformly over the period. For our case study, we used the constant $k = 1.175 \times 10^{-8}$ to reflect the 2-hour periods. The net head (in feet) and efficiency for each hydro facility i are listed in Table 16.1.

The objective coefficients cr_{it}, for each (i, t), were then computed via the equation

$$cr_{it} = (1.175 \times 10^{-8}) \times (\text{turbine efficiency of hydro } i)$$
$$\times (\text{net head of hydro } i) \times (\$ \text{ per megawatt-period at period } t)$$

where the product of the first three terms represents the power produced in megawatt-periods per unit release and the final term is $50, $40, or $30, depending on the period. For each time period-hydro pair (i, t), the value cr_{it} in dollars per 100,000 cubic feet of release is summarized in Table 16.2.

Table 16.1 Case study hydro and turbine specifications

	Saluda	Piedmont	Upper Pelzer	Lower Pelzer	Hollidays Bridge	Ware Shoals
head(ft)	40.7	26.0	25.0	40.0	41.5	58.0
efficiency	0.74	0.74	0.78	0.74	0.80	0.80

Table 16.2 cr_{it} in dollars per 100,000 cubic feet of release

	Saluda	Piedmont	Upper Pelzer	Lower Pelzer	Hollidays Bridge	Ware Shoals
on-peak	1.77	1.13	1.14	1.74	1.95	2.72
off-peak	1.41	0.90	0.91	1.39	1.56	2.18
weekend	1.06	0.67	0.68	1.04	1.17	1.63

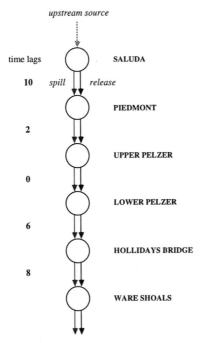

FIGURE 16.4 **Small-capacity hydros on the Saluda River**

The characteristics of each hydro i are given in Table 16.3. The lower and upper per-period turbine specifications, l_i and R_i, respectively, are expressed in thousands of cubic feet per period. The reservoir capacity D_i is given in cubic feet and the drainage area DA_i is shown in square miles.

Table 16.3   Case study input parameters

	Saluda	Piedmont	Upper Pelzer	Lower Pelzer	Hollidays Bridge	Ware Shoals
l_i (ft^3 × 1,000)	648	720	720	720	1,152	720
R_i (ft^3 × 1,000)	5,760	4,248	8,640	10,137	11,592	6,480
D_i (ft^3)	24,219	0	2,178	6,969	20,299	5,662
DA_i (mi^2)	290	85	35	1	70	100

Table 16.4   Drainage areas for the USGS gauges

	USGS gauge		
	G	P	W
drainage area (mi^2)	295	405	581

The exogenous inflows $\mathcal{I}_{it}$ into the system are not known and must be approximated based on the geographics of the individual facilities and the historic daily stream flow measurements provided by the USGS at nearby gauges. The inflows for the Saluda and Piedmont hydros were computed using the USGS gauge 02.1625.00 near Greenville (G), the inflows for the Upper and Lower Pelzer hydros used the USGS gauge 02.1630.00 near Pelzer (P), and the inflows for the Hollidays Bridge and Ware Shoals hydros used the USGS gauge 02.1635.00 near Ware Shoals (W). The drainage area for each of these gauges is found in Table 16.4.

We tested two different sets of inflow data, one relating to a dry period and one to a wet period. For the dry period, we used the historic flow from the first week in October of 1970 and for the wet period, we used the flow from the first week in April of 1970. The flows in cubic feet per 2-hour period from the USGS gauges for the dry and wet periods are recorded in Tables 16.5 and 16.6, respectively, with the seven table columns representing the individual days of the week. Since the USGS collects only one gauge measurement per day, we assumed that all 2-hour intervals within a given day had the same exogenous inflows.

Table 16.5   Flows in thousands of cubic feet per 2-hour period at USGS gauges

Gauge	First week of October, 1970 (dry period)						
G	1,303.2	1,281.6	1,303.2	1,267.2	1,209.6	1,209.6	1,231.2
P	1,605.6	1,663.2	1,663.2	1,605.6	1,692.0	1,519.2	1,605.6
W	1,980.0	1,857.6	1,454.4	1,195.2	2,268.0	2,210.4	1,958.4

Table 16.6 Flows in thousands of cubic feet per 2-hour period at USGS gauges

Gauge	First week of April, 1970 (wet period)						
G	6,098.4	7,488.0	10,224.0	8,136.0	5,947.2	6,400.8	5,644.8
P	8,136.0	10,728.0	11,880.0	10,224.0	8,208.0	7,041.6	7,200.0
W	7,920.0	14,040.0	12,672.0	11,592.0	9,000.0	7,106.4	7,416.0

Table 16.7 Estimated exogenous inflows in thousands of cubic feet per 2-hour period

Hydro	First week of October, 1970 (dry period)						
Saluda	1,281.1	1,259.8	1,281.1	1,245.7	1,189.0	1,189.0	1,210.3
Piedmont	375.4	369.2	375.4	365.1	348.5	348.5	354.7
U. Pelzer	138.7	143.7	143.7	138.7	146.2	131.2	138.7
L. Pelzer	3.9	4.1	4.1	3.9	4.1	3.7	3.9
Hollidays	238.5	223.8	175.2	144.0	273.2	266.3	235.9
Ware Sh.	340.7	319.7	250.3	205.7	390.3	380.4	337.0

The values $\mathcal{I}_{it}$ were computed from the information in Tables 16.3 through 16.6 as follows. For each of the hydro facilities, we adjusted for the difference between the drainage area of the associated USGS gauge and the drainage area of the hydro via the equation

$$\mathcal{I}_{it} = \frac{\text{drainage area of hydro } i}{\text{drainage area of the gauge}} \times \text{ flow at the gauge during period } t.$$

These exogenous inflows, in thousands of cubic feet per 2-hour period, for the dry and wet periods are found in Tables 16.7 and 16.8, respectively. As with Tables 16.5 and 16.6, the columns represent the days of the week, and all 2-hour intervals within a given day are assumed to have the same inflows.

The minimum stream flow regulations m_i were calculated as fixed percentages of the expected flows into the hydros. For each of the dry and wet periods, we calculated six different minimum stream flow scenarios

Table 16.8 Estimated exogenous inflows in thousands of cubic feet per 2-hour period

Hydro	First week of April, 1970 (wet period)						
Saluda	5,995.0	7,361.0	10,050.7	7,998.1	5,846.4	6,292.3	5,549.1
Piedmont	1,757.1	2,157.5	2,945.8	2,344.2	1,713.6	1,844.3	1,626.4
U. Pelzer	703.1	927.1	1,026.6	883.5	709.3	608.5	622.2
L. Pelzer	20.0	26.4	29.3	25.2	20.2	17.3	17.7
Hollidays	954.2	1,691.5	1,526.7	1,396.6	1,084.3	856.1	893.4
Ware Sh.	1,363.1	2,416.5	2,181.0	1,995.1	1,549.0	1,223.1	1,276.4

Table 16.9 Mean exogenous inflows in thousands of cubic feet per 2-hour period

	Saluda	Piedmont	Upper Pelzer	Lower Pelzer	Hollidays Bridge	Ware Shoals
mean inflow–dry	1,236.5	362.4	140.1	3.9	222.4	317.7
mean inflow–wet	7,013.2	2,055.5	782.9	22.3	1,200.4	1,714.8

by fixing the percentage used at values ranging from 0 to 50, in increments of 10. The expected flow into a hydro is defined as the sum of the mean exogenous inflow over the horizon and the minimum emission required from the facility immediately upstream (when one exists). For each of the six hydros, the week's mean inflow in thousands of cubic feet per period is given in Table 16.9, for each of the dry and wet periods. An example of the recursive calculations of m_i follows. Given a fixed percentage, say 10%, we begin with the facility furthest upstream and require the minimum emission at Saluda to be 10% of its exogenous inflow. For the dry period, we require Saluda to emit a minimum of 123.65 cubic feet per period. The regulation at Piedmont is then 10% of the sum of the regulation at the Saluda facility and Piedmont's historic mean inflow. For the dry period, the minimum emission at Piedmont is therefore 48.605 cubic feet per period. We repeat this logic for each hydro progressively further downstream until all facilities have been considered. The computed m_i values are summarized in Table 16.10.

Finally, three accommodations were made for modeling an ongoing process with a finite time horizon. First, the value of water remaining in the reservoir at the end of the planning horizon was assigned on-peak rates, as recommended by Duke Power. Second, we initialized the reservoirs in the first period at 50% of their capacity. Third, the nodes in "early" periods (those which received no flow from upstream facilities due to time lags) had their exogenous inflows supplemented. Since the total inflow to the system had to remain constant across all minimum stream flow scenarios so that the numerical results would be comparable, the supplemental amount to "early" nodes corresponding to each hydro i in the dry (wet) case was chosen to be the maximum among all minimum stream flow regulations enforced at hydro i in the dry (wet) scenario.

16.5.2 Computational Results and Conclusions

We conducted a number of test runs with various input data. Our intent was not only to inform the SCWRC as to the impact of environmental regulations on hydroelectric revenues, but also to provide them with a mathematical tool which they can use to perform additional testing.

Table 16.10   Minimum stream flow restrictions in thousands of cubic feet per 2-hour period

% mean inflow for m_i values		Saluda	Piedmont	Upper Pelzer	Lower Pelzer	Hollidays Bridge	Ware Shoals
dry	0	0.0	0.0	0.0	0.0	0.0	0.0
	10	123.6	48.6	18.8	2.2	22.4	34.0
	20	247.3	121.9	52.4	11.2	46.7	72.8
	30	370.9	220.0	108.0	33.5	76.7	118.3
	40	494.6	342.8	193.1	78.8	120.4	175.2
	50	618.2	490.3	315.2	159.5	190.9	254.3
wet	0	0.0	0.0	0.0	0.0	0.0	0.0
	10	701.3	275.6	105.8	12.8	121.3	183.6
	20	1,402.6	691.6	294.9	63.4	252.7	393.5
	30	2,103.9	1,247.8	609.2	189.4	416.9	639.5
	40	2,805.2	1,944.3	1,090.8	445.2	658.2	949.2
	50	3,506.6	2,781.0	1,781.9	902.1	1,051.2	1,383.0

All the runs were conducted by submitting our formulations in Mathematical Programming System (MPS) format to the mixed-integer optimization package MINTO [5], version 1.6a, on a Sun SPARCserver 1000 computer. The linear programming optimizer used by our installation of MINTO was CPLEX [2], version 4.0.7.

To accomplish our objectives, we considered two operating scenarios: a dry period during October of 1970 and a wet period during April of 1970. As mentioned in the previous subsection, we enforced different minimum stream flow regulations as a function of the mean hydro inflows during that period (found in Table 16.10), ranging from 0% to 50%, in increments of 10%. We also considered the scenario where the hydros are permitted no storage: that is, they must operate as run-of-river facilities. For each of these seven scenarios we computed the maximum system revenue obtainable in both the wet period and the dry period. As these flow requirements are progressively more restrictive, we expect the system revenue to decrease. We also intuitively expect more revenue to be generated in the wet period, when water is plentiful, than in the dry period. Our revenues for each of the dry and wet periods, under each of the seven sets of minimum flow restrictions, are summarized in Table 16.11.

Examining this table, we notice that the system revenue does not substantially decrease as the minimum stream flow regulations are systematically tightened. The percent decrease in revenue from operating with no environmental regulations to 50% minimum stream flow regulations is less than 1.2% in the dry period and less than 0.2% in the wet period. In fact, the only notable change in revenue for either case

Table 16.11   Maximum system
revenue under varying minimum
stream flow regulations

% mean inflow	Revenue	
for m_i values	dry	wet
0	$14,131	$55,950
10	$14,122	$55,950
20	$14,113	$55,948
30	$14,060	$55,919
40	$14,021	$55,892
50	$13,975	$55,862
run-of-river	$12,652	$53,714

occurs when the run-of-river scenario is enforced. The percent change in revenue from the current unrestricted policy to a totally restricted policy is roughly 10% in the dry period and 4% in the wet period. This somewhat surprising result can be explained in part by the fact that the reservoir storage capacities are fairly small relative to the exogenous inflows, so the ability to store water up to the reservoir capacity does not profoundly affect the revenues.

Turning our attention to the relative merits of Problems MI1 and 2-P, we submitted each of these formulations in MPS file format to MINTO for both the dry and wet periods, under all seven sets of minimum stream flow regulations. The results of the algorithm are summarized in Table 16.12, with the number of branch-and-bound nodes encountered within the enumerative scheme (# of nodes) and CPU seconds listed for each run. An asterisk (*) indicates that the code did not run to completion within our self-imposed limit of 48 hours CPU time or 200 megabytes of memory. Following this, in Table 16.13, we provide the objective function values to the continuous relaxations of Problems MI1 and 2-P for each of these same scenarios, along with the CPU time required to obtain these values.

A few notable results emerge when comparing the results of Tables 16.11, 16.12, and 16.13. First, the two relaxations provide identical objective function values in all but one test case, and they give very close approximations to the true discrete optimal solution value. As should be expected from the relative size of the linearizations, the times cited in Table 16.13 to solve the relaxations of 2-P exceed those of MI1. Second, the branch-and-bound algorithm was considerably more efficient for the wet period than the dry, irrespective of the formulation used. This is to be expected, as the greater flows permit the turbines to be almost continuously engaged. In fact, upon comparing the minimum

Table 16.12 Branch-and-bound statistics: formulations MI1, 2-P

| % mean inflow | Problem MI1 | | Problem 2-P | |
for m_i values	# of nodes	CPU seconds	# of nodes	CPU seconds
dry 0	*	*	33,251	45,712
10	*	*	91	410
20	651	358	26	359
30	*	*	1,925	2,360
40	28,448	10,736	19,051	2,447
50	19	57	19	386
run-of-river	714	698	1	160
wet 0	62	63	22	286
10	27	55	15	343
20	27	58	14	364
30	1	50	1	335
40	1	51	1	335
50	1	50	1	343
run-of-river	1	45	1	153

Table 16.13 Comparison of relaxations: formulations MI1, 2-P

| % mean inflow | Problem MI1 | | Problem 2-P | |
for m_i values	obj. value	CPU seconds	obj. value	CPU seconds
dry 0	14,144	2	14,144	56
10	14,123	3	14,123	53
20	14,113	3	14,113	48
30	14,061	4	14,060	45
40	14,022	5	14,022	43
50	13,976	4	13,976	36
run-of-river	12,652	2	12,652	22
wet 0	55,950	3	55,950	53
10	55,950	4	55,950	56
20	55,948	4	55,948	53
30	55,919	4	55,919	52
40	55,892	4	55,892	54
50	55,862	5	55,862	57
run-of-river	53,714	1	53,714	14

stream flow regulations of Table 16.10 with the turbines' lower specifications of Table 16.3, we see that various y_{it} values can be fixed to
the value 1 (see the discussion in Section 16.3 and Exercise 2 in Section
16.6) since the minimum stream flow regulations will exceed the turbine
lower bounds. Third, even though the continuous relaxations provided
values extremely close to the true discrete objective function values, the
branch-and-bound schemes often enumerated a number of nodes to verify optimality. This observation is particularly striking in the dry, 40%
case, where MINTO enumerated 28,448 and 19,051 nodes to obtain and
verify optimality for the formulations MI1 and 2-P, respectively. Indeed,
for the dry, run-of-river case, Problem MI1 enumerated 714 nodes simply
to verify the optimality of the original linear programming solution.

As far as an overall solution strategy is concerned, Table 16.12 demonstrates that neither formulation uniformly dominates the other. The
larger representations afforded by Problem 2-P do not appear attractive
in the simpler, wet cases, since the additional time required to solve the
associated relaxations is not justified in the enumerative phase. However, their utility manifests itself in the dry cases when the additional
inequalities derived in Section 16.4, and found in Problem 2-P but not
in MI1, serve to expedite the enumerative process.

16.6   Exercises and Projects

1. A hydro system has three facilities operating in series, beginning upstream with facility 1, and progressing downstream to facility 2 and then
3. The time lag for water flow between hydros 1 and 2, and hydros 2 and
3, is 12 hours each. Using time periods of 12 hours, formulate Problems
MI1 and CMI1, and solve over a one-week planning horizon, beginning
on Monday at 0:00. Assume that no minimum stream flow restrictions
are imposed. Initialize each reservoir at half capacity, and value water
remaining in the reservoirs at termination at on-peak rates. Report your
results, both in terms of operational strategy and maximum realizable
revenues. Is your answer sensitive to the time-period duration? For
example, how would your operational strategy change if 2-hour periods
are used instead? Why might a 12-hour period be unrealistic? (What
happens to the electric production at hydro 1 when 12-hour periods are
used?) The following system characteristics apply.

 a. The on-peak, off-peak, and weekend rates for electricity are $25,
    $20, and $15 per megawatt-hour, respectively. Here (for simplic-
    ity), on-peak hours occur from 0:00 to 12:00 on Monday through
    Friday, off-peak hours comprise the remainder of the weekdays,
    and weekend hours run all day Saturday and Sunday.

    b. Each turbine's lower and upper specifications are 105 and 250 cubic feet per second respectively.

    c. The net head for each hydro is 20 feet.

    d. All turbine efficiencies are 0.75. (In computing the objective function coefficients, use the constant $k = \frac{1.175}{6} \times 10^{-8}$.)

    e. The exogeneous inflows into each hydro are 50 cubic feet per second, throughout the week.

    f. The storage capacity of each hydro is 400,000 cubic feet.

2. Prove the following statement (mentioned in Section 16.3) which can be used, in certain instances, to reduce the number of variables and constraints in Problem MI1. Given any instance of Problem MI1 and any i such that $l_i \leq m_i$, an optimal solution will have $y_{it} = 1\ \forall t$.

3. Suppose turbines can be engaged with any amount of flow. Modify Problem MI1 to accommodate this assumption. Is your new formulation more or less difficult to solve computationally than MI1 and 2-P?

4. How would the model change if instead of small-capacity hydros we were concerned with large-capacity hydros? Which modeling assumptions made in this chapter would be most affected? What types of procedures might be employed to solve this formulation?

5. In our problem scenario, each hydro operated in the best interest of the system as a whole without regard to its own profitability. What would be the effect on the total revenue if the individual operators each worked in their own best interest? Model such a noncollaborative system. Computationally verify your answer.

6. For each of the formulations MI1 and 2-P, how do the following parameter changes affect its continuous relaxation:

    a. an increase in the reservoir storage capacities,

    b. a dry year as opposed to a wet year?

7. There are many different types of hydroelectric plants; some have the capability to pump water upstream. During periods when the value of electricity is low, water is pumped from the tail waters below the hydro to the head waters above the hydro. The water is then available to produce electricity during high-demand periods. While there is a cost associated with pumping the water, the difference in electric rates for these facilities is great enough to make pumping attractive. Construct a network flow diagram of a system where the hydros have such pumping capabilities. Use your representation to devise a mathematical formulation.

8. (Project) For discrete optimization programs in general, there is a wide body of literature devoted to obtaining formulations whose continuous relaxations provide good approximations to the original discrete problem. Construct additional linear inequalities for each of the two formulations that can be used to strengthen the continuous relaxations.

9. The 2-path formulation of Section 16.4 has the advantage over Problem MI1 of yielding a tighter continuous relaxation. Devise a model which uses 3-paths. How does the strength of this model compare with that of Problem 2-P? What happens to the size of the problem in terms of the number of variables and constraints?

10. In this chapter, we discussed, for a given discrete optimization problem, the existence of more than one mixed 0-1 linear formulation of the problem, with different formulations having different continuous relaxations. We also defined the convex hull of a set S to be the smallest convex set containing S. For each of the following sets, give a system of linear inequalities that defines the convex hull of the set.

   a. $S = \{(x_1, \ldots, x_3) \text{ binary} : x_1 + x_2 \leq 1, x_2 + x_3 \leq 1, x_1 + x_3 \leq 1\}$

   b. $S = \{(x_1, \ldots, x_4) \text{ binary} : x_1 + x_2 \leq 1, x_2 + x_3 \leq 1, x_3 + x_4 \leq 1, x_1 + x_4 \leq 1\}$

   c. $S = \{(x_1, \ldots, x_5) \text{ binary} : x_1 + x_2 \leq 1, x_2 + x_3 \leq 1, x_3 + x_4 \leq 1, x_4 + x_5 \leq 1, x_1 + x_5 \leq 1\}$

11. Complete the first part of the proof of Theorem 16.1 by showing that equations (16.1) can be computed as linear combinations of (16.6) – (16.12).

12. Complete the second part of the proof of Theorem 16.1 by showing that, given any (r, s, d, y) feasible to CMI1, these variables, together with the 2-path variables defined in the proof, satisfy equations (16.6) through (16.12) of CMI2.

13. Construct a small numeric example to show that inequalities (16.18) can tighten the feasible region to Problem CMI2.

Acknowledgments

The authors are grateful to the Department of Energy (DOE) for funding the first author's work under a DOE graduate fellowship, to the Air Force Office of Scientific Research for partially supporting the second author's research under grant number AFOSR F49620-96-1-274, and to Dr. A. W. "Bud" Badr of the South Carolina Water Resources

Commission for both introducing us to this problem and patiently answering our questions over a two-year period. We would also like to thank Konstantin Staschus, formerly of Pacific Gas and Electric Co., George Galleher of Duke Power Co., and Ralph Walker, Jr. of Consolidated Hydro Southeastern, Inc., who all generously shared with us their time and expertise.

16.7   References

[1] M. S. Bazaraa, J. J. Jarvis, and H. D. Sherali, *Linear Programming and Network Flows*, 2nd ed., Wiley, New York, 1990.

[2] CPLEX, ILOG CPLEX Division, Incline Village, NV
<http://www.cplex.com/>

[3] J. S. Gulliver and R. E. Arndt (Eds.), *Hydropower Engineering Handbook*, McGraw-Hill, New York, 1991.

[4] R. Lougee-Heimer, *Combinatorial Approaches to Energy, Economic, and Allocation Problems*, Ph.D. dissertation, Clemson University, Clemson, SC (1993).

[5] MINTO, Georgia Tech Research Institute, Atlanta, GA
<http://akula.isye.gatech.edu/~mwps/projects/minto.html>

[6] G. L. Nemhauser and L. A. Wolsey, *Integer and Combinatorial Optimization*, Wiley, New York, 1988.

[7] United States Department of the Interior Geological Survey, *Water Resources Data for South Carolina*, 1970, 1971.

Chapter 17

Vertical Stabilization of a Rocket on a Movable Platform

William F. Moss

Prerequisites: differential equations, linear algebra, numerical analysis

17.1 Introduction

Suppose that a rocket, mounted on a movable platform, may be slightly misaligned from the vertical position. It is desired to correct this misalignment just before the rocket is launched by applying a control force to the platform. In this chapter, we construct a simple model that can be used to determine such a control force in the case where the misalignment occurs in the direction parallel to the wheels of the platform. For our initial model, following Ogata [6], we will model the rocket as an inverted pendulum with its mass concentrated at the tip. We will assume that the platform rolls without resistance along the ground and that the mass and length of the rocket and the mass of the platform are known with a relative error less than 0.0001. Certain constraints will be imposed on the control force and the motion of the rocket; these will constitute the *design criteria*.

The following launch sequence is assumed. The rocket is initially at rest and constrained by a mechanical arm to be within 1 degree of vertical. Ten seconds before lift-off, the mechanical arm is removed and a control force is applied to the platform to bring the rocket into a vertical position within 4 seconds. During the time remaining before lift-off, subsystem go/no-go signals are analyzed and a launch go/no-go decision is made.

Beginning with our idealized physical model, we derive a mathematical, state-space model consisting of a system of four nonlinear ordinary differential equations. The state variables are the angle of the rocket from vertical, the position of the platform, and their time derivatives. We linearize these equations for small angular deviations of the rocket from vertical, arriving at a linear system of ordinary differential equations. We use state feedback to control the system, in which case the control force is represented as a linear combination of state variables. At this point, standard linear control theory can be applied. The response of the linear system is governed by the eigenvalues of the system matrix, and the eigenvalues of the system matrix determine the control force.

We use the following design procedure in order to find a suitable control force.

- Choose a set of eigenvalues for the linear system.

- Compute the corresponding control force.

- Simulate the nonlinear system using the computed control and evaluate the response.

- Repeat until all design criteria are satisfied.

In Section 17.2 the mathematical model is derived. Section 17.3 presents the relevant state-space control theory. In Section 17.4 we outline the algorithm of Kautsky, Nichols, and Van Dooren [4] for computing the control force from given system eigenvalues. Section 17.5 presents a step-by-step design procedure as a set of exercises and projects that use MATLAB [5].

17.2   Mathematical Model

Figure 17.1 shows the configuration of the platform and rocket system, in which the rocket is modeled as an inverted pendulum of length l with mass m concentrated at the end. The movable platform has mass M.

The x-coordinate of point P, the angle θ, and the control (force) u are functions of time t. The masses M and m as well as the length l are constant; that is, they do not depend on time. Let s' and θ' denote the derivatives of s and θ with respect to time. The *state* of the system is the vector-valued function of time defined by $x = [\theta, \theta', s, s']^T$. Our problem is to design a control $u = u(t)$ that will satisfy the design criteria.

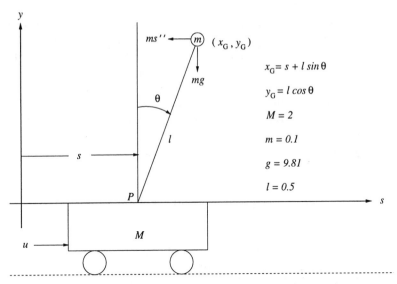

$$x_G = s + l\sin\theta$$

$$y_G = l\cos\theta$$

$$M = 2$$

$$m = 0.1$$

$$g = 9.81$$

$$l = 0.5$$

FIGURE 17.1   Rocket and platform configuration

In this section we outline the derivation of a nonlinear state-control model and its linearization for small motions about $\theta = 0$. Newton's laws of motion are used to derive a pair of nonlinear, second-order ordinary differential equations involving s, θ, and u. We put these nonlinear equations into standard form and then linearize them. We rewrite these linear and nonlinear, second-order systems as first-order systems of four equations in four unknowns. Using matrix notation, our linearized system has the form $x' = Ax + Bu$, where A is a real 4×4 matrix and B is a real 4×1 vector. In Section 17.5 this linear system is used to design a suitable control and the resulting design is tested by simulating the nonlinear system.

First, we analyze the motion of the system in the x-direction. From Newton's second law it follows that the total mass $M + m$ of the system times the acceleration of the center of mass equals the sum of the forces acting on the system:

$$(M + m)\left(\frac{Ms + mx_G}{M + m}\right)'' = u,$$

where $x_G = s + l\sin\theta$. After substituting for x_G, differentiating twice, and using the chain rule on the $\sin\theta$ term, we arrive at the nonlinear, second-order ordinary differential equation

$$(M + m)\,s'' - ml\,(\theta')^2\sin\theta + ml\,\theta''\cos\theta = u. \qquad (17.1)$$

Second, we analyze the motion of the inverted pendulum in a coordinate system fixed to the platform. The origin of this coordinate system is the point P in Figure 17.1. A gravitational force of magnitude mg in the negative y direction is applied to the pendulum tip mass. Because the coordinate system can accelerate, the pendulum tip mass is also acted on by an inertial force of magnitude ms'' in the negative x direction. From Newton's second law it follows that the rate of change of the angular momentum of the inverted pendulum is equal to the sum of the torques applied to it:

$$(\mathbf{r} \times \mathbf{p})' = \mathbf{r} \times \mathbf{f}, \tag{17.2}$$

where $\mathbf{r} = l\sin\theta\,\mathbf{i} + l\cos\theta\,\mathbf{j} + 0\mathbf{k}$, $\mathbf{p} = m\mathbf{r}'$, and $\mathbf{f} = -ms''\mathbf{i} - mg\,\mathbf{j} + 0\mathbf{k}$.

Computing the cross products, we find

$$\mathbf{r} \times \mathbf{p} = -ml^2\theta'\mathbf{k}, \quad \mathbf{r} \times \mathbf{f} = (-mgl\sin\theta + mls''\cos\theta)\,\mathbf{k}. \tag{17.3}$$

Substituting (17.3) into (17.2), rearranging terms, and equating $\mathbf{k}$ components gives

$$s''\cos\theta + l\,\theta'' = g\sin\theta. \tag{17.4}$$

Next, we put the nonlinear, second-order system (17.1), (17.4) into standard form. Solving (17.4) for s'' and substituting into (17.1) produces

$$\theta'' = \frac{(M+m)\,g\sin\theta - ml\,(\theta')^2\sin 2\theta/2 - u\cos\theta}{l\,(M + m\sin^2\theta)}. \tag{17.5}$$

Solving (17.4) for θ'' and substituting into (17.1) gives

$$s'' = \frac{ml\,(\theta')^2\sin\theta - mg\sin 2\theta/2 + u}{M + m\sin^2\theta}. \tag{17.6}$$

Next, we linearize the system (17.5), (17.6) for small motions about $\theta = 0$ to find a linear system that approximates the nonlinear one. We are motivated by the fact that analysis and control design are far easier for linear systems. Once a control has been found for the linear system, it must be tested on the nonlinear one. We will analyze the response of the nonlinear system to the control by simulating the nonlinear system using an ordinary differential equation solver.

To linearize (17.5) and (17.6) for small motions about $\theta = 0$, we can use $\cos\theta \approx 1$, $\sin\theta \approx \theta$, $(\theta')^2 \approx 0$, and $\sin^2\theta \approx 0$. Applying these approximations results in the linearized system

$$\theta'' = \frac{(M+m)\,g\theta - u}{Ml} \tag{17.7}$$

$$s'' = \frac{-mg\theta + u}{M}. \tag{17.8}$$

Finally, we convert our systems of second-order equations into systems of first-order equations. The state vector and its derivative are given by

$$x = \begin{bmatrix} x_1 \\ x_2 \\ x_3 \\ x_4 \end{bmatrix} = \begin{bmatrix} \theta \\ \theta' \\ s \\ s' \end{bmatrix}, \quad x' = \begin{bmatrix} x_1' \\ x_2' \\ x_3' \\ x_4' \end{bmatrix} = \begin{bmatrix} \theta' \\ \theta'' \\ s' \\ s'' \end{bmatrix}.$$

Substituting for θ'' from (17.5) and for s'' from (17.6), we find

$$x' = \begin{bmatrix} x_2 \\ \dfrac{(M+m)g \sin x_1 - mlx_2^2 \sin 2x_1/2 - u \cos x_1}{(M+m \sin^2 x_1)l} \\ x_4 \\ \dfrac{mlx_2^2 \sin x_1 - mg \sin 2x_1/2 + u}{M+m \sin^2 x_1} \end{bmatrix}. \tag{17.9}$$

Substituting for θ'' from (17.7) and for s'' from (17.8), we find

$$x' = \begin{bmatrix} x_2 \\ \dfrac{(M+m)gx_1 - u}{Ml} \\ x_4 \\ \dfrac{-mgx_1 + u}{M} \end{bmatrix}. \tag{17.10}$$

The matrix form of the linear system (17.10) is $x' = Ax + Bu$, where

$$A = \begin{bmatrix} 0 & 1 & 0 & 0 \\ \dfrac{(M+m)g}{Ml} & 0 & 0 & 0 \\ 0 & 0 & 0 & 1 \\ -\dfrac{mg}{M} & 0 & 0 & 0 \end{bmatrix}, \quad B = \begin{bmatrix} 0 \\ -\dfrac{1}{Ml} \\ 0 \\ \dfrac{1}{M} \end{bmatrix}. \tag{17.11}$$

17.3   State-Space Control Theory

In state-space control theory we encounter the linear, time-invariant system of ordinary differential equations

$$x' = Ax + Bu, \tag{17.12}$$

where x is an n-vector function of time, u is an m-vector function of time, A is a real $n \times n$ matrix, and B is a real $n \times m$ matrix, with $1 \le m \le n$ and $\operatorname{rank}(B) = m$. The variable x is called the *state*, while u is called the *control*. The matrix A is called the *open-loop matrix* and the matrix B is called the *control matrix*. If $m = 1$ the system is called *single input*; if $m > 1$ it is called *multiple input*.

One well-studied problem is to choose the control so that starting from a nonzero state, the system is driven rapidly toward $x = 0$. A popular method for solving this problem is to use *state feedback*

$$u = Fx, \tag{17.13}$$

where F is a real $m \times n$ matrix called the *gain matrix*. Substituting (17.13) into (17.12) produces the following *closed-loop system*:

$$x' = (A + BF)x. \tag{17.14}$$

The behavior of solutions to the closed-loop system are governed by the eigenvalues of the *closed-loop matrix* $A + BF$. Because the closed-loop matrix is real, the eigenvalues are real or complex conjugate pairs, in which case the eigenvalues are said to be closed under complex conjugation. The characteristics of the closed-loop system can be easily seen by examining the general solution to (17.14). Let us assume for the moment that the closed-loop matrix is *nondefective*: that is, there are linearly independent eigenvectors $c_1, \ldots, c_n$ corresponding to eigenvalues $\lambda_1, \ldots, \lambda_n$. Then the general solution to (17.14) is given by

$$x(t) = \alpha_1 c_1 e^{\lambda_1 t} + \cdots + \alpha_n c_n e^{\lambda_n t},$$

where $\alpha_1, \ldots, \alpha_n$ are scalars that can be determined from the state at the initial time. If all eigenvalues have negative real parts, then the state $x \to 0$ as $t \to \infty$ at a rate determined by the real parts. The more negative the real parts, the faster will be the approach to zero. The same conclusion holds even if the closed-loop matrix is defective but the general solution in this case is more complicated.

Solving the control problem using state feedback can be divided into two parts. First, choose eigenvalues, closed under complex conjugation, so that the closed-loop system has the desired response. Some experimentation is generally required here. Second, determine a gain matrix F so that the closed-loop matrix $A + BF$ will have these eigenvalues. The second problem is called the *eigenvalue assignment problem* by mathematicians and the *pole assignment problem* by engineers. Engineers use the term *pole* because the eigenvalues are the poles of the system transfer function. The eigenvalue assignment problem does not always have a solution as we shall now see.

The system $x' = Ax$ is called the *open-loop system*. Let μ be an eigenvalue of A. Then A has corresponding left and right eigenvectors. A *right eigenvector*, usually just referred to as an eigenvector, is a vector

$x \neq 0$ satisfying $(A - \mu I)x = 0$. A *left eigenvector* is a vector $y \neq 0$ satisfying $y^H(A - \mu I) = 0$; here y^H indicates the Hermitian transpose of the column vector y. Suppose μ_i, y_i, $i = 1, \ldots, n$ are the eigenvalues and left eigenvectors of A. In the pole assignment problem, we are trying to adjust the gain matrix so that μ_i moves to λ_i for $i = 1, \ldots, n$. Now suppose that $y_i^H B = 0$ holds for some i. Then $y_i^H(A - \mu_i I + BF) = 0$ for any F. This means that μ_i will be an eigenvalue of the closed-loop system for any choice of F. Consequently, it will not be possible to move the eigenvalue μ_i to a new location. The eigenvalue μ_i is called *uncontrollable*. In this case, it is necessary for μ_i to be one of the desired eigenvalues for the closed-loop system so that the eigenvalue assignment problem will have a solution.

The pair (A, B) is called *completely controllable* if $y^H B \neq 0$ for any left eigenvector y of A. The following theorems are well established [9].

THEOREM 17.1

The pole assignment problem has a solution for any set of eigenvalues, closed under complex conjugation, if and only if the pair (A, B) is completely controllable.

THEOREM 17.2

The pair (A, B) is completely controllable if and only if the controllability matrix

$$[B, AB, \ldots, A^{n-m}B]$$

has rank n.

Theorem 17.2 is easier to use than the definition when checking complete controllability.

Let us assume now that the pair (A, B) is completely controllable. In the single input case $(m = 1)$, the eigenvalue assignment problem has a unique solution F. In the multiple input case $(1 < m \leq n)$, there are infinitely many solutions, and it is standard practice to look for a solution F so that $A + BF$ is nondefective; otherwise, the closed-loop system may be overly sensitive to errors made in the measurement of system parameters such as M, m, and l in Figure 17.1. A necessary condition for $A + BF$ to be nondefective is that no eigenvalue is assigned more than m times; see [4]. A sufficient condition for $A + BF$ to be nondefective is that no eigenvalue is assigned more than m times and no assigned eigenvalue is an eigenvalue of A; see [1].

System parameters are generally not known exactly. Let $\widetilde{M}$, $\widetilde{m}$, and $\widetilde{l}$ denote the exact values of the parameters in Figure 17.1; also let $\widetilde{A}$ and $\widetilde{B}$ denote the corresponding open-loop and control matrices defined using equation (17.11). Then

$$\widetilde{M} = M + \Delta M, \ \widetilde{m} = m + \Delta m, \ \widetilde{l} = l + \Delta l, \ \widetilde{A} = A + \Delta A, \ \widetilde{B} = B + \Delta B.$$

Let us refer to the system corresponding to the pair (A, B) as the *design system* and the system corresponding to the pair $(\widetilde{A}, \widetilde{B})$ as the *physical system*. When the system is physically constructed, the control designed using the pair (A, B) will be applied to the system corresponding to the pair $(\widetilde{A}, \widetilde{B})$. The response of the physical system will be close to the response of the design system provided that the eigenvalues of

$$\widetilde{A} + \widetilde{B}F = A + \Delta A + (B + \Delta B) F = A + BF + \Delta A + (\Delta B)F$$

are close to the eigenvalues of $A+BF$. Suppose that $A+BF$ has distinct eigenvalues $\lambda_1, \ldots, \lambda_n$ and let x_i, y_i, $i = 1, \ldots, n$, denote corresponding unit right and left eigenvectors; that is,

$$(A - \lambda_i I) \, x_i = 0, \quad y_i^H \, (A - \lambda_i I) = 0,$$

where $\|x_i\|_2 = 1$, $\|y_i\|_2 = 1$. It is well known (see, for example [3]) that

$$\frac{\|\Delta A + (\Delta B)F\|_2}{|y_i^H x_i|} \tag{17.15}$$

is a good estimate for the perturbation in λ_i due to the perturbation $\Delta A + (\Delta B)F$ in $A+BF$. The expression $1/|y_i^H x_i|$ is called the *condition number* of λ_i.

17.4   The KNvD Algorithm

There are numerous algorithms for solving the eigenvalue assignment problem. In this section we outline the algorithm of Kautsky, Nichols, and Van Dooren [4]. This algorithm deals only with the nondefective case. It attempts to compute a set of n linearly independent eigenvectors for $A + BF$ before actually computing F. This is a bit of a trick. In the multiple input case there is some freedom in choosing the eigenvectors. They are generally chosen to minimize the sensitivity of the closed-loop system to errors. The KNvD algorithm assumes that no eigenvalue is assigned more than m times because of the necessary condition cited in Section 17.3. We will examine the single input case $(m = 1)$, so the eigenvalues to be assigned must be distinct.

ALGORITHM 17.7: KNvD algorithm, $m = 1$

1. *Perform the singular value decomposition on the control matrix B:*

$$B = U\Sigma V^T, \qquad \Sigma = \begin{bmatrix} S \\ 0 \end{bmatrix}, \qquad S = \begin{bmatrix} \sigma_1 & & 0 \\ & \ddots & \\ 0 & & \sigma_m \end{bmatrix},$$

   where U is an $n \times n$ orthogonal matrix, V is an $m \times m$ orthogonal matrix, and Σ is $n \times m$.
2. *Partition U as $U = [U_0, U_1]$, where U_0 is $n \times m$ and U_1 is $n \times (n - m)$. Then*

$$B = \begin{bmatrix} U_0 & U_1 \end{bmatrix} \begin{bmatrix} Z \\ 0 \end{bmatrix}, \qquad \text{where } Z = SV^T.$$

3. *A unit eigenvector c_i of $A + BF$, corresponding to the eigenvalue λ_i, is an element of the m-dimensional null space of $U_1^T(A - \lambda_i I_n)$. If $m = 1$ this basis consists of a single vector which we take to be c_i. If $m > 1$ this basis consists of more than one vector and a good choice for c_i is more complicated.*
4. *Let $\Lambda = \mathrm{diag}(\lambda_1, \ldots, \lambda_n)$ and $C = [c_1, \ldots, c_n]$. The gain matrix F can be found from*

$$F = Z^{-1} U_0^T (C\Lambda C^{-1} - A).$$

Algorithm 17.1 can be easily implemented using the MATLAB software package. Namely, the MATLAB `svd` function will generate the singular value factorization required in Step 1. An orthonormal basis for the null space described in Step 3 can be found using the MATLAB function `null`. The inversions required in Step 4 to find F can be done using the MATLAB slash notation; this is preferred to inverting C and then multiplying.

The validity of Algorithm 17.1 will now be established. First, note that since $\mathrm{rank}(B) = m$, we have $\sigma_1 \geq \cdots \geq \sigma_m > 0$ and thus S is nonsingular. Then Z is nonsingular because it is the product of nonsingular matrices. Also since the columns of U form an orthonormal set, it follows that

$$U_0^T U_0 = I_m , \qquad U_1^T U_0 = 0_{(n-m) \times m} . \tag{17.16}$$

In Step 3 of Algorithm 17.1, it is asserted that c_i is an element of the null space of $U_1^T(A - \lambda_i I_n)$. This result is easily derived. We know that

$$B = U_0 Z , \qquad (A + BF)c_i = \lambda_i c_i ,$$

giving $(A - \lambda_i I_n)c_i + U_0 Z F c_i = 0$. Multiplying both sides of this equality on the left by U_1^T produces $U_1^T(A - \lambda_i I_n)c_i + U_1^T U_0 Z F c_i = 0$ and thus $U_1^T(A - \lambda_i I_n)c_i = 0$, using (17.16).

Finally, we establish the validity of the expression for F in Step 4 of Algorithm 17.1. First, the n vector equations

$$(A + BF)c_i = \lambda_i c_i , \quad i = 1, \ldots, n$$

are equivalent to the matrix equation

$$(A + BF)C = C\Lambda .$$

We then have

$$
\begin{aligned}
(A + U_0 Z F)C &= C\Lambda \\
A + U_0 Z F &= C\Lambda C^{-1} \\
U_0 Z F &= C\Lambda C^{-1} - A \\
I_m Z F &= U_0^T(C\Lambda C^{-1} - A) \\
F &= Z^{-1} U_0^T(C\Lambda C^{-1} - A) .
\end{aligned}
$$

17.5   Exercises and Projects

In this section we present computational exercises to implement the KNvD algorithm and then to design a state feedback control that satisfies the following design criteria.

- Starting with a zero state except for a rocket angle 1 degree from vertical, the rocket must be brought into the vertical position with a settling time less than 4 seconds.

- Overshoot must be less than 0.85 degrees.

- The components of the gain vector F must be positive. The angle gain must be less than 75 and the other three gains must be less than 25.

- Assuming the masses of the rocket and platform and the length of the rocket are known with a relative error less than 0.0001, the maximum allowed uncertainty in the real parts of the closed-loop eigenvalues must be less than 10%.

The constraint on overshoot means that the oscillation of the rocket must be reasonably well damped. The constraint on the gain F limits the wattage of the feedback amplifier. The constraint on eigenvalue

uncertainty guarantees that errors in measured system parameters will
not lead to a large deviation from the desired system response.

In the following exercises, we will use MATLAB (version 5 or higher)
to build, one step at a time, a script file npend.m which implements the
KNvD algorithm, simulates the nonlinear system, and plots pendulum
angle and platform position versus time. We will then use npend.m to
design a suitable control. Maple [7] or Mathematica [8] could just as
easily be used here.

1. The script npend.m begins with several lines of comments followed
by lines specifying values for system parameters.

```
% Vertical stabilization of an inverted pendulum on a movable platform.
% The model is a set of four nonlinear ordinary differential equations.
% Full-state feedback is used to stabilize an approximate model obtained
% by linearization about the vertical position of the pendulum.
% Simulation of the nonlinear equations is used to evaluate performance.

delete npend.out
diary npend.out
global CL mass Mass l g F

% Define problem data.
n = 4;
m = 1;
mass = 0.1;
Mass = 2;
g = 9.81;
l = 0.5;
A = [0 1 0 0;g*(Mass+mass)/(Mass*l) 0 0 0; 0 0 0 1; -mass*g/Mass 0 0 0];
B = [0;-1/(Mass*l);0;1/Mass];
```

Choose the closed-loop eigenvalues to assign. Following Ogata [6], let
us start with a dominant complex conjugate pair of eigenvalues and two
real eigenvalues farther to the left in the complex plane.

```
% Choose the closed-loop eigenvalues to assign.
% Modified Ogata -- second-order dominant pair.
zeta = 0.5; % zeta = damping ratio
wn = 2.4;   % wn = undamped natural frequency
sigma = zeta*wn;
wd = wn*sqrt(1 - zeta^2); % wd = damped natural frequency
lambda = [-sigma+wd*i -sigma-wd*i -5*sigma -6*sigma]; % modified Ogata
```

We determine if the system is completely controllable using the following MATLAB commands.

```
% Determine if the system is completely controllable.
L = B;
for p = 1:n-m
  L = [B,A*L];
end
rnk = rank(L);
if rnk < n
  error('This system is not completely controllable.')
else
  disp('This system is completely controllable.')
end
disp(' ')
```

Now we implement the KNvD algorithm with the following MATLAB commands.

```
% Begin the KNvD algorithm
[U,Sigma,V] = svd(B);
S = Sigma(1:m,1:m);
U0 = U(:,1:m);
U1 = U(:,m+1:n);
Z = S(1:m,1:m)*V.';
for k = 1:n
  C(:,k) = null(U1.'*(A - lambda(k)*eye(n)));
end
Lambda = diag(lambda);
M = C*Lambda/C;
FF = Z\U0.'*(M - A);
F = real(FF);
disp('gain matrix, compare to [75 25 25 25]:')
disp(F)
if max(F >= [75 25 25 25])
   disp('Gain criterion not satisfied.')
else
   disp('Gain criterion satisfied.')
end
disp(' ')
CL = A + B*F; % the closed-loop matrix
```

2. We verify the computed gain matrix F as follows. We calculate the eigenvalues of the closed-loop matrix $A + BF$ and compare these with the assigned closed-loop eigenvalues.

```
disp('Comparison of eigenvalues')
disp('open-loop eigenvalues:')
disp(eig(A))
disp('assigned closed-loop eigenvalues:')
disp(lambda.')
disp('computed closed-loop eigenvalues:')
disp(eig(CL))
```

This computation provides a kind of quality control. However, it should be noted that some problems are sensitive to errors so that agreement may not be good in all cases.

3. If D and H are $m \times n$ matrices, then define

$$D = |H| \iff d_{ij} = |h_{ij}|, \ i = 1, \ldots, m, \ j = 1, \ldots, n$$
$$D \leq H \iff d_{ij} \leq h_{ij}, \ i = 1, \ldots, m, \ j = 1, \ldots, n.$$

From Taylor's theorem for functions of several variables it follows that

$$\Delta A \approx \frac{\partial A}{\partial M} \Delta M + \frac{\partial A}{\partial m} \Delta m + \frac{\partial A}{\partial l} \Delta l \tag{17.17}$$

$$\Delta B \approx \frac{\partial B}{\partial M} \Delta M + \frac{\partial B}{\partial l} \Delta l. \tag{17.18}$$

The following estimates for ΔA and ΔB can be derived using (17.17), (17.18), and the triangle inequality:

$$|\Delta A| \leq dA, \ \ |\Delta B| \leq dB,$$

where

$$dA = \begin{bmatrix} 0 & 0 & 0 & 0 \\ \frac{(M+m)g}{Ml} \left| \frac{\Delta l}{l} \right| + \frac{mg}{Ml} \left(\left| \frac{\Delta m}{m} \right| + \left| \frac{\Delta M}{M} \right| \right) & 0 & 0 & 0 \\ 0 & 0 & 0 & 0 \\ \frac{mg}{M} \left(\left| \frac{\Delta m}{m} \right| + \left| \frac{\Delta M}{M} \right| \right) & 0 & 0 & 0 \end{bmatrix} \tag{17.19}$$

$$dB = \begin{bmatrix} 0 \\ \frac{1}{Ml} \left(\left| \frac{\Delta M}{M} \right| + \left| \frac{\Delta l}{l} \right| \right) \\ 0 \\ \frac{1}{M} \left| \frac{\Delta M}{M} \right| \end{bmatrix} \tag{17.20}$$

Use the following MATLAB commands to define the matrices dA and dB in (17.19) and (17.20), where we are assuming that the relative errors in M, m, and l are at most 0.0001.

```
% perturbation analysis
remass = 0.0001;
reMass = 0.0001;
rel = 0.0001;
dA = zeros(n,n);
dB = zeros(n,m);
dA(2,1) = (g*(Mass+mass)/(Mass*l))*rel + (mass*g/(Mass*l))*(remass + reMass);
dA(4,1) = (mass*g/Mass)*(remass + reMass);
dB(2,1) = (1/(Mass*l))*(reMass + rel);
dB(4,1) = (1/Mass)*reMass;
```

From properties of matrix norms [3], it follows that

$$\|\Delta A + (\Delta B)F\|_2 \le \|\Delta A\|_2 + \|\Delta B\|_2\|F\|_2 \le \|\Delta A\|_F + \|\Delta B\|_F\|F\|_2$$
$$= \|\|\Delta A\|\|_F + \|\|\Delta B\|\|_F\|F\|_2 \le \|dA\|_F + \|dB\|_F\|F\|_2.$$
$$(17.21)$$

Here $\|W\|_F$ denotes the Frobenius norm [3] of the matrix W. Substituting (17.21) into (17.15) gives estimates for the eigenvalue perturbations which can be computed using the following MATLAB commands.

```
for k = 1:n
  y = null(CL' - conj(lambda(k))*eye(n));
  evcond(k) = 1./abs(y'*C(:,k));
end
normdCL = norm(dA,'fro') + norm(dB,'fro')*norm(F,2);
disp('Maximum eigenvalue displacements:')
delta_lambda = normdCL*evcond;
disp(delta_lambda)
disp('10% of the absolute value of the real part of eigenvalues:')
disp(0.1*abs(real(lambda)))
if max(delta_lambda >= 0.1*abs(real(lambda)))
    disp('Uncertainty (perturbation) criterion not satisfied.')
else
    disp('Uncertainty (perturbation) criterion satisfied.')
end
disp(' ')
```

4. We test the response of the nonlinear system to the control by using the MATLAB ordinary differential solver `ode45` to simulate the system from $t = 0$ to $t = 5$, and we plot the angle and position responses.

```
% Begin simulation of the nonlinear system.
tspan = [0 5];
theta0 = 1; % Initial theta.
thetap0 = 0; % Initial theta rate.
position0 = 0; % Initial cart position.
positionp0 = 0; % Initial cart position rate.
rpd = pi/180; % 1 degree in radians.
disp('Initial State [rad rad/sec m m/sec]')
x0 = [theta0*rpd thetap0*rpd position0 positionp0]; % Initial state.
disp(x0)
[t,x] = ode45('ndepole',tspan,x0);

% Analyze settling time and overshoot.
x(:,1) = (180/pi)*x(:,1); % theta in degrees
p = find(abs(x(:,1)) > 0.01*abs(x(1,1)));
Z = x(:,1);
Z(Z > 0) = [];
mxp = max(p);
```

```
if mxp < length(t)
  mxp = mxp + 1
end
disp(['Settling time = ', num2str(t(mxp))])
disp('settling time, should be < 4')
if t(mxp) >= 4
    disp('Settling time criterion not satisfied.')
else
    disp('Settling time criterion satisfied.')
end
disp(' ')
disp(['Overshoot = ',num2str(max(abs(Z)))])
disp('Overshoot, should be < 0.85')
if max(abs(Z)) >= 0.85
    disp('Overshoot criterion not satisfied.')
else
    disp('Overshoot criterion satisfied.')
end
disp(' ')

% Plot rocket angle and platform position.
subplot(1,2,1)
plot(t,x(:,1))
if theta0 == 0, theta0 = 1; end
axis([tspan -theta0 theta0])
axis([0 5 -1 1])
axis('square')
grid
title('Rocket Angle Response')
xlabel('Time (s)')
ylabel('Degrees')
subplot(1,2,2)
plot(t,x(:,3))
if position0 == 0, position0 = 0.04; end
axis([tspan -position0 position0])
axis('square')
grid
title('Platform Position Response')
xlabel('Time (s)')
ylabel('Position')
% print -deps2 plot1.eps % Create plot for inclusion in report.
```

Note that the *settling time* is the time it takes for the system to decay to 1% of its initial value, while the *overshoot* is the maximum amount the system overshoots its final value.

5. A separate MATLAB function file called ndepole.m must be written to define the right-hand side of the nonlinear first-order system (17.9). This file should contain the following MATLAB commands.

```
function xdot = ndepole(t,x)
global CL mass Mass l g F
xdot = zeros(4,1);
xdot(1) = x(2);
xdot(3) = x(4);
u = F*x;
s1 = sin(x(1));
s2 = sin(2*x(1));
m1 = Mass + mass*s1^2;
m2 = mass*l*x(2)^2;
xdot(2) = (-u*cos(x(1)) + (Mass+mass)*g*s1 - m2*s2/2)/(l*m1);
xdot(4) = (m2*s1 - mass*g*s2/2 + u)/m1;
```

6. The file npend.m is now complete. Next, we design a control that meets our design criteria. In practice the designer examines several different sets of closed-loop eigenvalues and determines the corresponding gain matrices. The response of each system is determined by simulation. The designer chooses the gain matrix that "best" meets the design criteria.

Let $\Delta\lambda$ denote the estimate of eigenvalue perturbation, let t_s denote the settling time, and let M_p denote the overshoot. We will refer to the initial choice of closed-loop eigenvalues as Design 1.

Now run npend.m. Table 17.1 summarizes the output. We see that the design criteria have not been met for our initial choice of closed-loop eigenvalues; in particular, the gain criterion and the perturbation criterion have not been satisfied. Make sure you see that the perturbation in each eigenvalue is greater than 10% of the absolute value of the real part of the eigenvalue.

Table 17.1   Design 1: $t_s = 3.9256$, $M_p = 0.49807$

	Design 1 Output			
λ	-1.20000 + 2.0785i	-1.20000 - 2.0785i	-6.0000	-7.2000
$\Delta\lambda$	0.1485	0.1485	2.3872	2.4651
F	113.9236	24.7596	25.3651	18.3193

7. It is well known [2, 6] that gains increase as the closed-loop eigenvalues are moved away from the open-loop eigenvalues. As can be seen from the output of npend.m, the open-loop eigenvalues are 0, 0, 4.5388, and -4.5388. Now modify one line of npend.m and call this Design 2.

```
lambda = [-sigma+wd*i -sigma-wd*i -4.5388 -6]; % Design 2.
```

Run `npend.m` again. The output is summarized in Table 17.2. Here the single stable open-loop pole is not being moved. The two real closed-loop poles are now closer to the origin and this has affected the system response. The components of the gain have decreased but the gain criterion is still not satisfied. We are also closer to satisfying the perturbation criterion. The settling time criterion is not satisfied.

Table 17.2 Design 2: $t_s = 4.0406$, $M_p = 0.52961$

	Design 2 Output			
λ	-1.20000 + 2.0785i	-1.20000 - 2.0785i	-4.5388	-6.0000
$\Delta\lambda$	0.1384	0.1384	0.9952	1.0716
F	86.8819	19.3640	15.9899	12.8504

8. Next, modify two lines of `npend.m` and call this Design 3.

```
wn = 2.8  % wn = undamped natural frequency
lambda = [-sigma+wd*i -sigma-wd*i -2.5388 -4.5388]; % Design 3.
```

Run `npend.m` again. The output is summarized in Table 17.3. Now we are really close. All the design criteria are satisfied except the perturbation criterion and it is nearly satisfied.

Table 17.3 Design 3: $t_s = 3.7766$, $M_p = 0.59835$

	Design 3 Output			
λ	-1.4000 + 2.4249i	-1.4000 + 2.4249i	-2.5388	-4.5388
$\Delta\lambda$	0.1512	0.1512	0.2900	0.3465
F	64.3859	14.3502	9.2091	8.9453

9. Finally, use the same damping ratio and undamped natural frequency as in Design 3 but move one real closed-loop eigenvalue closer to the origin. Call this Design 4.

```
wn = 2.8  % wn = undamped natural frequency
lambda = [-sigma+wd*i -sigma-wd*i -2 -4.5388]; % Design 4
```

Run `npend.m` again. The output is summarized in Table 17.4, and the response of the rocket angle θ and the platform position s are plotted in Figure 17.2. Now all the design criteria are satisfied.

Table 17.4   Design 4: $t_s = 3.8035$, $M_p = 0.59427$

	Design 4 Output			
λ	-1.4000 + 2.4249i	-1.4000 + 2.4249i	-2.0000	-4.5388
$\Delta\lambda$	0.1374	0.1374	0.1909	0.2236
F	59.4546	13.2471	7.2547	7.8167

We have taken a trial-and-error approach to finding a suitable design and have been successful using a dominant complex conjugate pair of eigenvalues and two real eigenvalues. Ogata [6] and Franklin, Powell, and Emami-Naeini [2] present various methods for eigenvalue assignment including the ITAE method. An ITAE solution is obtained using

```
wn = 3.3 % Franklin and Powell, Chapter 7.
lambda = wn*[-.424+1.263*i -.424-1.263*i -.626+.4141*i -.626-.4141*i];
```

The analysis of the effect of eigenvalue placement on system response is one of the main topics of linear control theory; see [2, 6].

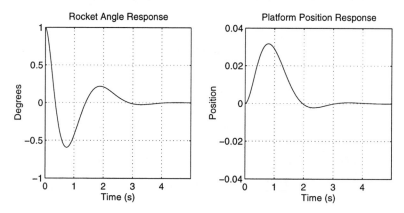

FIGURE 17.2   **Angle and position response for Design 4**

17.6   References

[1] J. W. Demmel, "On condition numbers and the distance to the nearest ill-posed problem," *Numerische Mathematik* **51**, 251–289 (1987).

[2] G. F. Franklin, J. D. Powell, and A. Emami-Naeini, *Feedback Control of Dynamic Systems*, 3rd ed., Addison-Wesley, Reading, MA, 1994.

[3] G. H. Golub and C. F. Van Loan, *Matrix Computations*, 2nd ed., Johns Hopkins University Press, Baltimore, 1989.

[4] J. Kautsky, N. K. Nichols, and P. Van Dooren, "Robust pole assignment in linear state feedback," *Int. J. Control* **41**, 1129–1155 (1985).

[5] The MathWorks, Inc., *Using MATLAB Version 5*, The Math-Works, Inc., Natick, MA, 1997. <http://www.mathworks.com/>

[6] K. Ogata, *Modern Control Engineering*, 2nd ed., Prentice-Hall, Englewood Cliffs, NJ, 1990.

[7] Waterloo Maple, Inc., *Maple V Learning Guide*, Springer-Verlag, New York, 1997. <http://www.maplesoft.com/>

[8] S. Wolfram, *The Mathematica Book*, 3rd ed., Cambridge University Press, Cambridge, 1996. <http://www.wri.com/>

[9] W. M. Wonham, *Linear Multivariable Control: A Geometric Approach*, Springer-Verlag, New York, 1985.

Chapter 18

Distinguished Solutions of a Forced Oscillator

T. G. Proctor

Prerequisites: differential equations

18.1   Introduction

Students in science and engineering study a differential equation that arises from modeling the vibration of a simple mechanical spring. This basic model, in expanded form, can be used to study the motion of bridges, buildings, and the surface of the earth when these structures are placed in duress by external forces such as wind, explosions, or earthquakes. The geometric shape (as a function of time) of the solution to systems that are driven by external forces depends on the shape of the forcing function. Knowledge of the possible solution shapes arising from periodic forcing should be part of a quantitative scientist's "vocabulary." The objective of this chapter is to show that, under several types of forcing functions, both linear and nonlinear mathematical models have solutions with sustained oscillations of a particular character.

For example, the solution of the linear differential equation

$$\frac{d^2y}{dt^2} + \omega^2 y = F_0 \cos \gamma t, \quad y(0) = \frac{dy}{dt}(0) = 0 \qquad (18.1)$$

is given by

$$y(t) = \frac{2F_0}{\omega^2 - \gamma^2} \sin\left(\frac{\omega + \gamma}{2} t\right) \sin\left(\frac{\omega - \gamma}{2} t\right) \qquad (18.2)$$

when $\omega \neq \gamma$. The solution $y(t)$ is graphed for the case $\omega = 2$ and $\gamma = 22/9$ in Figure 18.1, along with the enveloping curves.

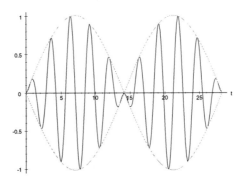

FIGURE 18.1   **Beats vibration with its envelope**

This special solution is called a *beats vibration* and is quite unusual. The homogeneous part of the basic linear mathematical model (18.1) has periodic solutions of period $2\pi/\omega$, whereas the forcing function has period $2\pi/\gamma$. The presence of this (apparently) periodic solution with a different period, although surprising, arises from the interaction of the forcing function with the solutions of the associated homogeneous problem. Moreover, the structure of this special solution is intriguing because (18.2) has the form of a relatively slowly varying amplitude $\sin \frac{1}{2}(\omega - \gamma)t$ multiplying a factor that varies at a faster rate. Note that the solutions are unchanged when ω and γ are interchanged and that when $|\omega - \gamma|$ is small, the amplitude is large.

A first step in analyzing the possible motion of a flexible structure such as a bridge is a simplified model in which the displacement of a particular position on the structure (for example, the center of the bridge) is modeled as a function of time. An equation for the vertical displacement $z(t)$ from equilibrium of such a position in a suspension bridge might assume the form

$$\frac{d^2z}{dt^2} + a\frac{dz}{dt} + bz + c(z) = w + f(t), \qquad (18.3)$$

in which $c(z) = gz$ if $z \geq 0$ and $c(z) = 0$ otherwise. Here, w is the effective gravitational force due to the weight of the structure, the term bz accounts for the force provided by the material of the bridge to restore the bridge to equilibrium, and $c(z)$ accounts for the pull of the cable which happens only when $z \geq 0$. The term $a\frac{dz}{dt}$ accounts for frictional forces in the bridge which tend to dampen the oscillations. The external forcing function $f(t)$ is provided by swirling winds and the resulting vortices which are peeled off of the structure. The resulting differential equation is a linear model when $z > 0$ and is a different linear model when $z < 0$. Solutions to these separate models can be pieced together at the time when $z = 0$.

After exploring the possible motions of the simplified system (18.3), more realistic models can be constructed; several extensions are discussed in Sections 18.4 and 18.5. The infamous collapse of the Tacoma Narrows Bridge in 1940 led to detailed analysis and explanation of the motion of the bridge under the influence of crosswinds present at that time. An alternative nonlinear analysis was subsequently carried out in [3]. The linear differential equation (18.1) was part of the explanation given in the 1940s; see [2]. This analysis was widely accepted since such models are introduced in elementary differential equations classes and have been successful in studying various applied problems. However, there is a basic nonlinear feature which occurs in important oscillation problems such as the motion of a suspension bridge. This chapter will show that other somewhat simpler forcing functions in the linear model also permit beats solutions and that piecewise linear equations can be used to construct solutions that resemble beats solutions for a nonlinear suspension bridge model under periodic forcing (thus simulating possible disturbances).

18.2   Linear Model with Modified External Forcing

The external forcing term $F_0 \cos \gamma t$ applied to the system (18.1) has period $2\pi/\gamma$ and average value zero. If we replace $F_0 \cos \gamma t$ with another periodic function $f(t)$ with similar properties, is there a solution which resembles the special solution displayed in Figure 18.1? As an example, consider the function

$$f(t) = F_0 \sum_{n=0}^{\infty} (-1)^n \delta\left(t - \frac{n\pi}{\gamma}\right)$$

consisting of a sum of Dirac delta functions. (It can be easily demonstrated that the solution $y(t)$ resulting from this forcing function is virtually indistinguishable from that arising from a forcing function consisting of a sequence of positive and negative step functions.) The use of Laplace transforms [1] yields the following solution to (18.1):

$$y(t) = \frac{F_0}{\omega} \sum_{n=0}^{\infty} (-1)^n \sin \omega \left(t - \frac{n\pi}{\gamma}\right) u\left(t - \frac{n\pi}{\gamma}\right),$$

where the function $u(t) = 0$ when $t < 0$, and $u(t) = 1$ when $t \geq 0$. A plot of $y(t)$ for $\omega = 2$ and $\gamma = 22/9$ is shown in Figure 18.2 for $0 \leq t \leq 9\pi \approx 28.274$. Note that the input (forcing) function changes sign every 1.285 units, so we must sum at least 22 terms to graph $y(t)$ over the interval $[0, 9\pi]$. Observe that this graph generally has the same shape as that arising from the forcing function $f(t) = \cos \gamma t$. However, there are some discrepancies occurring for positive t near $t = 0$ where the slope of the graph is 2 rather than 0 as in the cosine case; also flat spots are apparent near $t = 9\pi/2$ and near $t = 9\pi$. The solution for both the classical and modified systems has period 9π.

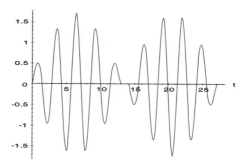

FIGURE 18.2   **Response of oscillator to periodic impulses**

The remarkable similarities between the solutions for cosine and impulse forcing is also demonstrated in Figure 18.3 for the case $\omega = 2.4$ and $\gamma = 2$. In this case, both the cosine and impulse forcing solutions have period 5π.

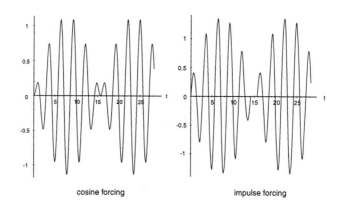

cosine forcing   impulse forcing

FIGURE 18.3   **Solutions for cosine and impulse forcing**

The reader should be warned that changing the parameters ω and γ by small amounts may result in a significant change in the time scales and amplitudes of the beats solution. This is illustrated for the case $\omega = 9/5$ and $\gamma = 2$ in Figure 18.4, which shows the solution for impulse forcing (the plot for cosine forcing is similar). This behavior can be contrasted with that exhibited by previous examples.

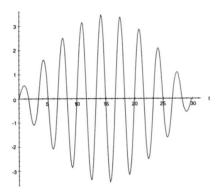

FIGURE 18.4   **Solution for impulse forcing: $\omega = 9/5$, $\gamma = 2$**

The examples discussed so far have had relatively small $|\gamma - \omega|$. When there is a significant difference in the values of γ and ω, the similarities may disappear. To see a case with a peculiar response, we examine $\omega = 7/2$ and $\gamma = 1/2$ and then reverse the value assignments. (Recall that in the case of cosine forcing, the response is not changed by interchanging ω and γ.) Figure 18.5 shows the response $y(t)$ under impulse forcing for the parameters $\omega = 7/2$ and $\gamma = 1/2$. (In this case, only four terms were needed in the response function.) It is seen that several beats occur with one amplitude, and then the system is pulsed to a new level, and so forth. Recall that an impulse changes the first derivative so the system has not returned to exactly the same state when the next impulse arrives to the system.

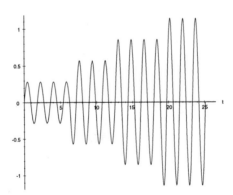

FIGURE 18.5   **Solution for impulse forcing: $\omega = 7/2$, $\gamma = 1/2$**

By contrast, if the values of ω and γ are reversed, then the response graph of Figure 18.6 is obtained for impulse forcing. For completeness the graph resulting from cosine forcing is shown in Figure 18.7.

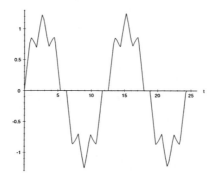

FIGURE 18.6   **Solution for impulse forcing: $\omega = 1/2$, $\gamma = 7/2$**

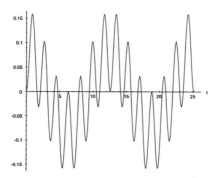

FIGURE 18.7   **Solution for cosine forcing: parameters $1/2$, $7/2$**

Now suppose we force the oscillator with a periodic piecewise linear function given by

$$f(t) = \left(1 - \frac{t}{a}\right)[u(t) - u(t - 2a)]$$
$$+ \left[-1 + \frac{1}{a}(t - 2a)\right][u(t - 2a) - u(t - 4a)]$$
$$+ \left[1 - \frac{1}{a}(t - 4a)\right][u(t - 4a) - u(t - 6a)]$$
$$+ \left[-1 + \frac{1}{a}(t - 6a)\right][u(t - 6a) - u(t - 8a)] + \cdots$$

Such a forcing function has a sawtooth-shaped graph, as shown in Figure 18.8. The solution to (18.1) can again be obtained by the technique of Laplace transforms. As before, the response exhibits behavior similar to that for the usual cosine forcing; see [5] for further details. In summary, the results of this section show that "beats solutions" are present when the linear oscillator is forced with a variety of periodic functions.

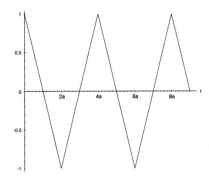

FIGURE 18.8   **Sawtooth forcing function**

18.3   Nonlinear Oscillator Periodically Forced by Impulses

Recall our simplified mathematical model (18.3) for the vertical displacement from equilibrium $z(t)$ of a particular position in a suspension bridge

$$\frac{d^2z}{dt^2} + a\frac{dz}{dt} + bz + c(z) = w + f(t),$$

in which $c(z) = gz$ for $z \geq 0$ and $c(z) = 0$ otherwise. We introduce the variable $x = z - \frac{w}{b+g}$, representing the distance from equilibrium when no external forcing is present, and study as a first approximation the simpler system

$$\frac{d^2x}{dt^2} + G(x) = p(x) + f(t), \tag{18.4}$$

where

$$G(x) = \begin{cases} k^2x & \text{if } x < -\nu \\ K^2x & \text{if } x \geq -\nu \end{cases} \quad \text{and } p(x) = \begin{cases} g\nu & \text{if } x < -\nu \\ 0 & \text{if } x \geq -\nu \end{cases}$$

with $0 < k^2 = b < K^2 = b + g$ and $\nu \geq 0$. Such a system might model a spring in which the restoring force is different when the spring is under extension than when under compression. In our nonlinear bridge model, $\nu = \frac{w}{b+g} > 0$. The case $\nu = 0$ will be studied first to steer our approach. We are presently ignoring the damping term proportional to dx/dt; at a later time this complication will be reintroduced.

First, we study the system (18.4) when $\nu = 0$ and $f(t) = 0$. Notice that the solution of

$$\frac{d^2x}{dt^2} + \sigma^2 x = 0$$

with specified $x(\tau)$ and $x'(\tau) = \frac{dx}{dt}(\tau)$ is given by

$$x(t) = x(\tau)\cos\sigma(t-\tau) + \frac{x'(\tau)}{\sigma}\sin\sigma(t-\tau). \qquad (18.5)$$

Suppose we start a solution of (18.4) at $x = 0$ and with $x' = s > 0$. The resulting problem may be solved on a sequence of time subintervals. The form of the equations of motion will change whenever the solution $x(t)$ changes sign, and we use (18.5) in the intervening intervals. The resulting solution is given by

$$x(t) = \begin{cases} \frac{s}{K}\sin Kt & \text{for } 0 \le t \le \frac{\pi}{K} \\ -\frac{s}{k}\sin k\left(t - \frac{\pi}{K}\right) & \text{for } \frac{\pi}{K} \le t \le \frac{\pi}{K} + \frac{\pi}{k} \end{cases}$$

The solution $x(t)$ is graphed for the parameters $k = 3/2$, $K = 3$, $s = 1$ in Figure 18.9. Clearly this is a nonlinear system since the negative of the solution shown above is not a solution.

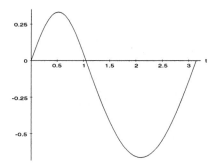

FIGURE 18.9   **Unforced response for (18.4)**

For the forced case, we "hit" the system with an alternating sequence of pulses. First consider the pattern of impulses given by

$$f(t) = F_0 \sum_{n=0}^{\infty} \left\{ \delta(t - nT) - (1 + \alpha)\delta(t - (n + \theta)T) \right\}.$$

Here the positive and negative "impulses" have respective strengths F_0 and $(1 + \alpha)F_0$, and they are applied to the system at the times $t = 0, \theta T, T, (1 + \theta)T, 2T, \ldots$ The relevant question is whether we can select θ and α so that a "beats solution" will occur.

This problem may be treated on a sequence of time subintervals. The form of the equations of motion will change whenever the solution $x(t)$ changes sign or when an impulse is delivered to the system.

Case 1: Suppose $\tau = nT$ where n is a positive integer. We get a pulse of strength F_0 at $t = \tau$. If $x(\tau) > 0$, then the change of variable $u(t) = x(t + \tau)$ gives $u'' + K^2 u = F_0 \delta(t)$, $u(0) = x(\tau)$, $u'(0) = x'(\tau)$. Solving this problem and transforming back to $x(t)$ produces

$$x(t) = x(\tau) \cos K(t - \tau) + \frac{1}{K}(x'(\tau) + F_0) \sin K(t - \tau)$$

for $t > \tau$. A similar analysis for $x(\tau) < 0$ gives

$$x(t) = x(\tau) \cos k(t - \tau) + \frac{1}{k}(x'(\tau) + F_0) \sin k(t - \tau)$$

for $t > \tau$. Likewise, if $\tau = (n + \theta)T$, we get the two cases listed above with $-(1 + \alpha)F_0$ replacing F_0.

Case 2: Suppose $t \neq nT$ and $t \neq (n + \theta)T$ and $x(\tau) = 0$. If $x'(\tau) > 0$, the solution is

$$x(t) = x(\tau) \cos K(t - \tau) + \frac{1}{K}x'(\tau) \sin K(t - \tau)$$

for $t > \tau$, and if $x'(\tau) < 0$ the solution is

$$x(t) = x(\tau) \cos k(t - \tau) + \frac{1}{k}x'(\tau) \sin k(t - \tau)$$

for $t > \tau$.

Example 18.1

Suppose $k = K = 2$ (so we have symmetric forcing) and the input period is $T = \pi/2$ ($\gamma = 4$). The solution given above for $F_0 = 1$, $\theta = 0.5$, $\alpha = 0$ is graphed in Figure 18.10 and shows an output period π.

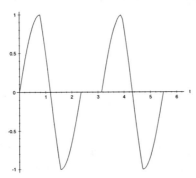

FIGURE 18.10 Response with symmetric forcing: $k = K = 2$

If we take the same γ (i.e., input period) as in Example 18.1 and $0 < k < K$, we get a response that somewhat resembles the one shown in Figure 18.10. The first negative pulse occurs after the first peak and before the first positive zero. Call such a point t_1 and the accompanying displacements and velocities u_1 and v_1 (taken at the left limit). Call the next point of interest, that is the first positive zero, t_2 and the accompanying velocity v_2. The next point of interest occurs when the second positive impulse is applied. Call this time t_3, with accompanying displacement and velocity u_3 and v_3 (left limit). The final time of interest occurs when the second negative impulse is applied. Call this point t_4, with accompanying displacement and velocity u_4 and v_4 (left limit). At the final time we desire $u_4 = 0$ and $v_4 - (1 + \alpha) = 0$. The output $x(t)$ for a typical θ, α trial is shown in Figure 18.11.

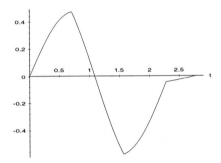

FIGURE 18.11   **Near beats response**

Suppose the function $P(\theta, \alpha)$ is defined to be the vector having components u_4 and $v_4 - (1 + \alpha)$. We want to choose θ and α so that $P(\theta, \alpha) = 0$. The author evaluated the resulting P function over a θ, α grid and was not successful in locating a good starting value for using Newton's method [4] for zeros. Such a starting value would be corrected by choosing increments $\Delta\theta$ and $\Delta\alpha$ given by the components of the vector $-J(\theta, \alpha)^{-1}P(\theta, \alpha)$, where J is the Jacobian matrix

$$J(\theta, \alpha) = \begin{bmatrix} \frac{\partial u_4}{\partial \theta} & \frac{\partial u_4}{\partial \alpha} \\ \frac{\partial v_4}{\partial \theta} & \frac{\partial v_4}{\partial \alpha} - 1 \end{bmatrix}.$$

Moreover the matrix J is poorly conditioned. Hence this attempt to get a beats output from the input period $2T$ is unsuccessful for this pattern of input impulses.

However, a second and more successful strategy for obtaining a generalized beats phenomenon is as follows. Suppose that $0 < k < K$ is given with $\frac{1}{K} + \frac{1}{k} = 1$ and $T = \frac{\pi}{2}$. We pulse a system at rest at $t = 0$

and apply a pulse of strength -1 at $t_1 = \frac{T}{K}$ so that $x_1(t) = \frac{1}{K}\sin Kt$ for $0 \le t \le \frac{T}{K}$. At this time, $x = \frac{1}{K}$, $x' = 0$. In the next interval $x_2(t) = -\frac{1}{K}\sin K(t - t_1) + \frac{1}{K}\cos K(t - t_1)$. At $t_2 = t_1 + \frac{T}{2K} = \frac{3T}{2K}$ we have $x = 0$, $x' = -\sqrt{2}$. In the next interval we have $x_3(t) = -\frac{\sqrt{2}}{k}\sin k\left(t - \frac{3T}{2K}\right)$. At $t_3 = \frac{3T}{2K} + \frac{T}{2k}$, we have $x = -\frac{1}{k}$ and $x' = -1$. At this time we apply a impulse of strength 1 to get the solution $x_4(t) = -\frac{1}{k}\cos k(t - t_3)$ for $t_3 \le t \le t_3 + \frac{T}{k} = t_4$, at which time we have $x = 0$, $x' = 1$. An impulse of strength -1 restores the system to rest. At $t = 2T$ the cycle begins anew.

Example 18.2

Consider a system with $T = \frac{\pi}{2}$, $k = \frac{3}{2}$, $K = 3$ and which is pulsed as described above. The resulting output $x(t)$ is shown in Figure 18.12.

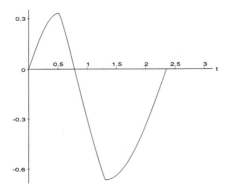

FIGURE 18.12   **Beats response for $k = \frac{3}{2}, K = 3$**

This input of period $2T$ generates a beats vibration response having period $2T$ and with the property that as $k \to K$, the resulting input period reduces to T (the output still has period $2T$). Thus, as the symmetry of the unforced oscillator is broken by allowing $k \ne K$, the period of the required input doubles. Figure 18.13 schematically depicts the instants at which ± 1 impulses are delivered for the two cases.

FIGURE 18.13   **Input impulse timing**

18.4   A Suspension Bridge Model

We turn now to the bridge model (18.4) introduced in Section 18.3. Suppose w and ν are positive numbers with $\frac{w}{\nu} > g > 0$ and that

$$\frac{d^2x}{dt^2} + \left(\frac{w}{\nu} - g\right)x = g\nu + f(t) \text{ for } x < -\nu, \qquad (18.6)$$

$$\frac{d^2x}{dt^2} + \frac{w}{\nu}x = f(t) \text{ for } x \geq -\nu. \qquad (18.7)$$

Here the forcing function consists of a series of positive and negative impulses. Clearly, if the strength of the forcing function is so weak that the amplitude of the response never achieves $x = -\nu$, we have a response as given in Section 18.2. However, if strong forcing is present, it will be necessary to consider both regions $x < -\nu$ and $x \geq -\nu$. We will look for such a forcing function that will result in a "beats-like" response. The sequence of impulses in our forcing function will be intimately tied to the times at which the response achieves the level $x = -\nu$.

Suppose we first consider an interval containing $t = \tau$ in which there is no external forcing. Then the solution of (18.6) in terms of $x(\tau)$, $x'(\tau)$ is given by

$$x(t) = \left(x(\tau) - \frac{g\nu^2}{w - \nu g}\right)\cos\left((t - \tau)\sqrt{\frac{w}{\nu} - g}\right)$$
$$+ \frac{x'(\tau)}{\sqrt{\frac{w}{\nu} - g}}\sin\left((t - \tau)\sqrt{\frac{w}{\nu} - g}\right) + \frac{g\nu^2}{w - \nu g}. \qquad (18.8)$$

Likewise the solution of (18.7) in terms of $x(\tau)$, $x'(\tau)$ is given by

$$x(t) = x(\tau)\cos\left((t - \tau)\sqrt{\frac{w}{\nu}}\right) + \frac{x'(\tau)}{\sqrt{\frac{w}{\nu}}}\sin\left((t - \tau)\sqrt{\frac{w}{\nu}}\right). \qquad (18.9)$$

Consider the *phase plane* (x, y), defined relative to $y = \frac{dx}{dt} = x'$. Let $\alpha = \frac{g\nu^2}{w - \nu g}$, $K = \sqrt{\frac{w}{\nu}}$, $k = \sqrt{\frac{w}{\nu} - g}$. Then for $x < -\nu$,

$$x - \alpha = \left(x(\tau) - \alpha\right)\cos k(t - \tau) + \frac{x'(\tau)}{k}\sin k(t - \tau)$$

$$\frac{y}{k} = \frac{x'(\tau)}{k}\cos k(t - \tau) - \left(x(\tau) - \alpha\right)\sin k(t - \tau)$$

giving the ellipse

$$(x - \alpha)^2 + \left(\frac{y}{k}\right)^2 = \left(x(\tau) - \alpha\right)^2 + \left(\frac{y(\tau)}{k}\right)^2 = \beta^2, \qquad (18.10)$$

for some positive constant β. Similarly, for $x \geq -\nu$, we obtain the ellipse

$$x^2 + \left(\frac{y}{K}\right)^2 = x(\tau_1)^2 + \left(\frac{y(\tau_1)}{K}\right)^2 = \gamma^2, \qquad (18.11)$$

for some positive constant γ.

In an interval containing t, τ, τ_1 and during which there is no external forcing, the solution $(x(t), y(t)) = (x(t), x'(t))$ travels on the ellipses (18.10) and (18.11). If the interface $x = -\nu$ is crossed, ellipse (18.10) must match up with ellipse (18.11), which gives a relationship between β and γ. That is,

$$k^2\left(\beta^2 - (\nu + \alpha)^2\right) = K^2(\gamma^2 - \nu^2)$$

from which it follows that

$$\beta = \sqrt{(\nu + \alpha)^2 + \frac{K^2}{k^2}(\gamma^2 - \nu^2)}. \qquad (18.12)$$

Example 18.3

Suppose that $K = 3, k = 1.5, \gamma = 5, \nu = 1, \alpha = 2$. Then application of (18.12) gives $\beta = \sqrt{105} \approx 10.247$ and the first ellipse (18.10) is given by

$$(x - 2)^2 + \frac{y^2}{2.25} = 105.$$

This ellipse can be parametrized as

$$x = 2 + 10.247 \sin z, \quad y = 15.37 \cos z$$

for $0 \leq z \leq 2\pi$. The second ellipse (18.11) is given by

$$x^2 + \frac{y^2}{9} = 25$$

so that

$$x = 5 \sin z, \quad y = 15 \cos z$$

for $0 \leq z \leq 2\pi$. The two ellipses are displayed in Figure 18.14. Notice that they cross at the abscissa value $x = -\nu = -1$.

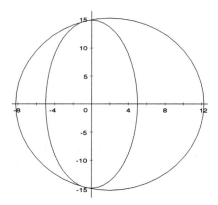

FIGURE 18.14   **Phase plane ellipses for Example 18.3**

In order to design a forcing function for the nonlinear system (18.6)–(18.7), we first reconsider the beats solution of

$$\frac{d^2x}{dt^2} + \omega^2 x = F_0 \sum_{n=0}^{\infty} (-1)^n \delta(t - \frac{n\pi}{2}).$$

In between pulses, the solution of the differential equation satisfies

$$(\frac{dx}{dt})^2 + \omega^2 x^2 = Z^2,$$

where Z is a constant. Thus, in the phase plane (x, x') traverses along an ellipse parametrized by Z between pulses. At pulse time t_n, the new value of Z is given by

$$Z = \sqrt{[x'(t_n^-) + (-1)^n F_0]^2 + [\omega x(t_n)]^2}.$$

For example, consider the case when $\omega = F_0 = 1.8$. Table 18.1 shows the corresponding pulse times and the resulting ellipse parameter that applies until the next pulse. In the phase plane we have a trajectory which is composed of partial ellipses. Figure 18.15 shows the first half of the trajectory described above. The second half is symmetric but taken in reverse order. The solution trajectory travels along the ellipse (shown as a solid curve) until an impulse is delivered to the system and the trajectory jumps to a new ellipse. The time along each ellipse for this linear system is $\frac{n\pi}{2}$. The x versus t graph for this solution is shown in Figure 18.4.

Table 18.1 Position and derivative data at pulse times

n	$t_n = \frac{n\pi}{2}$	$x(t_n)$	$x'(t_n^-)$	Pulse	Z until next pulse
0	0	0	0	1	1.8
1	$\frac{\pi}{2}$	0.309	-1.712	-1	3.556
2	π	-0.897	3.168	1	5.224
3	$\frac{3\pi}{2}$	1.706	-4.226	-1	6.763
4	2π	-2.657	4.872	1	8.136
5	$\frac{5\pi}{2}$	3.657	-4.782	-1	9.309
6	3π	-4.608	4.226	1	10.252
7	$\frac{7\pi}{2}$	5.417	-3.168	-1	10.943
8	4π	-6.005	1.712	1	11.365
9	$\frac{9\pi}{2}$	6.313	0	-1	11.506
10	5π	6.314	-1.8	1	11.365
11	$\frac{11\pi}{2}$	6.005	3.512	-1	10.943
12	6π	-5.417	-4.968	1	10.252
13	$\frac{13\pi}{2}$	4.608	6.026	-1	9.309
14	7π	-3.567	-6.582	1	8.136
15	$\frac{15\pi}{2}$	2.657	6.763	-1	6.763
16	8π	-1.706	-6.026	1	5.224
17	$\frac{17\pi}{2}$	0.897	4.968	-1	3.556
18	9π	-0.309	-3.512	1	1.8
19	$\frac{19\pi}{2}$	0	1.8	-1	0

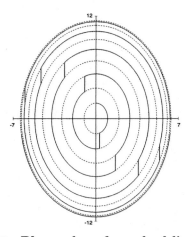

FIGURE 18.15 **Phase plane for pulsed linear system**

If we construct appropriate phase plane trajectories for the unforced nonlinear system, each consists of the two ellipses (18.10) and (18.11) joined together. Then we apply a sequence of pulses which drives the trajectory from one such path to another along vertical segments at the time of the pulses. The total trajectory has the appearance of Figure 18.4, and we have a forcing function that will produce a beats solution. Note that the time for a solution of the unforced system to travel any closed trajectory differs from curve to curve if such trajectories cross the line $x = -\nu$. Consequently, the time between pulses for the pulsed system will vary and the forcing function will have only the period given by the time the system takes to traverse the envelope of the beat.

If damping terms are included in the model, such as the $a\frac{dz}{dt}$ term in the model described in Section 18.3, then the beats solution we constructed will gradually disintegrate into a periodic solution with the same period as the forcing period. Since the system is nonlinear, the form of the periodic response may depend on the size of the pulses imposed on the system. Further details are provided in [1].

18.5   Model Extension to Two Spatial Dimensions

Nonlinear oscillations of the type considered previously have been discussed in a series of papers by McKenna and his collaborators; see [3] and [5] for an expanded list of references. An elementary explanation of the model is given in [1]. An extension of the model we have considered concerns motion that differs at different positions x along the bridge and could assume the form

$$\frac{\partial^2 u}{\partial t^2} + EI\frac{\partial^4 u}{\partial x^2} + a\frac{\partial u}{\partial t} + c(u) = W(x) + f(x,t), \qquad (18.13)$$

where $u(x,t)$ is the displacement at position x along the bridge at time t and EI is a constant representing the effects of bridge materials and structure. Appropriate boundary conditions are posed in [5]. Further extensions of the model would concern functions $u(x,y,t)$, with u being the vertical displacement at position (x,y) on the bridge at time t. Simple models of this type would take x to be a position measured horizontally along the bridge. Even more realistic models would take x to be a measurement along the arc length of the bridge and allow for the total length to change semi-elastically. Photographs taken of the collapsing Tacoma Narrows Bridge show violent twisting and longitudinal motion, so the displacement u in that failure did exhibit a dependence on x and y. For further details concerning such models, the reader should consult [5]. One approach in the investigation of solutions of (18.13) is to search for solutions that have the form $u(x,t) = X(x)T(t)$.

18.6   Exercises and Projects

1. Verify that the expression for $y(t)$ in (18.2) satisfies the linear differential equation and initial conditions given in (18.1).

2. Use the substitution $x = z - \frac{w}{b+g}$ in (18.3) to obtain the system (18.4), where $G(x)$ and $p(x)$ are as specified in Section 18.3.

3. Derive the expression (18.10) from equation (18.8). Similarly, derive the expression (18.11) from equation (18.9).

As a prelude to the next exercise, consider the solution of

$$\frac{d^2x}{dt^2} + 9x = \sum_{n=0}^{\infty} (-1)^n \delta\left(t - \frac{n\pi}{4}\right), \quad x(0) = \frac{dx}{dt}(0) = 0.$$

This equation results from (18.4) when $K = k = 3$, $T = \pi/2$, $\nu = 0$. The solution is given by

$$x(t) = \frac{1}{3} \sum_{n=0}^{\infty} (-1)^n \sin 3\left(t - \frac{n\pi}{4}\right) u\left(t - \frac{n\pi}{4}\right).$$

The corresponding response graph (with vertical scale multiplied by 3) is shown in Figure 18.16. A summary of the values of $x(t)$ and $x'(t)$ at the times when the pulses are applied is given in Table 18.2.

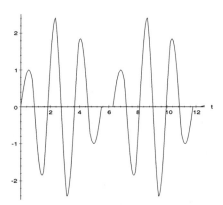

FIGURE 18.16   **Beats-like response for $k = K = 3$**

In this case, we note the following:

- In all, there are 8 pulses before the cycle begins again.

- The negative pulse at $\frac{3\pi}{4}$ occurs just when the curve has peaked.

Table 18.2   System data when pulses are applied

	Pulse	t	$3x(t)$	$x'(t^-)$
t_0	positive	0	0	0
t_1	negative	$\frac{\pi}{4}$	$\sqrt{2}/2$	$-\sqrt{2}/2$
t_2	positive	$\frac{2\pi}{4}$	$-(1+\sqrt{2}/2)$	$\sqrt{2}/2$
t_3	negative	$\frac{3\pi}{4}$	$1+\sqrt{2}$	0
t_4	positive	$\frac{4\pi}{4}$	$-(1+\sqrt{2})$	-1
t_5	negative	$\frac{5\pi}{4}$	$1+\sqrt{2}/2$	$1+\sqrt{2}/2$
t_6	positive	$\frac{6\pi}{4}$	$-\sqrt{2}/2$	$-(1+\sqrt{2}/2)$
t_7	negative	$\frac{7\pi}{4}$	0	1

- The positive pulse at $\frac{4\pi}{4}$ occurs when the value of x is exactly the negative of the value at the previous pulse, and the effect of the pulse is to give a slope of 0 (exactly the value of the slope when the previous pulse was delivered).

- The effect of a pulse of strength ± 1 at time τ for a spring with restoring constant σ is given by

$$x(t) = x(\tau)\cos\sigma(t-\tau) + \frac{1}{\sigma}\left(x'(\tau^-) \pm 1\right)\sin\sigma(t-\tau), \quad t > \tau,$$

so the slope gains or loses value 1 from this pulse.

4. Devise an input pattern for k and K such that $k < 3 < K$ and $T = \pi/2$ ($\gamma = 4$). This will generalize the beats vibration with $\omega = 3$, $\gamma = 4$. Use the constant $K = 3.2$ when $x > 0$ and the constant $k = 2.8$ when $x < 0$, and treat the nonsymmetric problem on subintervals of time. Alter the timing of the positive and negative pulses so that they are applied exactly when the values of x' shown above are achieved. For example, in the first segment $x = \frac{1}{K}\sin Kt$, so $t_1 = \frac{3\pi}{4K} \approx 0.7363$. The next time of interest is when $x = 0$ (because the spring constant becomes k) and this occurs at $t \approx 0.8590$. When $x' = -\sqrt{2}/2$ we apply a positive pulse: namely, at $t \approx 1.560277$. Continue to compute times when impulses are applied and show that

$$t_1 \approx 0.7363, \quad t_2 \approx 1.560277, \quad t_3 \approx 2.33165, \quad t_4 \approx 3.120555,$$

$$t_5 \approx 3.92699, \quad t_6 \approx 4.6808, \quad t_7 \approx 5.5223308,$$

at which time we apply a negative pulse driving the system to $x = 0$, $x' = 0$ for $t_7 < t < 2\pi$. This sequence of pulses is then repeated anew. Consequently, a beats-like solution results from a slight variation of the periodically applied pulses over the interval $[0, 2\pi]$. Moreover the response does have an exact period 2π. Notice that the size of F_0 is not crucial to the construction given here.

5. (Project) Let $g = \nu = 1$, $w = 3.5$, and $F_0 = 1.8$. Construct a table of values of the polar angles for the points $(x(t_n), x'(t_n^-))$ given in Table 18.1. Then starting at $x(0) = x'(0) = 0$, apply a pulse of strength F_0. Follow the curve in the phase plane until the angle from the origin is equal to arctan $x(t_1)/x'(t_1^-)$ and apply a pulse of strength $-F_0$, then follow the new energy curve until the angle from the origin is arctan $x(t_2)/x'(t_2^-)$, etc. Continue this process, recording the times of the pulses, thereby constructing a solution x versus t graph. Notice that the time interval between pulses will vary; however, the resulting solution graph of x versus t will again resemble Figure 18.4.

18.7   References

[1] P. Blanchard, R. L. Devaney, and G. R. Hall, *Differential Equations*, PWS Publishing, Boston, 1995.

[2] M. Braun, *Differential Equations and Their Applications: An Introduction to Applied Mathematics*, Springer-Verlag, New York, 1983.

[3] Y. S. Choi, H. O. Her, and P. J. McKenna, "Galloping: nonlinear oscillation in a periodically forced loaded hanging cable," *Journal of Computational and Applied Mathematics* **52**, 23–34 (1994).

[4] P. Deuflhard and A. Hohmann, *Numerical Analysis: A First Course in Scientific Computation*, W. de Gruyter, Berlin, 1995.

[5] A. C. Lazer and P. J. McKenna, "Nonlinear flexings in a periodically forced floating beam," *Mathematical Methods in the Applied Sciences* **14**, 1–33 (1991).

[6] T. G. Proctor, "Distinguished oscillations of a forced harmonic oscillator," *The College Mathematics Journal* **26**, 111–117 (1995).

Chapter 19

Mathematical Modeling and Computer Simulation of a Polymerization Process

Hong Zhang
Yuan Tian

Prerequisites: differential equations, scientific computing

19.1 Introduction

Polymers are widely used chemical materials. Many commercially important polymers (such as polyvinyl chloride, polymethyl methacrylate, and polystyrene) are produced by *radical chain polymerization methods*. Recent developments in using electron spin resonance methods to study radicals in a free radical polymerization system have provided new insights into the mechanism of diffusion-controlled kinetical processes [7]. However, the analytical experimentations involved are expensive, time consuming, and offer only snapshots — so it is difficult to deduce a dynamic and microscopic picture of the production process. On the other hand, the current development of mathematical modeling techniques, advanced computers, and in addition highly accurate and efficient numerical software now offer the potential of making computer simulations feasible for many engineering applications. In order to better understand the mechanisms of the molecular process of radical growth, and to facilitate further research and experimentation, it is desirable to formulate and numerically solve mathematical models for such chemical reactions.

Radical polymerization is a chain reaction consisting of three sequential steps — *initiation*, *propagation*, and *termination*. The initiation

step is considered to involve two reactions. The first is the generation
of primary radicals R_1 by the homolytic dissociation of an initiator I:

$$I \xrightarrow{k_d} 2R_1,$$

where k_d is the rate constant for the initiator dissociation. The second
part of the initiation involves the addition of this primary radical to a
monomer molecule to produce the chain initiating species R_2:

$$R_1 + M \xrightarrow{k_p} R_2,$$

where M represents (the concentration of) a monomer molecule and k_p
is the rate constant for propagation. Propagation results in the growth
of R_2 by the successive addition of large numbers of monomer molecules.
Each addition creates a new radical which has the same chemical structure as the previous one, except that it is longer by one monomer unit.
The successive additions are represented by

$$R_i + M \xrightarrow{k_p} R_{i+1}, \quad i = 2, \ldots, N-1, \tag{19.1}$$

where R_i is (the concentration of) the radical of chain length i. At some
point, the propagating polymer chain stops growing and terminates as
a result of reacting with another radical, forming nonreactive polymer
molecules. In general, we express the termination step by

$$R_i + R \xrightarrow{k_{t_i}} \text{dead polymer}, \quad i = 1, \ldots, N, \tag{19.2}$$

where $R = \sum_{j=1}^{N} R_j$ is the sum of (the concentrations of) the growing
radicals R_j and k_{t_i} is the termination rate constant for radical R_i. A detailed introduction to this process can be found in textbooks on polymer
chemistry; for example, see [4].

Before presenting mathematical equations for the polymerization process described above, we briefly explain a general model for such chemical reactions. Suppose that two chemicals, W and V, are present in a
solution, and that they interact to produce a chemical U at a reaction
rate k:

$$W + V \xrightarrow{k} U.$$

Thus, W decreases and U increases at rate kVW. If there are no other
changes that take place, the reaction can be quantified by

$$\frac{dW}{dt} = -kVW, \qquad \frac{dU}{dt} = kVW,$$

where t represents time.

Application of this general model to the polymer propagation step (19.1) and the termination step (19.2) then gives

$$\frac{dR_i}{dt} = k_p M R_{i-1} - k_p M R_i - k_{t_i} R R_i, \quad i = 2, \ldots, N. \tag{19.3}$$

The model suggested by (19.3) can be modified to account for the spatial difficulty that the monomer has in encountering a radical to foster the interaction. When the amount of monomer is large and the amount of R_i is small, there is no difficulty in the interaction; however, when the amount of monomer is small and the amount of R_i is large, the interaction is less likely to occur. Thus the term describing the interaction is altered to $-k_p R_i(M - \epsilon\phi_m R)$. This gives the model

$$\frac{dI}{dt} = (-k_d + k_p R\epsilon\phi_m)I$$

$$\frac{dX}{dt} = k_p R(1 - X)$$

$$\frac{dR_1}{dt} = 2fk_d I + (-k_p M + k_p R\epsilon\phi_m - k_{t_1} R)R_1 \tag{19.4}$$

$$\frac{dR_i}{dt} = k_p M R_{i-1} + (-k_p M + k_p R\epsilon\phi_m - k_{t_i} R)R_i, \quad i = 2, \ldots, N$$

$$I(0) = I_0, \quad X(0) = 0; \quad R_i(0) = 0, \quad i = 1, \ldots, N,$$

where t is the polymerization time and X is the fraction of monomer that has been converted. Other important variables and parameters are:

- the monomer concentration M, given by $M = M_0(1 - X)$, where M_0 is the initial monomer concentration;

- the initiation efficiency f, given by $f = \frac{D_m}{\gamma_0 k_0 + D_m}$, where γ_0 and k_0 are constants and the monomer diffusion function $D_m(X)$ is to be discussed later;

- the monomer volume fraction $\phi_m = \frac{1-X}{1-\epsilon X}$, with $\epsilon = \frac{d_p - d_m}{d_p}$, where d_p is the polymer density and d_m is the monomer density;

- the rate constant for initiator dissociation $k_d = e^{(33.1 - 1.48 \times 10^4/T)}$, where T represents absolute temperature (Kelvin).

For the polymerization system (19.4), the values of all parameters involved (except the termination rate constants k_{t_i}) are either determined or can be obtained from experimental data. For example, Table 19.1 provides experimental data values obtained from [5]; Table 19.2, at the end of this chapter, lists values for various parameters appearing in the

Table 19.1   Experimental data X, R, k_p for $I_0 = 0.0217$

t	X	dX/dt	R	k_p
min	(%)	(10^2 min^{-1})	(10^7 mol/l)	(10^{-4} l/mol $\cdot$ min)
5	1.3	0.26	0.60	4.390
10	2.6	0.26	0.65	4.110
15	3.8	0.26	0.70	3.860
20	5.2	0.26	0.73	3.760
25	6.5	0.26	0.77	3.610
30	7.8	0.26	0.80	3.520
50	13.2	0.26	0.84	3.750
70	18.5	0.26	0.96	3.320
80	20.8	0.27	1.00	3.410
90	23.5	0.30	1.20	3.300
100	27.0	0.40	2.00	2.740
110	31.0	0.45	3.50	1.850
120	37.0	0.66	6.00	1.750
125	40.5	0.85	8.00	1.790
130	46.0	1.29	11.0	2.170
132	50.0	1.75	13.0	2.690
135	54.5	2.00	15.0	2.930
136	57.0	2.05	16.5	2.890
137	59.0	2.10	17.2	2.970
138	62.0	2.10	18.0	3.070
139	64.0	2.05	19.0	3.000
140	66.0	2.00	20.0	2.940
141	68.0	1.90	20.5	2.890
142	70.0	1.75	21.2	2.750
143	71.0	1.70	22.2	2.640
144	72.0	1.60	23.0	2.480
145	73.2	1.40	24.0	2.180
146	74.5	1.30	25.0	2.040
147	75.3	1.15	26.0	1.790
148	76.1	1.05	27.0	1.630
149	77.0	0.95	27.5	1.500
150	77.5	0.85	28.5	1.330
152	79.0	0.65	30.0	1.030
154	80.0	0.60	31.0	0.968
156	80.5	0.50	31.5	0.814
158	81.3	0.40	32.0	0.436
160	81.6	0.35	32.0	0.594
167	83.0	0.25	31.0	0.474

model (19.4). In addition, the order N of (19.4), which represents the number of radicals participating in a reaction, is extremely large.

The lack of an accurate model for k_{t_i} as well as the large magnitude of N represent major obstacles for a computer solution of (19.4). One approach taken by polymer engineers is to assume that the radical termination coefficients are independent of the chain length: namely, it is assumed that

$$k_{t_i} = k_t, \qquad i = 1, \ldots, N. \tag{19.5}$$

The large and complex mathematical model (19.4) can then be transformed into the small and computable system

$$\frac{dI}{dt} = (-k_d + k_p R \epsilon \phi_m) I$$

$$\frac{dX}{dt} = k_p R (1 - X) \tag{19.6}$$

$$\frac{dR}{dt} = 2 f k_d I + (k_p \epsilon \phi_m - k_t) R^2$$

$$I(0) = I_0, \quad X(0) = 0, \quad R(0) = 0,$$

with the empirical models

$$k_p = k_p^0 \left(1 - \frac{X}{X_\infty} \right), \qquad X_\infty = \lim_{t \to \infty} X(t), \tag{19.7}$$

and

$$\frac{1}{k_t} = \frac{1}{k_t^0} + \theta \frac{R}{D_M}. \tag{19.8}$$

Here k_p^0, k_t^0, and θ are constants, and the polymer diffusion function $D_M(X)$ is to be discussed later. The assumption $R_N \equiv 0$ is used.

It is known that the reactivity of the radicals is strongly influenced by their chain lengths during the termination process. The assumption (19.5) obviously fails to describe such sensitive behavior. Another consequence of this artificial assumption is that the resulting simplified model (19.6) no longer provides the R_i, the concentrations of growing radicals with different chain lengths. A crucial piece of microscopic information for the study of the dynamic polymerization process is thereby lost.

19.2   *Formulating a Mathematical Model*

Our goal is to build a model that describes the diffusion-controlled behavior of growing radicals. Two criteria have to be met. Physically, the model should be realistic; mathematically, it should be solvable.

Considerable synergism — involving both the collection of data from chemical analytical experiments and the development of new mathematical techniques — is required to build a realistic and computable model. Frequent communications and discussions with polymer engineers identified the following information as critical to this work (see [6] for details):

- A set of experimental data (see Table 19.1 for a sample data set).

- The order of (19.4) can be as large as one million.

- The concentration R_i of chain length i radicals varies proportionally to the decay rate of k_{t_i} and converges to 0 as $i \to \infty$.

- The solution of (19.4) is not sensitive to k_p. The empirical model (19.7) is acceptable in general.

- The solution of (19.4) is expected to be very sensitive to k_{t_i}. The termination rate constant k_{t_i} can be modeled using

$$k_{t_i} = k_{DM}\left(i^{-1/2} + \frac{1}{R}\sum_{j=1}^{N} j^{-1/2}R_j\right), \qquad (19.9)$$

where k_{DM} is a diffusion-controlled rate depending on X and R.

With the collected information, the specific tasks of our model building are to (a) reduce the order of (19.4) sufficiently so that the resulting system admits a feasible computer solution; (b) build a model for k_p that is better than (19.7); and (c) model the diffusion-controlled rate k_{DM} in (19.9).

19.2.1   Reducing the Order of the Polymerization System

For a representable model, the order N of (19.4) needs to be large enough so that $R_N(t) \approx 0$. Experimental and computational observations have indicated that N can be as large as one million. From the viewpoint of physics, an infinite number of radicals actually participate in a polymerization process, so that $N = \infty$. However, computer solutions are possible only for systems of finite order, and any system with size one million is virtually uncomputable. Fortunately, R_i varies slowly with respect to the chain length i, especially for larger i. Thus, an approximation can be made by grouping the radicals as

$G_1 = \{R_1\}; \; G_i = \{R_{r_{i-1}+1}, \ldots, R_{r_i}\}, i = 2, \ldots, n; \; G_{n+1} = \{R_{r_n+1}, \ldots\},$

with $r_1 = 1$. Also let w_i denote the number of R_i in G_i: namely,

$$w_1 = 1; \quad w_i = r_i - r_{i-1}, \quad (i = 2, \ldots, n); \quad w_{n+1} = \infty.$$

Let $\overline{R}_i$ be the representative of group G_i,

$$\overline{R}_i = \frac{1}{w_i} \sum_{j=r_{i-1}+1}^{r_i} R_j,$$

and let $\overline{k}_{t_i}$ be an approximation to the termination rate constants for group G_i,

$$\overline{k}_{t_i} = k_{DM}(r_i^{-1/2} + \frac{1}{R} \sum_{j=1}^{n} r_j^{-1/2} w_j \overline{R}_j).$$

Assuming $R_{r_i} \approx \overline{R}_i$ and $k_{t_j} \approx \overline{k}_{t_i}$ for $j = r_{i-1} + 1, \ldots, r_i$, then the system (19.4) can be approximated by

$$\frac{dI}{dt} = (-k_d + k_p R \epsilon \phi_m) I$$

$$\frac{dX}{dt} = k_p R (1 - X)$$

$$\frac{d\overline{R}_1}{dt} = 2 f k_d I + (-\frac{1}{w_1} k_p M + k_p R \epsilon \phi_m - \overline{k}_{t_1} R) \overline{R}_1 \qquad (19.10)$$

$$\frac{d\overline{R}_i}{dt} = \frac{1}{w_i} k_p M \overline{R}_{i-1} + (-\frac{1}{w_i} k_p M + k_p R \epsilon \phi_m - \overline{k}_{t_i} R) \overline{R}_i, \quad i = 2, \ldots, n,$$

$$I(0) = I_0, \quad X(0) = 0; \quad \overline{R}_i(0) = 0, \quad i = 1, \ldots, n,$$

where

$$R = \sum_{i=1}^{n} w_i \overline{R}_i.$$

The variation of the R_i with respect to i is governed by the radical termination coefficient k_{t_i}. It is reasonable to group the R_i according to the decay rate of k_{t_i}, which is proportional to $i^{-1/2}$; see (19.9). Therefore, the groups G_i can be formed using the criterion

$$R_{j_1}, R_{j_2} \in G_i \quad \text{whenever} \quad \left| j_1^{-1/2} - j_2^{-1/2} \right| \leq \delta$$

for a given small positive number δ. For instance, we set $r_1 = 1$ and $\delta < 1$. While $r_i^{-1/2} > \delta$, the next r_{i+1} is chosen as

$$r_{i+1} = \min\{j : j \geq r_i + 1 \text{ and } r_i^{-1/2} - j^{-1/2} > \delta\}, \quad i = 1, \ldots, n-1,$$

until $r_n^{-1/2} \leq \delta$. As an example, the system (19.4) of order $N = 10^6$ can be approximated by (19.10) with $n = 187$ when $\delta = 10^{-3}$.

19.2.2   Modeling the Propagation Rate

The empirical model (19.7) for k_p might be improved by using the observed data values for X and k_p, such as those given in Table 19.1. Numerical tests have revealed that least squares data-fitting gives good simulations in general. For the example represented by Table 19.1, the regressions of k_p on X produced the linear least squares model

$$k_p = 647.425 - 574.568X \tag{19.11}$$

and the cubic least squares model

$$k_p = 10^3(0.7908 - 2.4614X + 5.2182X^2 - 3.9441X^3). \tag{19.12}$$

It has been observed that the solution of (19.10) is not sensitive to k_p; consequently either (19.7) or the least squares regressions (19.11) and (19.12) can be used with quite satisfactory results.

19.2.3   Modeling the Diffusion-Controlled Termination Rate

According to the free volume theory of polymer diffusion [1], the monomer diffusion function D_m and the polymer diffusion function D_M are determined by

$$D_m = 1.35e^{-1850/T}e^{-A/B}, \qquad D_M = 1.35e^{-1850/T}e^{-A/(\xi B)},$$

where A and B are proportional to $X(t)$ and $1-X(t)$ and can be modeled as $A = 0.868[1 - X(t)] + 0.433X(t)$ and $B = [5.29(T - 32.9)(1 - X(t)) + 2.82(T - 301)X(t)]/10^4$. The parameters $T = 333.16$ and $\xi = 0.295$ are appropriate for the experimental data in Table 19.1.

The value of the diffusion-controlled termination rate constant k_{DM} involved in the model (19.9) for k_{t_i} cannot be directly obtained using current experimental techniques. However, physically, we know that k_{DM} is dominated by the polymer diffusion function D_M during the early period of the radical propagation, and is controlled by the monomer diffusion function D_m in the late phase of the radical termination. The experimental data for X listed in Table 19.1 can be used to plot D_m and D_M in Figure 19.1. Shifting the D_M-curve upward as shown by Figure 19.2, the fact described above suggests that k_{DM} can be approximately modeled as a curve that traces the shifted D_M (dotted line) in the early phase of the reaction, and then switches to the tail of the D_m-curve (dash-dot line), as shown by the solid line. The geometric transformation performed can be expressed as

$$k_{DM} = \theta(\theta_0 D_M(1 - \alpha) + D_m\alpha), \tag{19.13}$$

where α is a weight and where θ and θ_0 represent certain proportionality constants.

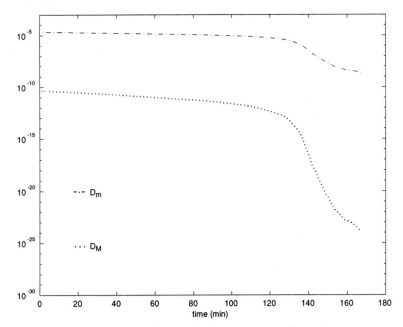

FIGURE 19.1 D_m **and** D_M **vs. reaction time**

The transition of k_{DM} from D_M-dominance to D_m-control is smooth in real chemical reactions, indicating that it should occur near a point X_0 at which the monomer conversion X accelerates. Thus, θ_0 is chosen as a multiplier of D_M that renders $\theta_0 D_M$ and D_m equal at $X = X_0$:

$$\theta_0 = \frac{D_m}{D_M} \qquad \text{at } X = X_0.$$

Based on experimental observations, we choose $X_0 = 0.45$. We also choose the weight α using

$$\alpha = \frac{R}{R_\infty}, \qquad R_\infty = \lim_{t \to \infty} R(t),$$

to guarantee a smooth and clear transition of k_{DM} from $\theta_0 D_M$ to D_m at X_0, since $R/R_\infty \approx 0$ for $X < X_0$ and it quickly increases to $R/R_\infty \approx 1$ for $X \geq X_0$.

Using (19.13) for k_{DM}, computational results have revealed that k_{DM} is often relatively large for $X < X_0$, indicating that (19.13) does not put enough weight on D_m. This observation is consistent with the actual radical propagation process, during which the monomer diffusion constant D_m also makes a nonnegligible contribution. Therefore, we inserted a heuristic linear function $c_1 + c_2X$, with $c_1 + c_2X_0 = 1$, into k_{DM} when $X < X_0$, which simulates a moderate impact of D_m during the early radical propagation process. The model for k_{DM} is then

$$k_{DM} = \begin{cases} \theta(\theta_0 D_M(c_1 + c_2X)(1 - \alpha) + D_m\alpha) & \text{if } X < X_0 \\ \theta(\theta_0 D_M(1 - \alpha) + D_m\alpha) & \text{otherwise.} \end{cases} \qquad (19.14)$$

Now, the only unidentified parameter is θ. Comparing the numerical solution of (19.10) with the available experimental data, the value of θ is found to be in the range of

$$\theta = (1.2 \pm 10\%)10^{12}. \qquad (19.15)$$

The variation is due to the error in the experimental data. The indicated 10% relative error is usually acceptable.

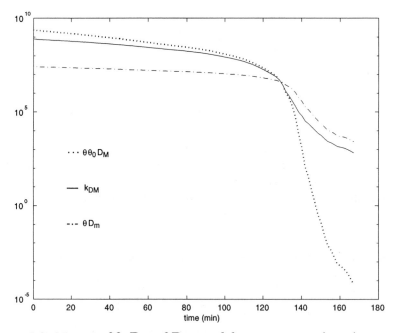

FIGURE 19.2 $\theta\theta_0 D_M$, θD_m, and k_{DM} vs. reaction time

19.3   Computational Approach

We have translated the essence of the polymerization process into a solvable mathematical model (19.10). The model is built upon both physical and mathematical considerations. In this section, we shall discuss related numerical algorithmic issues and carry out a computer simulation of the chemical process.

19.3.1   Optimizing θ

The only nonuniquely determined parameter in the model (19.10) is θ, the constant of proportionality involved in k_{t_i}; see (19.9) and (19.14). Since the solution of (19.10) is highly sensitive to k_{t_i}, the selected value of θ is critical to the accuracy of the computer simulation. The values given by (19.15) are found to be the best, based upon all the available experimental data. When a particular set of experimental data X and R is provided, θ can be customized so that the solution of (19.10) more closely fits the given experimental data. This process is equivalent to solving the constrained optimization problem

$$\min\{\psi(\theta) : \; \theta \in [1.1 \times 10^{-12}, 1.3 \times 10^{-12}]\}, \qquad (19.16)$$

where ψ measures the error in fitting the observed data with the model (19.10), relative to the value θ specified in (19.14). Since the monomer conversion fraction X is informative and less sensitive than the total concentration R of the radical chain, we recommend that X be used in the objective function ψ for data fitting. For example, the error function can be defined as

$$\psi(\theta) = \sum_i w(i)\,|X_e(i) - X(t_i, \theta)|, \qquad (19.17)$$

where w is a vector of weights determined from physical considerations, $X_e(i)$ is the observed data at t_i, and $X(t_i, \theta)$ is the computed value obtained from (19.10). In this work, we use

$$w(i) = \begin{cases} X_e(i) & \text{if } 0 < X_e(i) < 0.7 \\ 0 & \text{otherwise.} \end{cases} \qquad (19.18)$$

The monomer conversion fraction X changes rapidly from 0.4 to 0.7, and most of the radical interactions occur during this period. The accuracy of the simulation at the final stage of the polymerization is not a practical concern, and the experimental data in that range could involve large data errors; see Section 19.3.3. Since the error function defined in this way is unimodal in a neighborhood of its optimal solution, established stable computational routines that use optimal search plan schemes, such as FMIN [2], can satisfactorily solve the optimization problem (19.16).

19.3.2   Scaling Considerations

The magnitude of the R_i is of the order $O(10^{-7})$. A small perturbation during computation could cause unwanted oscillations of the solution and result in failure of the numerical simulation. A remedy for this problem is to scale the R_i by a constant s, such as $s = 10^7$. The effect of such scaling is illustrated by comparing Figure 19.3 (without scaling) to Figure 19.4 (scaling with $s = 10^7$). The solid lines in these figures show the numerically obtained values of X and R, while the circles designate the experimental data. Clearly, the scaling has a beneficial effect on increasing the accuracy of the numerical solution.

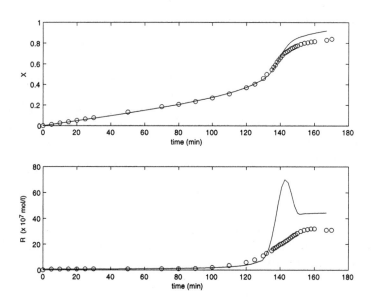

FIGURE 19.3   **Numerical and observed X and R, no scaling**

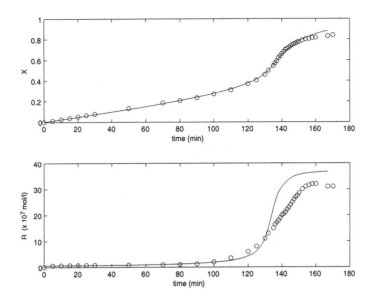

FIGURE 19.4   **Numerical and observed X and R, with $s = 10^7$**

19.3.3   *Computer Simulation*

Using MATLAB [3], we solved the model (19.10) having order $n = 188$. This order represents radicals with chain lengths ranging from 1 up to 10^8. Since (19.10) is a *stiff ODE system* [2], the MATLAB function ode15s was used. Figures 19.5 and 19.6 compare the computed monomer conversion fraction X and the total concentration R of the radical chains (solid lines) with the experimental data (in circles). Two different initiator concentrations I_0 are illustrated, where $I_0 = 0.0217$ in Figure 19.5 and $I_0 = 0.0434$ in Figure 19.6. Note that the monomer conversion fraction X starts accelerating at approximately the value 0.4, illustrating what is known as the *gel effect* — the polymerization system solidifies at that time, and the polymerization stops within minutes. This gel effect is characterized by a significant increase in the radical concentration R, as shown in Figures 19.5–19.6. The computed solution of (19.10) fits the experimental data very well except at the final stage of the polymerization, which is still considered satisfactory because the measurement of radical concentration using an electron spin resonance spectrometer can involve relatively large data errors at the final stage. The good agreement between the calculated values and the experimental data suggests that the proposed diffusion-controlled kinetic model

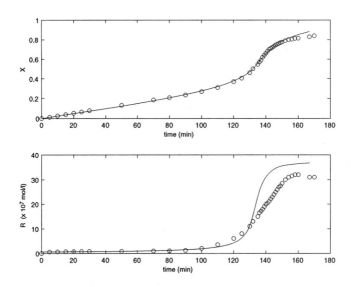

FIGURE 19.5   **Plots of X and R vs. time for $I_0 = 0.0217$**

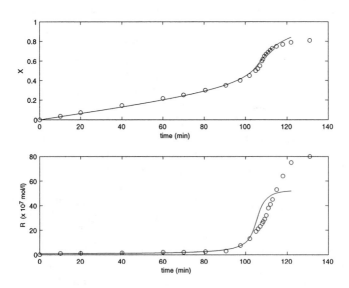

FIGURE 19.6   **Plots of X and R vs. time for $I_0 = 0.0434$**

(19.10) qualitatively and quantitatively describes the mechanism of the radical polymerization, and it adequately captures the gel effect.

Figures 19.7–19.8 indicate the distribution of the calculated radical sizes R_i. Specifically, the plotted curves correspond (from top to bottom) to the eight chain lengths

$$i = 1;\ 1{,}100;\ 4{,}400;\ 18{,}000;\ 64{,}000;\ 240{,}000;\ 870{,}000;\ 6{,}300{,}000.$$

Again, two different initiator concentrations I_0 are illustrated, where $I_0 = 0.0217$ in Figure 19.7 and $I_0 = 0.0434$ in Figure 19.8. It is interesting to observe that, at the very beginning of the reaction, only small or primary radicals are present in the system. These radicals grow quickly, and their distribution curves propagate to high molecular weight within a very short time period. These microscopic chemical activities which we have discovered through computer simulation would be extremely difficult (in fact, practically impossible) to be captured experimentally.

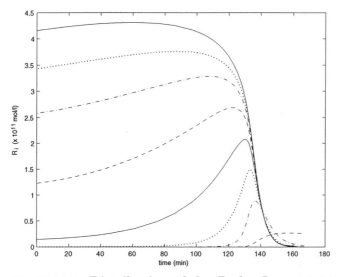

FIGURE 19.7   Distribution of the R_i for $I_0 = 0.0217$

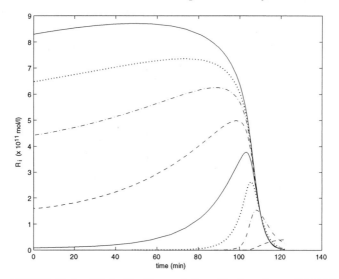

FIGURE 19.8   **Distribution of the R_i for $I_0 = 0.0434$**

Figure 19.9 shows a simulation of the kinetics of the polymerization process with the initiator concentration $I_0 = 0.0217$. The results are divided into six time intervals, so that Figure 19.9 consists of six graphs, (a)–(f), that illustrate the distribution of the R_i over the respective time intervals (in minutes) of $[0, 7.4 \times 10^{-5}]$, $[7.4 \times 10^{-5}, 3.6 \times 10^{-4}]$, $[3.6 \times 10^{-4}, 1.2 \times 10^{-3}]$, $[1.2 \times 10^{-3}, 6.4 \times 10^{-3}]$, $[6.4 \times 10^{-3}, 18]$, and $[18, 167]$. Figure 19.9 clearly describes the complete dynamic behavior of the growing radicals. All of these graphs are very helpful for understanding the underlying molecular process. For detailed interpretation and empirical verification of the simulation results, the reader can consult [6]. Table 19.2 provides a summary of the parameters used for the data set in Table 19.1.

19.4   *Conclusion*

We have introduced a mathematical model that predicts and describes the steps of radical growth in an important chemical process. The approach has been to construct a mathematical model that is computationally tractable, while still being realistic. This is achieved by appropriately aggregating groups of radical chains in the original model (19.4). In addition, several other models for the rate constants k_p and k_{t_i} are required in order to successfully simulate the behavior of the chemical

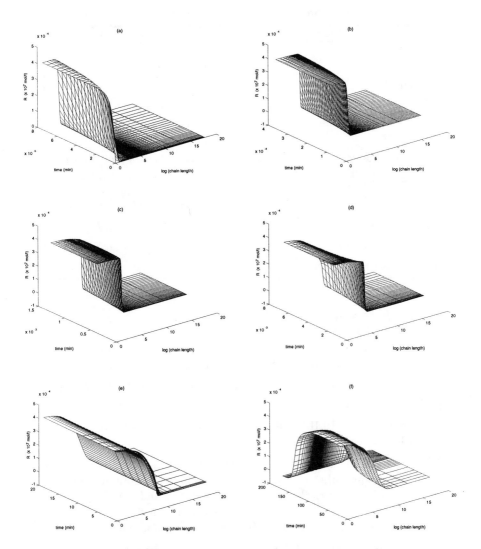

FIGURE 19.9 R_i over successive time intervals

system. The resulting model and proposed computational approach are able to accurately and efficiently represent the polymerization process. The modeling and simulation presented in this chapter provide polymer engineers with a methodology for studying and understanding the molecular process of radical growth.

Table 19.2 Summary of parameters used for simulation

Parameter	Parameter
$T = 333.16$	$\xi = 0.2950$
$X_0 = 0.45$	$X_\infty = 0.85$
$\epsilon = 0.2027$	$s = 10^7$
$M_0 = 8.54$	$R_{max} = R_\infty = 120$
$I_0 = 0.0217$ (mol/liter)	$\gamma_0 k_0 = 2.8 \times 10^{-5}$
$k_p^0 = 735$	$k_t^0 = 6.38 \times 10^7$
$k_p = k_p^0 - X k_p^0 / X_\infty = 735 - 864.706X$	$\phi_m = (1 - X)/(1 - \epsilon X)$
$k_{t_i} = k_{DM}\left(\frac{1}{\sqrt{i}} + \frac{1}{R}\sum_{j=1}^{N}\frac{R_j}{\sqrt{j}}\right)$	$k_d = e^{(33.1 - 1.48 \times 10^4/T)}$
$A = 0.868(1 - X) + 0.433X$	$D_m = 1.35 e^{-1850/T} e^{-A/B}$
$B = \frac{5.29(T - 32.9)(1 - X) + 2.82(T - 301)X}{10^4}$	$f = D_m/(\gamma_0 k_0 + D_m)$
$D_M = 1.35 e^{-1850/T} e^{-A/(\xi B)}$	$\theta_0 = D_m/D_M$ at X_0
$\theta = $ output from `FMIN`	

19.5   Exercises and Projects

1. Derive (19.3) by applying the general model of chemical change to the radical chemical propagation and termination steps (19.1)–(19.2).

2. Assume (19.5) holds and that $\lim_{N\to\infty} R_N(t) = 0$ for $t > 0$. Show that the system (19.4) of order $N \approx \infty$ is equivalent to (19.6).

3. Show in model (19.10) that $\overline{R}_{n+1}$ automatically satisfies $\overline{R}_{n+1}(t) = 0$ for all $t > 0$.

4. A real-valued function $F(x)$ defined on $[a, b]$ is called *unimodal* if there is a unique value x^* such that (a) $F(x^*)$ is the minimum of $F(x)$ on $[a, b]$; (b) $F(x)$ strictly decreases for $x \leq x^*$; and (c) $F(x)$ strictly increases for $x \geq x^*$. Show that the error function defined by (19.17) and (19.18) is unimodal in a neighborhood of the optimal solution.

5. Using the experimental data from Table 19.1, compare the following models for the propagation rate constant k_p:

   a. the empirical model (19.7);

   b. the linear least squares model (19.11);

   c. the cubic least squares model (19.12).

Carry out this comparison for the three additional data sets provided in Tables 19.3, 19.4, and 19.5.

Table 19.3   Data set (a)

t	X	R	k_p	t	X	R	k_p
10.0	3.6	1.0	3.73	108.0	60.0	25.5	3.04
20.0	7.2	1.1	3.53	108.5	62.0	27.0	2.97
40.0	14.4	1.5	2.80	109.2	65.0	29.0	2.71
60.0	21.6	2.0	2.30	110.0	67.0	32.0	2.37
70.0	25.2	2.0	2.64	111.0	69.0	38.0	1.91
80.5	30.0	2.5	2.54	112.0	71.0	41.0	1.60
90.5	35.0	3.0	3.33	113.0	73.0	45.0	1.36
97.5	40.0	7.5	2.00	115.0	75.0	53.0	0.98
102.0	45.0	13.0	2.17	118.0	77.0	64.0	0.68
105.0	50.0	19.0	2.16	122.0	79.0	75.0	0.44
106.0	52.0	21.0	2.48	131.0	81.0	80.0	0.44
107.0	55.0	23.0	2.85				
$\xi = 0.303$							

Table 19.4   Data set (b)

t	X	R	k_p	t	X	R	k_p
5.0	3.0	1.1	4.82	74.5	59.0	66.0	1.37
15.0	7.7	1.3	4.28	75.0	61.0	68.0	1.38
25.0	13.0	1.5	3.94	75.5	63.0	69.0	1.39
34.0	18.0	1.7	3.69	76.0	65.0	71.0	1.41
45.0	23.3	2.2	3.22	76.3	66.0	72.0	1.33
50.0	26.0	3.0	2.84	77.0	68.0	73.5	1.15
55.0	29.5	4.0	2.52	78.0	70.0	76.0	0.965
60.0	34.0	7.0	2.16	79.5	72.0	79.5	0.809
65.0	39.0	16.0	1.65	81.0	74.0	84.0	0.687
67.0	43.0	28.0	1.35	82.5	75.0	89.0	0.584
71.0	48.0	51.0	1.28	84.5	76.0	94.0	0.510
72.0	50.0	55.0	1.19	87.0	77.0	99.0	0.439
72.5	52.0	58.0	1.24	89.0	78.0	102.0	0.401
73.5	55.0	62.0	1.33	92.0	84.0	104.0	0.401
74.0	57.0	64.0	1.34				
$\xi = 0.3135$							

6. (Project) Using different values for θ in (19.8), compute the numerical solution of (19.6). Discuss the sensitivity of the solution of (19.6) to the termination coefficient k_t.

7. (Project) Compare numerical solutions of (19.10) for different orders n. The computation time is expected to increase rapidly as n becomes larger. Do you have any suggestions for making the computation more efficient?

Table 19.5   Data set (c)

t	X	R	k_p	t	X	R	k_p
5.0	3.5	2.0	3.63	50.0	44.0	27.0	1.98
10.0	7.0	2.0	3.76	50.5	47.0	34.0	2.08
15.0	10.5	2.0	3.91	51.0	50.0	37.0	2.30
20.0	14.0	2.0	4.07	52.0	54.0	42.0	2.30
25.0	17.5	2.15	3.95	52.5	57.0	46.0	2.26
30.0	21.0	3.0	2.95	53.0	60.0	50.0	2.20
35.0	24.5	3.5	2.76	54.0	63.0	60.0	1.80
40.0	28.0	5.0	2.22	54.5	65.0	63.0	1.47
42.0	30.0	6.0	2.26	55.5	67.0	66.0	1.33
45.0	33.0	8.5	2.11	57.5	70.0	73.0	1.03
48.0	38.0	11.0	2.71	59.0	72.0	77.0	0.88
49.0	40.0	23.0	1.63	61.0	74.0	80.0	0.88
49.5	42.0	24.0	1.83				

$\xi = 0.3202$

19.6   References

[1] J. L. Duda, J. S. Vrentas, S. T. Ju, and H. T. Liu, "Prediction of diffusion coefficients for polymer-solvent systems," *Amer. Inst. Chem. Eng. J.* **28**, 279–285 (1982).

[2] D. Kahaner, C. Moler, and S. Nash, *Numerical Methods and Software*, Prentice-Hall, Englewood Cliffs, NJ, 1989.

[3] The MathWorks, Inc., *Using MATLAB Version 5*, The MathWorks, Inc., Natick, MA, 1997. <http://www.mathworks.com/>

[4] G. Odian, *Principles of Polymerization*, Wiley, New York, 1981.

[5] J. Shen, Y. Tian, G. Wang, and M. Yang, "Modeling and kinetic study of radical polymerization of methyl methacrylate: propagation and termination rate coefficients and initiation efficiency," *Makromol. Chem.* **192**, 2669–2680 (1991).

[6] Y. Tian and H. Zhang, "Computational modeling of radical polymerization of methyl methacrylate," *Computer Modeling and Simulation in Engineering*, 1999, to appear.

[7] Y. Tian, S. Zhu, A. E. Hamielec, and D. R. Eaton, "Conformation, environment and reactivity of radicals in copolymerization of methyl methacrylate/methylene glycol dimethylmethacrylate," *Polymer* **33**, 384–395 (1992).

Appendix A

The Clemson Graduate Program in the Mathematical Sciences

K. T. Wallenius

A.1 Introduction

This appendix focuses on the development of the mathematical sciences program at Clemson University. It is offered with the author's gratitude and deep respect for Clayton Aucoin, the architect and builder of that program. It takes one set of skills (foresight) to accurately anticipate called-for curricular reforms 20 years in advance, another (ingenuity) to design a plan of action to prepare for the predicted events, yet another (conviction) to believe strongly enough in the plan to take career chances to pursue the vision, still another (leadership) to convince others to join in the implementation of the plan, and maybe one more (equanimity) to tolerate skeptics, detractors, and bungling university administrators along the way. Clayton Aucoin has all these attributes with the possible exception of the last. But four out of five is better than your average mathematician. This appendix is not biographical in nature beyond using key events in Clayton's professional life as they pertain to the evolution of ideas and the creation of the program. It is about his vision in the early 1960s of how the mathematical sciences[1] could best contribute to solving problems in a society of ever-increasing technological

---

[1]For the purposes of this appendix, the definition of *mathematical sciences* given by the Board on Mathematical Sciences of the National Research Council in [1] will be adopted: "pure mathematics, applied mathematics, statistics and probability, operations research, and scientific computing."

complexity. It is about modeling in two related senses: Clayton's intuitive model of the importance of the mathematical sciences in the workplace of the 1980s and 1990s, as well as the role of mathematical and statistical modeling in academic curricula to meet the needs of that workplace. Clayton insisted that modeling play a central role in every phase of creating the program and building a faculty to deliver it. Thus it is appropriate that material focusing on the Clemson program be included in this volume of models.

A.2   Historical Background

To give some perspective on conditions surrounding the development of the Clemson program and the current state of education in the mathematical sciences, a brief historical summary will be given borrowing heavily from references [1, 6].

World War II created unprecedented opportunities for the application of the mathematical sciences in support of the war effort. Areas of application included weaponry, electronics, cryptography, and quality control. Urgency of the war effort shaped national priorities and gave importance to relatively new areas such as operations analysis. Research grew at a phenomenal rate aided by new sources of government support such as the Office of Naval Research (ONR) and, later, the National Science Foundation (NSF). The successful conclusion of the war brought appreciation for the contributions of science and mathematics. Increases in funding for research, support for graduate assistants, travel, conferences, etc. changed the focus at many institutions and gave momentum to a movement emphasizing academic research. Academicians who could obtain grants were in great demand by the late 1950s. Curricular development and teaching were not high priority concerns. Graduate assistants relieved senior faculty of considerable responsibilities relative to undergraduate teaching.

The 1950s and 1960s were decades of growth. Annual production of Ph.D.s surged from about 200 in 1950–51 to 1070 in 1968–69. Employment opportunities were unlimited during this period of expansion. The situation turned around in the 1970s and a decade of contraction followed the twenty years of growth. Overall Ph.D. production fell to about 791 in 1979–80. The decline was concentrated in the pure areas, which experienced a decrease of 55%, whereas statistics and operations research actually increased slightly. The decline was amplified by the emergence of computer science as a separate discipline with its own identity, drawing students and resources in the process. Market conditions forced graduates trained for academic research to seek employment in nonacademic positions and in educational institutions where teaching was stressed. These were jobs for which graduates were ill-prepared. Paradoxically, the declining demand in the 1970s for graduates trained for academic research was accompanied by a new respect for and reliance upon mathematical methods for solving problems in an increasingly technological society. One might have expected some serendipitous balance to have resulted from the two offsetting trends. But not so. Employers in Business, Industry, and Government (BIG) were reluctant to hire the traditionally trained mathematician whom they tended to view as an awkward luxury. Academic training, excellent by traditional standards, did not prepare graduates to communicate with clients, coworkers trained in science and engineering, or with management. Many graduates expected to be allowed to do independent research often tangential to what employers considered relevant [5]. This may not be entirely fair — no one would deny that there were many exceptions — but the image was widespread [3]: "Much of industry and business still regards mathematicians with suspicion. Few industries have career paths for mathematicians; contributions of a mathematical nature are often not recognized as such because they are made by physicists, engineers, and computer scientists."

The 1980s were relatively stable compared with the previous three decades of ups and downs. Ph.D. production continued to decline slowly, reaching a low of 726 in 1984–85 and then climbed again to 929 in 1989–90. While Ph.D. production was increasing, the number of degrees awarded to U.S. citizens actually decreased due to a dramatic change in the mix of international and U.S. students. The proportion of Ph.D.s awarded to U.S. students dropped from 68% in 1980–81 to 43% ten years later. This trend was and still is a source of concern throughout the mathematical sciences community, especially in major research institutions where graduate students are used to teach undergraduate service courses. Employment patterns and unemployment rates are also sources of concern. While the number of new Ph.D.s employed in the U.S. went up, most of the increase occurred in nondoctoral-granting academic institutions and in nonacademic positions. The percent of new Ph.D.s still looking for jobs six months or more after receiving the degree averaged nearly 10% in the late 1980s.

The slow upward trend of Ph.D. production in the 1980s accelerated into the 1990s reaching 1202 in 1992–93, an all-time high. But, a majority of these, fifty-six percent, were foreign nationals. Employment patterns of the late 1980s also continued into the early 1990s. Among those whose employment status was known as of late September 1993, 12.4% were still seeking employment. This rate nearly equaled the record rate of 12.7% in 1992. Only 252 or 21% of new Ph.D.s obtained positions in doctoral-granting institutions. It could be argued that some of the problems were due to a national economy in recession, but these statistics have persisted into the late 1990s, a time when the economy has rebounded dramatically. There is a general agreement that national needs are not reflected in most graduate programs and that revision of traditional training is long overdue. In short, we are producing too many square pegs in a market dominated by round holes.

Clayton Aucoin was part of the growth scene of the late 1950s and early 1960s described above, receiving his Ph.D. in algebraic topology in 1956. Never one to follow traditional paths, he took a position as Senior Systems Engineer with a defense contractor after graduation. He recognized that even those trained in the applied mathematics[2] of the day were seriously deficient in problem-solving abilities involving modeling, analyzing real data, constrained optimization, and computing. Traditional mathematics programs simply did not prepare their students very well for nonacademic careers. There were many strong mathematical statistics programs. Operations research and computing were developing, mostly in engineering school settings. It was not impossible, of course, for a student to obtain breadth of training across departmental lines, but breadth was not the paradigm of the day. The fact that component areas were thriving as parochial academic units augured against the creation of programs requiring cooperation across traditional departmental lines. Funding agencies were not supporting curricular development.

A short stint back in academe convinced Clayton that supporters of the conventional paradigm were too comfortable with business as usual to take risks attendant with curricular experimentation so he opted for a personal departure. Upon receiving an NSF Science Faculty Fellowship, he headed for Stanford University to broaden his own training at an institution known for its excellent programs in statistics and an emerging graduate program in operations research. It was here that Clayton and the author, a first-year graduate student in statistics, met and became good friends.

A.3   Transformation of a Department

Not long after completing the year at Stanford, Clayton moved to Clemson and soon became head of a small service-oriented department. In a short period of time, Aucoin was able to sell his concept to a rather

---

[2]Applied mathematics in the early 1960s was nearly synonymous with mathematical physics.

forward-looking administration willing to commit resources and take chances. Reform started with an innovative undergraduate major featuring courses in computing, operations research, probability, and statistics along with more traditional undergraduate mathematics courses. This program also required students to select an "option" (e.g., biology, communications, computing, statistics) that stressed modeling in an applications area. Before embarking on reform, there had to be something to reform. Traditional master's and doctoral programs were created to get a foothold in the graduate degree business in South Carolina. With a foot in the door, it was possible to justify the hiring of additional faculty to staff the new courses and start the transformation. Aucoin began assembling a faculty who shared his philosophy and would play key roles in developing the program. Diminished academic job opportunities in the 1970s made it possible to be rather selective in hiring new faculty. Despite heavy recruiting competition in the key areas of operations research, statistics, and computational mathematics, the novelty of the program concept and the enthusiasm of the faculty seemed to infect visitors, including the author of this paper who joined the faculty in 1968.

With the addition of faculty in computer science, computational mathematics, operations research, and statistics, more breadth was added in the graduate degree programs. In 1975, the department put together a successful curricular reform proposal to create an applied master's program in mathematical sciences. The resulting grant from the NSF program *Alternatives in Higher Education* gave visibility, legitimacy, and support to efforts to create such a program to prepare graduates for BIG careers. A distinguished Board of Advisors provided valuable guidance during the planning phase of the applied master's program. The Board members included H. T. Banks (Brown University), Don Gardiner (Oak Ridge National Labs), Stu Hunter (Princeton University), Bill Lucas (Cornell University), Bob Lundegard (Office of Naval Research), Don McArthur (Milliken Corporation), Jim Ortega (ICASE), and Bob Thrall (Rice University). The consensus opinion was that a level of training between the baccalaureate and the Ph.D., something more ambitious than the traditional master's degree, would be attractive to students and fill a significant national need. The applied master's program was formally created in December 1975. The NSF grant also called for the development of a Ph.D. program that would build on the breadth-and-depth philosophy of the M.S. degree. The resulting Ph.D. program was approved by the faculty in the spring of 1979.

A.4   The Clemson Program

This section starts with a brief description of the four graduate degree programs offered by the Department of Mathematical Sciences at Clemson University. The first two degrees are administered wholly within the department. The second two are jointly administered with other academic units. These programs involve roughly 50 faculty members and 100 graduate students in the Department of Mathematical Sciences along with colleagues and students from cooperating departments. Some measures of success will be noted. Attention will be given to elements of the programs that prepare graduates for careers in teaching in colleges and universities and for careers in business, industry, and government.

The applied M.S. program was based on the following premises stated in [7]:

a. The major source of employment for mathematical scientists in the future will be nonacademic agencies.

b. Most such employers will require more than a B.S. degree but less than a Ph.D. degree in the mathematical sciences.

c. Employers will prefer personnel who possess not only a concentration in a particular area of the mathematical sciences, but also a diversified training in most of the other areas.

d. Graduates should have more than superficial education in applying mathematical techniques to solve problems in areas other than the mathematical sciences. Inherent in such training is the ability to communicate, both orally and in writing, with persons from these areas.

e. It is desirable to obtain such broad-based education in the mathematical sciences prior to specializing for the Ph.D. degree.

The initial plan called for 14 three-hour courses (42 semester hours) plus a one-hour master's paper to be completed over two academic years and an intervening summer. This was a substantial increase from typical requirements in which the master's degree was often regarded as a consolation prize to Ph.D. candidates unable to pass their qualifying examinations. Requirements for the M.S. degree were broken down into two categories: breadth and concentration.

The breadth requirements consisted of eight courses:

- two "core" courses (e.g., analysis and advanced linear algebra);

- two computing courses (e.g., digital models and computational mathematics);

- one statistics course (e.g., data analysis);

- one operations research course (e.g., mathematical programming);

- one additional statistics/operations research course (e.g., selected from applied multivariate analysis, linear statistical models, network flows, advanced linear programming, stochastic processes);

- one applied models course taught outside the department (e.g., a course in science, engineering, or economics, employing mathematical models to study various system behaviors).

The seven departmental breadth courses and the external applied models course were designed to meet the requirements of premises (c) and (d), respectively. With the aid of a faculty advisor, each student selects six additional courses in one of five areas of concentration: discrete mathematics, applied analysis, computational mathematics, operations research, or statistics. Some flexibility was allowed to tailor programs to special career objectives. A concentration could be a blend of advanced courses in more than one area and could include courses taught in other departments if such a blend achieved sufficient depth and better met a student's objective. An important feature of the M.S. program was the mix of applied courses and theory/methodology courses, about seven of each. This is not to say that applications are not considered in theoretical courses or that applied courses are not based on sound theory or gloss over rigor. The common thread is that modeling is stressed in both. In some cases, two courses pertaining to the same subject area appear in the catalog, such as Decision Theory (a mathematical statistics course presented from a measure-theoretical viewpoint) and Applied Statistical Decisions (a course that models rational behavior leading to the Bayesian paradigm).

As might be expected, the initial program requirements described above have undergone minor changes over the years. For example, after many successful "external models" courses, a few bad experiences led to the abandonment of this requirement in favor of the inclusion of more modeling content in the breadth courses. The one-hour paper requirement was replaced by a variable credit master's project culminating with both written and oral presentations in conjunction with the master's examination. Despite these minor changes, the guiding philosophy of breadth and depth remains intact.

The present Ph.D. program, inaugurated in 1979, was developed after experience was gained with the master's program. It was fashioned along the same general philosophy of breadth and depth, with a heavy emphasis on modeling. Breadth requirements are met by taking at least two graduate courses in each of five areas: algebra/discrete mathematics, analysis, scientific computing, operations research, and statistics. *Breadth* is assessed by performance on three preliminary examinations selected from among the five areas listed above plus stochastic processes. These three prelims plus a fourth comprehensive examination administered by the student's advisory committee comprise the Ph.D. qualifying examination required for formal admission to degree candidacy. The comprehensive examination addresses *depth* of understanding and serves three purposes: to assess the student's readiness to perform independent research; to assess the student's competency in advanced graduate material relevant to the student's chosen research area; and to provide a forum for members of the advisory committee to learn about and provide input into the student's proposed research program. A thesis proposal is not a required part of this fourth examination although such a proposal is frequently discussed during the oral part of the examination. Weaknesses identified during the comprehensive examination can be remedied through additional course work and directed reading. A successful thesis defense is the final requirement.

A second Ph.D. program, this in Management Science, was created in 1971 through a joint effort of the faculties of the departments of Management and Mathematical Sciences. Program requirements consist of twelve courses considered fundamental (called "core" courses) plus an additional six advanced courses (called "concentration" courses) approved by the student's advisory committee. Core courses include training in both departments while the concentration might be in a single area such as production/operations management, stochastic models, applied statistics, etc. Admission to candidacy is based on successful completion of a comprehensive examination composed of one-day written examinations in applied statistics (statistical inference and data analysis) and in operations research (mathematical programming, stochastic models, and operations management) plus a week-long case study; the latter requires a definitive analysis and evaluation of a management science application, including financial and mathematical modeling, formulation, analysis, and final recommendations in the form of a written study citing appropriate references. The program culminates with a defense of the student's dissertation.

In addition, a Master of Education degree in mathematics education is offered by the College of Education. It is administered by a joint committee and requires a minimum of 18 hours of graduate credit in the mathematical sciences. The mathematical sciences community views the decline in mathematics preparation of incoming freshmen with considerable concern. We share in that concern and view our participation in the training of mathematics teachers for elementary and secondary education as an extremely important professional service. However, the training of school teachers is outside the focus of this appendix and will not be discussed further.

A.5   Communication Skills

In addition to breadth of academic training, virtually every study and paper dealing with graduate student preparation for today's job market stresses the importance of communication skills in preparation for college teaching as well as nonacademic jobs. Various opportunities to acquire and improve these skills exist on all campuses: the passive observation of what makes for good teaching by simply attending lectures and evaluating one's instructors, the requirement to orally defend one's Ph.D. dissertation or master's paper, and (for some) the opportunity to consult, tutor, or teach. These opportunities are often not highly exploited because of the emphasis placed on learning subject matter. One cannot expect communication skills to be absorbed by osmosis, so a program must facilitate their acquisition through conscious planning.

The most obvious opportunity is in the role of classroom instructor since, at most graduate institutions, teaching assistants (T.A.s) must be used to shoulder some of the heavy undergraduate service teaching load. The necessity to take on this kind of responsibility is viewed as an opportunity for students to gain poise and self-confidence in front of an audience, whether it be a group of students in the classroom or a future group of clients or business associates in a board room. The ability to prepare and deliver an organized lecture and to think on one's feet in response to questions in the classroom is extremely valuable. The importance of these skills is taken seriously at Clemson, as manifested by specific requirements. All new graduate T.A.s are required to attend an intensive week-long orientation prior to commencement of classes. They attend lectures on "The Role and Responsibilities of the Graduate Teaching Assistant," "Methods for Active Teaching and Learning," "Questioning and Discussion Techniques," "Planning for Instruction: the First Day," "Dealing with Potential Problems in the Classroom and Office," among others. They are given a ninety-two page *Guidebook for Clemson University Teaching Assistants* containing sections on preparation and teaching techniques along with other information on the use of audiovisual aids, computing facilities, administrative information, cheating, plagiarism, sexual harassment, etc. Students are divided into groups of four or five for practice teaching sessions. Each student is required to prepare a five-minute lecture (on an assigned topic), given to members of the group and recorded on videotape. After listening to the lecture and reviewing the tape, group members and a faculty advisor critique the lecturer's performance. During the course of actual classroom teaching, a faculty mentor is assigned to each T.A. The mentor is required to visit the classroom on several occasions and make a written evaluation of teaching performance after each visit. The mentor also checks on test preparation and evaluation and, in general, is a source of experience and support. T.A.s understand the importance of good recommendations by faculty mentors in letters of reference accompanying job applications.

In addition to the above initial training and mentoring given all T.A.s, special requirements have been established for all advanced Ph.D. students (i.e., students who have passed prelims). With support from the Fund for the Improvement of Post Secondary Education (FIPSE), Clemson was one of eight universities cooperating with the AMS-MAA-SIAM Committee on Preparation for College Teaching in developing specific practices aimed at advanced graduate students. (The other participants were: The University of Cincinnati, Dartmouth College, The University of Delaware, Harvard University, Oregon State University, The University of Tennessee, and Washington University.) These activities, directed by a faculty member who had been recognized for excellence in research and teaching with a University Alumni Professorship, focused on the following goals:

- helping students become more effective teachers;

- helping students become more aware of the components and expectations of the profession;

- broadening the student's mathematical sciences perspectives;

- accomplishing this with minimum expense (time and money) to the students.

These goals were met through the careful design of a professional seminar carrying three semester hours of graduate credit, the purchase of a small amount of video equipment, and the establishment of a special departmental library for the project. Details on all eight projects plus interesting survey statistics are listed in the Committee's report [2].

A.6 Program Governance

Implementing and operating a program requiring a diverse faculty that typically reside in separate academic units can present a challenging

administrative problem. People in the mathematical sciences are noto-
riously parochial and are sometimes accused of lacking an abundance
of interpersonal skills. It is natural for each faculty member to feel his
specific area of expertise should occupy a place of prominence among
the component disciplines. Governance, tenure and promotion policies,
procedures for determining raises, defining and filling new faculty positions, etc. are all potentially divisive issues when real or perceived
inequities exist between component areas. Since differences in opinion
are bound to occur within any organizational structure, it is important
to have clearly-stated and effective departmental governance practices
in place.

A structure built around *subfaculties* has been found to work reasonably well at Clemson. Each subfaculty has developmental responsibilities relative to academic matters associated with its interest area. The
subfaculties are Analysis, Computational Mathematics, Discrete Mathematics, Operations Research, Statistics and Probability, and Undergraduate Education. Each faculty member is encouraged to join several
subfaculties but can be a voting member of only one. Subfaculties elect
a representative to the Mathematical Sciences Council (MSC) and to
each of three rather standard standing committees (Research, Graduate
Affairs, Undergraduate Affairs). The MSC advises the department chair
on long-range planning, curricular issues, and all matters brought before
it by the faculty or the department chair. The MSC also functions as
the Personnel Committee relative to the recruitment of new faculty.

A.7   Measures of Success

It is difficult to define "success" let alone quantify the degree to which
it has been achieved. Traditional measures include success in recruiting
top students for graduate work, success of graduates in competing for rewarding positions, success of the faculty in obtaining tenure, promotion,
and in competing for research funding. Considerable effort has been expended in program development at Clemson. We are proud of the recognition accorded these efforts and the honors earned by our graduates.
Ten Ph.D. programs in the mathematical sciences, including Clemson's,
were selected for study in a 1992 report [1] by the Board on Mathematical Sciences of the National Research Council (NRC). These ten
programs were examined during site visits and recommended as models
of "successful" programs. The Clemson program was also one of some
twenty-seven programs in Departments, Institutes, and Centers listed
and described in the proceedings of the conference *Graduate Programs
in the Applied Mathematical Sciences II* [4]. The master's program has

received national recognition and it attracts outstanding students who have had great success competing with graduates trained in traditional programs at better-known universities.

Counter to national trends in which U.S. citizens comprise a minority of the candidates for graduate degrees in the mathematical sciences, of the eighty-two graduate students enrolled in the Clemson program in the spring of 1994, seventy (85%) were U.S. citizens. Of these, twenty-six were females and five were black or Hispanic. International students in the program represented China (5), Germany, India (2), Mexico, Poland, Turkey, and Togo. This trend has continued to the present time (1999), in which 81% of the sixty-three graduate students are U.S. citizens and the twelve foreign students represent nine different countries. Also, during 1999 females comprised 37% of the total number of graduate students and 48% of those enrolled in the master's program. The uniqueness of the program, the quality and enthusiasm of the faculty, and the success of graduates in obtaining exciting employment goes a long way in attracting high quality applicants, many of whom come from small, elite four-year colleges. Another measure of success, one of the most gratifying, is follow-on applications from institutions after one of their former students completes the program. For example, the first student from St. Olaf College in Northfield, Minnesota enrolled in the master's program in 1985. She had an outstanding undergraduate education and was well prepared. Her success set the tone and since then, there has been a steady flow of excellent students from St. Olaf College, ten in all, several of whom have completed the Ph.D. program. Besides the success of graduates in finding good initial positions, return visits by BIG recruiters after having hired a previous graduate indicates the customer is happy with the product.

Funded research has dramatically increased. There are some notable examples of subfaculty research synergism involving major funding. The department was successful in competing for NSF funding under the Experimental Program to Stimulate Competitive Research (EPSCoR). The five-year program, entitled *Research in Discrete Structures*, involved faculty members in algebra, operations research, and computational mathematics. (The Algebra subfaculty subsequently adopted the name Discrete Mathematics reflecting the broadened interest in combinatorics, graph theory, and computing.) Another successful joint effort, entitled *Distributed Computing* and funded for three years by ONR, involved the analysis, operations research, and computational mathematics subfaculties. More recently, a multidisciplinary team of researchers in analysis, operations research, and statistics (together with faculty from finance and accounting) are participating in a multi-year grant from ONR to study *Capital Budgeting and Affordability Models*.

A.8 Conclusion

The Clemson program in the mathematical sciences anticipated a national need at a time when traditional mathematics programs flourished, producing graduates trained to assume positions on faculties of doctoral-granting institutions. The Clemson model, based on the foresight of Clayton Aucoin, focused on producing graduates having a combination of depth and breadth of training. This approach was selected by the National Research Council as one of ten successful model programs in the nation [1]. Today's technology revolution requires an ever-increasing number of broadly trained mathematical scientists. The Clemson program is helping to meet that need. For further information on current developments in the program, the interested reader can visit the website <http://www.math.clemson.edu>.

A.9 References

[1] Board on Mathematical Sciences, *Educating Mathematical Scientists: Doctoral Study and the Postdoctoral Experience in the United States*, National Research Council, National Academy Press, Washington, D.C., 1992.

[2] Committee on Preparation for College Teaching, AMS-MAA-SIAM, *You're the Professor, What Next: Ideas and Resources for Preparing College Teachers*, Bettye Anne Case (Ed.), MAA Notes Number 35, Mathematical Association of America, 1994.

[3] Conference Board of the Mathematical Sciences, *Graduate Education in Transition: Report of a Conference*, CBMS, Washington, D.C., 1992.

[4] R. E. Fennell and R. D. Ringeisen (Eds.), *Conference on Graduate Programs in the Applied Mathematical Sciences II*, Clemson University, Clemson, SC, 1993.

[5] R. E. Gaskell and M. S. Klamkin, "The industrial mathematician views his profession: a report of the committee on corporate members," *American Mathematical Monthly* **81**, 699–716 (1974).

[6] D. E. McClure, "Report on the 1993 survey of new doctorates," *Notices of the AMS* **40**, 1164–1196 (1993).

[7] T. G. Proctor, "Graduate programs in the applied mathematical sciences: perspectives and prospects," Technical Report # 274, Department of Mathematical Sciences, Clemson University, Clemson, SC, 1978.

Index